AF764217

Cardiovascular Nanomedicine

Nanoscience & Nanotechnology Series

Editor-in-chief:
Nguyễn T. K. Thanh, University College London, UK

Series editors:
Shinya Maenosono, *Japan Advanced Institute of Science and Technology, Japan*
Stefanos Mourdikoudis, *University of Chemistry and Technology Prague, Czech Republic*
Zeev Rosenzweig, *University of Maryland Baltimore County, USA*
Olivier Sandre, *Université de Bordeaux (CNRS), France*

Titles in the series:
1: Nanotubes and Nanowires
2: Fullerenes: Principles and Applications
3: Nanocharacterisation
4: Atom Resolved Surface Reactions: Nanocatalysis
5: Biomimetic Nanoceramics in Clinical Use: From Materials to Applications
6: Nanofluidics: Nanoscience and Nanotechnology
7: Bionanodesign: Following Nature's Touch
8: Nano-Society: Pushing the Boundaries of Technology
9: Polymer-based Nanostructures: Medical Applications
10: Metallic and Molecular Interactions in Nanometer Layers, Pores and Particles: New Findings at the Yoctolitre Level
11: Nanocasting: A Versatile Strategy for Creating Nanostructured Porous Materials
12: Titanate and Titania Nanotubes: Synthesis, Properties and Applications
13: Raman Spectroscopy, Fullerenes and Nanotechnology
14: Nanotechnologies in Food
15: Unravelling Single Cell Genomics: Micro and Nanotools
16: Polymer Nanocomposites by Emulsion and Suspension
17: Phage Nanobiotechnology
18: Nanotubes and Nanowires, 2nd Edition
19: Nanostructured Catalysts: Transition Metal Oxides
20: Fullerenes: Principles and Applications, 2nd Edition
21: Biological Interactions with Surface Charge Biomaterials
22: Nanoporous Gold: From an Ancient Technology to a High-Tech Material
23: Nanoparticles in Anti-Microbial Materials: Use and Characterisation
24: Manipulation of Nanoscale Materials: An Introduction to Nanoarchitectonics
25: Towards Efficient Designing of Safe Nanomaterials: Innovative Merge of Computational Approaches and Experimental Techniques
26: Polymer–Graphene Nanocomposites
27: Carbon Nanotube-Polymer Composites
28: Nanoscience for the Conservation of Works of Art
29: Polymer Nanofibers: Building Blocks for Nanotechnology
30: Artificial Cilia
31: Nanodiamond
32: Nanofabrication and its Application in Renewable Energy

33: Semiconductor Quantum Dots: Organometallic and Inorganic Synthesis
34: Soft Nanoparticles for Biomedical Applications
35: Hierarchical Nanostructures for Energy Devices
36: Microfluidics for Medical Applications
37: Nanocharacterisation, 2nd Edition
38: Thermometry at the Nanoscale: Techniques and Selected Applications
39: Nanoceramics in Clinical Use: From Materials to Applications, 2nd Edition
40: Near-infrared Nanomaterials: Preparation, Bioimaging and Therapy Applications
41: Nanofluidics, 2nd Edition
42: Nanotechnologies in Food, 2nd Edition
43: ZnO Nanostructures: Fabrication and Applications
44: Diatom Nanotechnology: Progress and Emerging Applications
45: Nanostructured Materials for Type III Photovoltaics
46: Chemically Derived Graphene: Functionalization, Properties and Applications
47: Graphene-based Membranes for Mass Transport Applications
48: Carbon Nanostructures for Biomedical Applications
49: Surface Chemistry of Colloidal Nanocrystals
50: Reducing Agents in Colloidal Nanoparticle Synthesis
51: Carbon Nitride Nanostructures for Sustainable Energy Production and Environmental Remediation
52: Nanotubes and Nanowires, 3rd Edition
53: Bionanodesign: Old Forms for New Functions, 2nd Edition
54: Photothermal Nanomaterials
55: Concepts and Design of Materials Nanoarchitectonics
56: Anticorrosive Nanomaterials: Future Perspectives
57: Sustainable Nanotechnology
58: Anisotropic 2D Materials and Devices
59: Nanotechnology for Diabetes Management
60: Graphene-based Nanocomposite Sensors: Detection to Diagnosis
61: Nanoparticles as Sustainable Environmental Remediation Agents
62: Smart Nanomaterials for Infectious Diseases
63: Magnetic Nanoparticles: Materials Engineering, Properties and Applications
64: Cardiovascular Nanomedicine

How to obtain future titles on publication:
A standing order plan is available for this series. A standing order will bring delivery of each new volume immediately on publication.

For further information please contact:
Book Sales Department, Royal Society of Chemistry, Thomas Graham House, Science Park, Milton Road, Cambridge, CB4 0WF, UK
Telephone: +44 (0)1223 420066, Fax: +44 (0)1223 420247
Email: booksales@rsc.org

Visit our website at books.rsc.org

Cardiovascular Nanomedicine

Edited by

Roger Narayan
University of North Carolina
Email: roger_narayan@unc.edu

and

Tanveer Tabish
University of Oxford, UK
Email: tanveer.tabish@cardiov.ox.ac.uk

ROYAL SOCIETY
OF **CHEMISTRY**

Nanoscience & Nanotechnology Series No. 64

Print ISBN: 978-1-83767-058-1
PDF ISBN: 978-1-83767-111-3
EPUB ISBN: 978-1-83767-112-0
Print ISSN: 1757-7136
Electronic ISSN: 1757-7144

A catalogue record for this book is available from the British Library

© The Royal Society of Chemistry 2024

All rights reserved

Apart from fair dealing for the purposes of research for non-commercial purposes or for private study, criticism or review, as permitted under the Copyright, Designs and Patents Act 1988 and the Copyright and Related Rights Regulations 2003, this publication may not be reproduced, stored or transmitted, in any form or by any means, without the prior permission in writing of The Royal Society of Chemistry or the copyright owner, or in the case of reproduction in accordance with the terms of licences issued by the Copyright Licensing Agency in the UK, or in accordance with the terms of the licences issued by the appropriate Reproduction Rights Organization outside the UK. Enquiries concerning reproduction outside the terms stated here should be sent to The Royal Society of Chemistry at the address printed on this page.

Whilst this material has been produced with all due care, The Royal Society of Chemistry cannot be held responsible or liable for its accuracy and completeness, nor for any consequences arising from any errors or the use of the information contained in this publication. The publication of advertisements does not constitute any endorsement by The Royal Society of Chemistry or Authors of any products advertised. The views and opinions advanced by contributors do not necessarily reflect those of The Royal Society of Chemistry which shall not be liable for any resulting loss or damage arising as a result of reliance upon this material.

The Royal Society of Chemistry is a charity, registered in England and Wales, Number 207890, and a company incorporated in England by Royal Charter (Registered No. RC000524), registered office: Burlington House, Piccadilly, London W1J 0BA, UK, Telephone: +44 (0) 20 7437 8656.

For further information see our website at www.rsc.org

Printed in the United Kingdom by CPI Group (UK) Ltd, Croydon, CR0 4YY, UK

Preface

The book aims to give full coverage of the topic of cardiovascular nanomedicine, in which nanoparticles of different sizes, shapes, and dimensions can be bioconjugated with antibodies at appropriate ratios to encapsulate cardiovascular drugs, biomolecules, and gaseous molecules to deliver them selectively to the site of action in a safe, targeted, and cost-effective manner. Nanoparticles have received significant attention for storing large amounts of drugs and releasing therapeutically relevant amounts of drugs very selectively in response to external (*e.g.*, light, ultrasound, and magnetic field) or internal (*e.g.*, exposure to water, changes in pH, changes in redox states, and exposure to enzymes) stimuli. Integrating nanoparticles with 'classical' drugs to create biocompatible delivery platforms is a current area of research activity to treat cardiovascular diseases; innovations in this area can significantly impact patient welfare. This book seeks to expand our understanding and provide an in-depth and insightful discussion. Traditional drug delivery systems are not selective and targeted and hence induce severe collateral damage to surrounding non-diseased healthy cells/tissues.

Despite the significant interest and the rapid development of cardiovascular drug delivery nanomedicine, there is currently no adequate and accessible book on this topic, which motivated and inspired us to edit this book. The primary motivation is to give mechanistic, preclinical, and clinical insights into cardiovascular nanomedicine. Nanotechnology has mainly been explored and used for cancer and infection; it has not been developed to a similar extent for cardiovascular diseases. Hence, this book will discuss core chemical concepts *via* the synthesis of novel nanotechnology-based drug formulations and cover the latest drug delivery advances and innovative therapeutic targets in targeting cardiovascular lesions at an early, curable, and reversible stage. Such an approach can lead to a more targeted, safe,

Nanoscience & Nanotechnology Series No. 64
Cardiovascular Nanomedicine
Edited by Roger Narayan and Tanveer Tabish
© The Royal Society of Chemistry 2024
Published by the Royal Society of Chemistry, www.rsc.org

and cost-efficient form of drug development. We focus on the therapeutic applications of drug-carrying nanoparticle-based formulations, particularly emphasizing the chemistry and design of nanomedicines for treating all types of cardiovascular diseases. We aim to cover practical research studies on the mechanisms of action of nanomedicines and their interaction with endothelial and cardiac cells. Furthermore, this book covers topics such as novel therapeutic targets in cardiovascular diseases for ultimate drug-delivery therapeutics and the pharmaceutical basics and safety considerations of cardiovascular nanomedicine not described in other books. The book focuses on the general concept, design, and guidelines for preparing (including scale-up) nanomedicines and their interactions with diseased cells. We combine different approaches for utilizing the chemical sciences as a source for cardiovascular science.

We prepared this book to meet the needs and pique the interest of university students and researchers in this sector. We have also taken care to meet the needs of teachers and scientific specialists in this field by carefully consolidating the projected future research and development opportunities. We aimed to meet the needs of the scientific professionals and teachers in this field by carefully consolidating the prospects of research and development that are envisaged in this area. We want to express our gratitude to every contributor who made this book possible. We are grateful to the Royal Society of Chemistry team, Katie Morrey, Amina Headley, and Matt Caffrey, for their unwavering support and helpful input, without whom the book would not have been possible. Finally, we hope you, the reader, will find this book enjoyable and valuable.

Roger J. Narayan and Tanveer A. Tabish

Contents

Nanoscience & Nanotechnology Series No. 64
Cardiovascular Nanomedicine
Edited by Roger Narayan and Tanveer Tabish
© The Royal Society of Chemistry 2024
Published by the Royal Society of Chemistry, www.rsc.org

CHAPTER 1

Cardiovascular Nanomedicine: From Targeted Delivery to Theranostics

MARYAM REZVANI*[a,b] AND NEJAT DÜZGÜNEŞ[c]

[a]Department of Life and Environmental Sciences, Drug Science Division, University of Cagliari, 09124 Cagliari, Italy; [b]Advanced Nanobiotechnology and Nanomedicine Research Group (ANNRG), Iran University of Medical Sciences, Tehran, Iran; [c]Department of Biomedical Sciences, Arthur A. Dugoni School of Dentistry, University of the Pacific, San Francisco, CA 94103, USA

*E-mail: rezvani.mary@yahoo.com

1.1 Introduction

Cardiovascular diseases (CVDs) are ranked as the leading cause of mortality worldwide, imposing a large economic burden on society for healthcare. Epidemiological studies reported a variety of risk-enhancing factors for CVDs, including high blood pressure and cholesterol, overweight, smoking, lack of physical exercise and type 2 diabetes.[1] Various medical and pharmaceutical strategies have been studied for the treatment of CVDs, and efforts are ongoing to discover new techniques. In recent decades, nanomedicine has shown promising potential for multi-target cardiovascular applications by exploiting the special properties of nanomaterials in extra- and intravascular mobility to deliver drugs and/or diagnostic imaging agents. This emerging approach has been used for managing various cardiovascular

Nanoscience & Nanotechnology Series No. 64
Cardiovascular Nanomedicine
Edited by Roger Narayan and Tanveer Tabish
© The Royal Society of Chemistry 2024
Published by the Royal Society of Chemistry, www.rsc.org

disorders and risk factors, including myocardial infarction (MI), atherosclerosis (AS), hypertension (HP) and pulmonary arterial hypertension (PAH).[2] For this purpose, various nanodelivery systems have been designed and developed. Nanovesicles (liposomes, niosomes, *etc.*), polymeric nanoparticles (NPs) [poly(lactic-*co*-glycolic acid) (PLGA) copolymers, poly(ε-caprolactone), *etc.*], lipid-based NPs [nanostructured lipid carriers (NLCs), solid lipid NPs (SLNs), *etc.*] and micelles are some of these nanosystems.[3–6] Therapeutic nanodelivery systems can localize in pathological sites through the enhanced permeability and retention effect. They can also be actively transferred to a target area by using ligands with high specific affinity to particular receptors. These nanosystems can reduce side effects, increase the accuracy and efficiency of treatment and/or improve diagnosis by targeting high-risk areas with cardiovascular disorders. Furthermore, their biomimetic characteristics, stimuli responsiveness and theranostic aspects could promote the treatment of CVDs.[7] This chapter reviews recent advances in nanomedicine applications for the treatment and diagnosis of cardiovascular disorders by addressing some of the most common diseases and risk factors.

1.2 Nanomedicine-mediated Therapeutic Approaches

1.2.1 Nanomedicine for Treating MI

MI refers to the irreversible damage to the myocardium caused by the reduction in blood flow in coronary arteries or their blockage. MI can be followed by heart failure and death.[8] The most common clinical strategies for treating MI include percutaneous coronary intervention, thrombolysis and coronary artery bypass grafting. Despite the advantage of these methods in decreasing the mortality rate, some of their unpredictable complications, such as ischemia–reperfusion injury, hemorrhage and coronary restenosis, necessitate developing more effective approaches for hypoxic-ischemic tissue rescue, blood flow restoration, cardiac myocyte regeneration and tissue repair with minimum cardiac damage.[9] Nanomedicine has opened a new horizon in this field, as NPs have demonstrated good potential in the targeted delivery of therapeutic agents with their controlled release and/or direct impacts on the cardiac cells and prevention of post-MI remodeling. A combination of pro-angiogenesis, anti-apoptosis and anti-inflammatory approaches has been proposed as an effective treatment strategy for MI.[10] Tanshinone IIA is a phytochemical containing an estrogen ring with inhibitory effects on inflammation and cardiac hypertrophy. Its incorporation into monomethoxy-poly(ethylene glycol)-poly (lactic acid)-D-*α*-tocopherylpolyethylene glycol 1000 succinate NPs and intravenous injection into mice with left coronary artery ligation prevented the phosphorylation of IκB (inhibitor of NF-κB) and activation of NF-κB (nuclear factor kappa-light-chain-enhancer of activated B cells). Treatment with these NPs suppressed the pro-inflammatory cytokine expression, reduced cardiomyocyte apoptosis and fibrotic process, inhibited left ventricle dilation and limited the infarct expansion.[11] Angiogenesis plays a vital role in MI treatment and can alleviate patient symptoms. Vascular endothelial growth factor (VEGF), basic fibroblast growth factor, interleukin-8 (IL-8) and other angiogenic agents can stimulate vessel formation. The short half-life of the growth factors, however, is considered to be a drawback to their effectiveness. In one study, VEGF electrostatically complexed with polyglutamic acid polypeptides was loaded into a hyaluronic acid hydrogel, and the formulation was proposed as a delivery platform for growth factors to ischemic myocardium, providing sustained release of these angiogenic cargoes.[12] The application of the bioactive peptide nanofibers has been reported as an efficient approach for post-MI healing. The injection of glycosaminoglycan-mimetic peptide scaffolds in a cardiac infarct rat model increased VEGF-A expression, vascular cell recruitment and angiogenesis in the cardiovascular tissue after MI, providing more cardiac muscle preservation and better cardiac function.[13] Blood flow restriction and oxygen deficiency in MI can lead to cardiac cell death. Hence, oxygen delivery to the infarcted site using nanosystems is another effective strategy for restoring cardiac function and cell rescue. Intravenous injection of oxygen-releasing NPs and their accumulation in the infarcted area led to suppressed fibrosis and enhanced neovascularization and cell survival.[14] Apoptosis has been reported to have a pivotal role in cardiomyocyte death, determining the infarct size in MI and heart failure. *In vivo* studies have revealed the beneficial effects of apoptosis inhibition on cardiac functions and infarct size reduction in MI models. It has been proposed that tissue necrosis in the infarcted site at the early stage of MI can be decreased by blocking myocardial apoptosis. Nanoprotein complexes can prevent remodeling of the left ventricle in MI patients. These complexes could decrease the myocardial apoptosis-related protein expression in MI rat models and prevent MI cell apoptosis.[15] MI also involves the infiltration of inflammatory leukocytes into the myocardial area, cytokine production and the triggering of inflammatory responses. Therefore, another suggested strategy to decrease the post-MI myocardial injury is anti-inflammatory therapy. Elevated levels of IL-6, IL-1 and tumor necrosis factor-alpha (TNF-α) have been reported as the contributing factors for myocardial injury and myocyte death.[16,17] Calcium carbonate NP-based delivery systems containing colchicine reduced the pro-inflammatory cytokine expression in myocardial tissues, the levels of IL-1β and TNF-α in serum, myocardial fibrosis and infarct size after MI.[18] MiR-199a-3p showed a regulatory effect on cardiomyocyte biology and could promote cardiac function and decrease the infarct size in MI mice models.[19] NPs coated with macrophage membrane and containing miR-199a-3p could bind to pro-inflammatory cytokines (TNF-α, IL-6 and

IL-1β), improve cell proliferation, inhibit hypoxia-induced apoptosis and cardiac fibrosis and suppress inflammation in the MI mice model.[20]

1.2.2 Nanomedicine Therapeutic Approaches for AS

AS is an inflammatory disease characterized by atherosclerotic plaque formation in specific sites of arteries due to the accumulation of lipids, cholesterol and fibrous elements. Plaque rupture can result in thrombosis and may be followed by life-threatening consequences, such as MI and stroke.[21] The main challenges for the management of AS are the development of an efficient targeting nanosystem and the improvement of plaque stability. Oxidative stress and inflammation are among the main triggers for enhancing AS. Targeted delivery of IL-10-encapsulated cyclic arginylglycylaspartic acid (cRGD)-conjugated pluronic-based nanocarriers to atherosclerotic plaques resulted in the decreased generation of IL-1β in AS lesions and a reduction in the plaque size.[22] Targeted delivery of the designed tetrapod PdH nanozymes to arterial plaques led to the scavenging of reactive oxygen species (ROS), activation of autophagy, reduction of oxidative stress, suppression of inflammation and alleviation of AS progression.[23] Platelet endothelial cell adhesion molecule-1 (PECAM-1) is one of the adhesion molecules highly expressed during the AS process due to the endothelial inflammatory condition. Anti-PECAM-1 surface-functionalized nanocapsules with the core of the docosahexaenoic acid showed good potential for AS regression.[24] Biomimetic rapamycin-loaded PLGA NPs coated with macrophage membrane were developed for the effective and targeted treatment of AS (Figure 1.1). Integrin α4β1, a protein on the macrophage surface, could bind with the VCAM-1, an antibody on the human umbilical vein endothelial cells, and ensure the active targeting of the dysfunctional endothelium. *In vitro* studies revealed the sustained release kinetics of rapamycin and internalization of the NPs by umbilical vein endothelial and RAW264.7 cells. The results showed the accumulation of these nanosystems in atherosclerotic lesions and their inhibitory effects on the progression of AS in the mouse model.[25]

The promotion of plaque stability is a critical consideration in AS management. Angiogenesis in atherosclerotic lesions leads to plaque growth and instability.[26] Ginsenoside and catalase-co-loaded porous PLGA NPs with surface modification with U937 cell membranes targeted atherosclerotic plaques and reduced the progression of AS through a process involving anti-angiogenesis; scavenging of ROS; downregulation of IL-1β, TNF-α and intercellular adhesion molecule-1 (ICAM-1) and anti-oxidant and anti-inflammatory activities.[27] A biomimetic nanodelivery system developed by platelet membrane cloaking around selenium and ginsenoside Rb1 NPs led to anti-inflammatory, anti-oxidant and anti-angiogenic effects.[28] As the AS progression is essentially affected by cholesterol crystals in plaques, designing and developing nanosystems for cholesterol crystal removal can

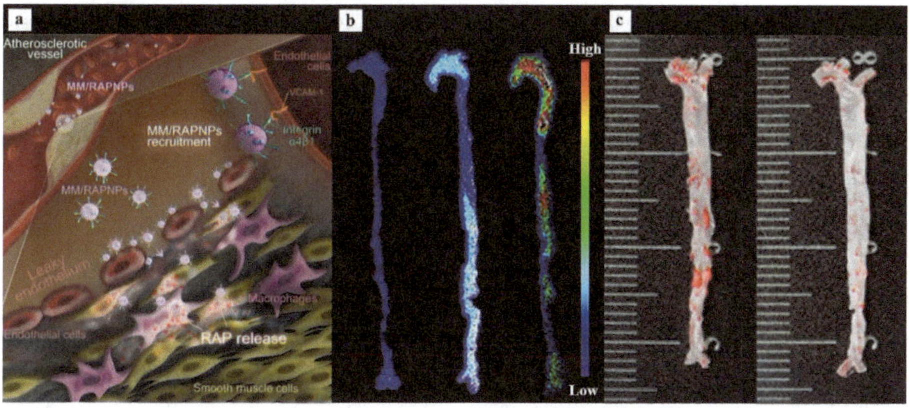

Figure 1.1 Effect of rapamycin-loaded PLGA NPs coated with macrophage membrane (MM/RAPNP) on targeted treatment of atherosclerosis: (a) mechanism of action; (b) targeting plaques in the aortic arc by the fluorescently labeled NPs coated with macrophage membrane (right) compared with the non-coated fluorescently labeled NPs (middle) and control (left); (c) inhibitory effect of MM/RAPNP on atherosclerosis progression (right) compared to control (left) in the $ApoE^{-}/^{-}$ mouse model. Reproduced from ref. 25, https://doi.org/10.7150%2Fthno.47841, under the terms of the CC BY 4.0 license, https://creativecommons.org/licenses/by/4.0/.

be a promising approach to inhibit AS-related cardiovascular diseases. Several studies have indicated the role of liposomes as cholesterol-lowering agents in the treatment of AS.[29,30] Nano-sponge-like liposomes containing ginsenosides Rb1 and surface-decorated with Annexin V showed anti-AS activity. They accumulated in atherosclerotic plaques, eliminated intra- and extra-cellular cholesterol, promoted cholesterol efflux and reduced apoptosis and inflammation in plaques. The plaque targeting was mediated by the ability of Annexin V to recognize phosphatidylserine (PS), which is upregulated in the plaques. The incorporation of ginsenosides Rb1 into the lipid bilayer increased the bilayer affinity to the cholesterol and cholesterol crystal solubilization.[31] Targeted delivery of coding or non-coding nucleic acid-loaded NPs is another nanomedicinal strategy to treat AS. Several studies have reported promising results on vascular wall targeting by microRNAs, such as miR-33, miR-21, miR-143 and miR-145.[32] Non-cationic nanostructures with a core of polyethylene glycol-coated superparamagnetic iron oxide NPs and a shell of phosphorothioate-modified miR-146a oligonucleotides were developed. Repeated intravenous injection of these nanostructures into the $ApoE^{-/-}$ mice alleviated AS by reducing and stabilizing the atherosclerotic plaques. This therapeutic approach was attributed to the effect of miR-146a-superparamagnetic iron oxide NPs in downregulating the expression of the genes related to vascular inflammation and immune responses. These NPs targeted the scavenger receptors of the endothelial cells and macrophages and enhanced the delivery of these nanostructures to the plaques.[33]

1.2.3 Nanomedicine-mediated Anti-arrhythmic Approaches

Nanomedicine has demonstrated great potential in normalizing abnormal heart rhythms by promoting the benefits of anti-arrhythmic agents, such as beta-blockers and sodium, potassium and calcium channel blockers.[34] Guanfu base A is a sodium and potassium channel antagonist. The solid nanolipids loaded with Guanfu base A caused a longer blood circulation time of the payload and a better anti-arrhythmic effect, especially in ventricular tachycardia and ectopia, than Guanfu base A HCl solution in rodent models.[35] Chitosan NPs containing botulinum toxin showed an antagonistic effect on the arrhythmia caused by the activation of cardiac ion channels.[36] Several nanosystems were developed to alleviate a common type of cardiac arrhythmia, called atrial fibrillation.[37] In this case, the localized release of the anti-arrhythmic and anti-inflammatory agents can act more effectively than intravenous injections. Parylene-C-nanostructured films loaded with amiodarone and dexamethasone efficiently released their payloads, prevented fibrotic tissue adhesion and reduced inflammation and atrial fibrillation.[38]

1.2.4 Nanomedicine Therapeutic Approaches for HP/PAH

Hypertension, the elevation of blood pressure to 140/90 mm Hg, is considered a crucial risk factor for cardiovascular diseases.[39] Heart failure, carotid atherosclerosis, left ventricular hypertrophy and peripheral vascular diseases are among the complications caused by high blood pressure.[40] Angiotensin-converting enzyme inhibitors, beta-blockers, angiotensin receptor blockers and calcium channel blockers are the main therapeutic agents for treating hypertension. These agents can be administered in combination according to the guidelines. Combining angiotensin receptor blockers with calcium channel blockers has been effective in reducing blood pressure and controlling other cardiovascular events.[41] One of the approaches for HP management is the oral administration of angiotensin-converting enzyme inhibitors. The application of nanocarriers has facilitated the controlled delivery and release of these compounds. Some bioactive peptides possess angiotensin-converting enzyme-inhibitory activity, but their poor bioavailability, short half-life and high reactivity with other components of the formulation hinder their oral administration. Loading such peptides into nanosystems has been suggested as a strategy to overcome the obstacles to their oral delivery.[42,43] Transdermal delivery of ethosomal hydrogels loaded with carvedilol, a beta-blocker, provided a proper anti-hypertensive effect with the controlled release of the payload. The tail-cuff measurement method showed that these nanosystems gradually reduced systolic blood pressure within 24 h in sodium chloride-induced hypertensive rat models.[44] Transdermal delivery of nanoethosomal formulations loaded with valsartan, an angiotensin II receptor antagonist, in medroxyprogesterone acetate-induced hypertensive rats effectively controlled the blood pressure for up to 48 h. Controlled release of valsartan from these formulations caused a gradual reduction in blood pressure.[45] The major goals of nanomedicine in the regulation of hypertension are increased bioavailability and controlled release of anti-hypertensive agents. For example, nisoldipine, a calcium channel antagonist, has a poor oral bioavailability of 5% due to pre-systemic metabolism and its role as a substrate for cytochrome P4503A4 enzymes.[46] Several studies have suggested the use of nano-based methodologies to improve its bioavailability. NLCs and SLNs were developed for the delivery of nisoldipine, and NLCs improved the oral bioavailability of the payload by 1.09- and 2.46-fold compared to SLN and nisoldipine suspension, respectively.[47] PLGA NPs loaded with nisoldipine and piperine reduced the systemic blood pressure in rats by more than 28% in comparison to nisoldipine suspension. Piperine acted as a cytochrome P4503A4 enzyme inhibitor and caused a 4.9-fold enhancement in nisoldipine bioavailability.[46]

Restriction of the cross-sectional area for blood flow through pulmonary arterial circulation results in PAH, increasing the workload on the right side of the heart. Conventional therapeutic approaches for PAH have exploited targeting the prostacyclin, nitric oxide (NO) and endothelin pathways.[48] Nebulization of sildenafil-laden NLCs and SLNs for PAH treatment in rats protected the normal lung parenchyma and minimized excessive vasodilation and intra-alveolar bleeding. Utilization of NLCs led to low interaction with mucin, efficient nebulization and a higher cell viability than SLNs.[49] Nanosized metal–organic frameworks have been proposed for the delivery of anti-PAH agents, as they are biocompatible and have a high drug-loading capacity, tunable size, shape and chemical nature and controllable drug-binding and release kinetics. NanoMIL-89, a nano-sized, iron-based metal–organic framework, loaded with sildenafil, exhibited sustained vasodilation in the mouse aorta, with an initial lag phase, which was in line with the payload release.[50] Injection of nanospheres containing ONO1301, a prostacyclin agonist, into the Sugen/hypoxia rat models led to a reduction in IL-1β, IL-6, transforming growth factor-β (TGF-β) and platelet-derived growth factor-β levels. Furthermore, it caused an increase in the expression of hepatocyte growth factor and an improvement in the percent medial wall thickness of the pulmonary vasculature and right ventricle pressure/left ventricle pressure in rats.[51] The pressure and hypertrophy of the right ventricle and muscularization of the pulmonary artery in PAH rat models were reduced by the intratracheal administration of beraprost NPs. The NPs incorporating this prostacyclin analog exhibited a sustained anti-proliferative effect on the smooth muscle cells of the pulmonary arteries.[52] NO possesses vasodilatory, anti-coagulant and antiproliferative properties, and the NO nanoformulations released NO with a kinetics consisting of an initial peak followed by a sustained release. These nanosystems induced the relaxation of the pulmonary arteries of hypoxia-induced PAH mice in a concentration-dependent manner.[53] Oral delivery of NPs containing bosentan, an endothelin receptor antagonist, prevented oxidative stress and cell proliferation and exhibited protective effects on smooth muscle cells of the pulmonary arteries.[54] Nanoencapsulation of bosentan can overcome the usual restricting aspects for its oral delivery, including low active time and bioavailability, the need for frequent administration, side effects and liver toxicity. Intratracheal administration of cerivastatin nanoliposomes prevented the smooth muscle cell proliferation in pulmonary arteries, exhibited sustained release kinetics and provided safe drug delivery for PAH treatment, without cytotoxicity and with less lactone metabolite formation compared to free cerivastatin.[55]

1.2.5 Stimuli-responsive Nanomedicine Approaches for Treating CVDs

The use of stimuli-responsive nanomedicines has opened a new horizon in overcoming the limitations of conventional approaches in CVD treatment. Such nanomedicines can adjust the release of a therapeutic agent in response to endogenous stimuli, including changes in pH and ROS levels at the pathological site, and/or exogenous stimuli, including magnetic fields, ultrasound and light. To treat inflammation and oxidative stress associated with AS, ROS-responsive nanomicelles composed of poly(ethylene glycol)-poly(propylene sulfide) and containing the botanical anti-inflammatory agent, andrographolide, were designed and developed. These polymeric nanomicelles reduced the ROS level and the expression of monocyte chemoattractant protein-1 and IL-6 (pro-inflammatory cytokines). Plaque-targeted release was achieved through the reaction between poly(propylene sulfide) and ROS, which caused micellar decomposition.[56] The stimulus-responsive drug release nanosystems can exploit the ROS-laden and acidic environment of the MI site as a trigger. A multifunctional hydrogel containing recombinant humanized collagen type III and curcumin-loaded PLGA NPs was developed to obtain the properties of direct injectability into the myocardial wall and on-demand release in response to the microenvironment of the MI site. This hydrogel accelerated angiogenesis, decreased apoptosis, prevented inflammatory responses, healed the damaged heart and promoted cardiac function.[57] The pH-responsive tannic acid/europium coordination nanocomplex showed a sustained release behavior in the acidic microenvironment, increased anti-oxidation and neovascularization, prevented inflammatory responses and decreased the infarct size.[58]

Nanomaterial-mediated photothermal therapy (PTT) benefits from an increase in temperature (hyperthermia) in the target area by converting light energy into heat. Most of the studies on PTT have been performed using near-infrared (NIR) light.[59] NPs containing nanophotosensitizers, with passive or active targeting, can accumulate in thrombi or atherosclerotic plaques, and through hyperthermia, they can cause loosening of the blood clot and plaque reduction by the destruction of the interactions between fibrins and ablation of the macrophages, respectively.[60] Immunoliposomes conjugated with single-chain variable fragment antibody (scFv) and loaded with IR780 dye showed efficient PTT for acute thrombosis, and their use was proposed as a superior, alternative approach to the plasminogen activator-based thrombolysis with bleeding complications. Thanks to the presence of the scFv antibody and IR780, these nanoliposomes could target activated platelets and induce hyperthermia with an average local temperature increase of 12 °C under NIR irradiation, leading to the reduction in the clot area in an $FeCl_3$-induced thrombosis mouse model. Co-loading of these nanosystems with low-dose, single-chain urokinase plasminogen activator (scuPA) caused a synergistic effect of hyperthermia-induced liposome penetration into clots and fibrinolytic effect *in vitro*. Co-loading of scuPA, however, did not significantly improve thrombolysis *in vivo*, confirming the potent photothermal effect for sufficient thrombolysis without the need for other fibrinolytic agents.[61]

In photodynamic therapy (PDT), photosensitizers accumulate in the target site, act as energy transfer catalysts, absorb light and result in cytotoxic ROS release.[62] Under NIR irradiation, RGD-modified mesoporous carbon nanospheres containing porphyrin-like metal centers caused site-specific thrombolysis *via* the synergistic effects of hyperthermia and ROS.[63] These nanosystems overcame the hemorrhagic risk of systematic fibrinolytic agents by targeting the glycoprotein IIb/IIIa receptors on the thrombus. The combination of PDT with PTT inhibited the secondary embolism caused by post-photothermal fragments. PDT-induced ROS, generated by NPs, caused lipid peroxidation and damaged platelet factor 3 (Figure 1.2).

ROS generation *via* the activation of sonosensitive materials under low-frequency ultrasound, as the stimuli, has been used in sonodynamic therapy (SDT), which showed a higher efficacy endowed by the potent penetration of ultrasound.[64] SDT utilizing curcumin nanosuspensions resulted in proper anti-AS effect and plaque stability. This technique increased the level of ROS in RAW264.7 cells, promoted macrophage apoptosis, decreased the levels of low-density lipoprotein and total cholesterol and stimulated macrophage transformation from M1 (pro-inflammatory phenotype) to M2 (reparative phenotype).[65]

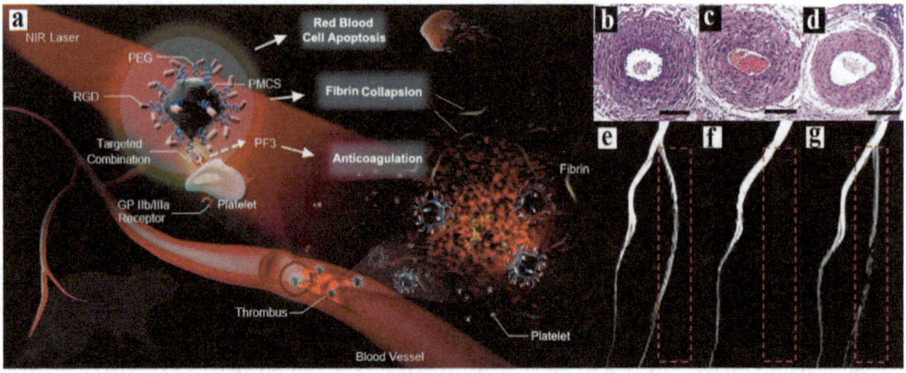

Figure 1.2 Site-specific thrombolytic effect of arginylglycylaspartic acid-modified mesoporous carbon nanospheres containing porphyrin-like metal centers under near-infrared irradiation (RGD-PMCS + NIR): (a) mechanism of action; (b and e) normal groups; (c and f) embolized groups; (d) no thrombus was observed in the vessels of rats treated by RGD-PMCS + NIR in the histopathology assessment and (g) MRI showed vascular recanalization in the RGD-PMCS + NIR group. The red dashed lines and scale bars indicate the site of the thrombus and 100 μm, respectively. Reproduced from ref. 63, https://doi.org/10.1002/advs.201901378, under the terms of the CC BY 4.0 license, https://creativecommons.org/licenses/by/4.0/.

1.3 Targeted Nanomedicine Approaches for CVD Diagnosis

Extending the patient's life, preventing the occurrence of fatal conditions and ensuring effective and successful treatment require the early diagnosis of CVD. Nanomedicine has facilitated the early detection of CVDs by using nanomaterials as contrast agents, including gold nanoshells, colloidal nanobeacons, quantum dots and iodine-based and iron oxide NPs. These nanosystems can be used in different diagnostic techniques, such as optical coherent tomography (OCT), photoacoustic imaging (PAI), fluorescence tomography, computed tomography (CT), magnetic resonance imaging (MRI) and positron emission tomography (PET).[66] The cLABL peptide-functionalized gold nanoshells exhibited binding capability for the ICAM-1 molecule, a molecule overexpressed in the earliest stage of AS, and acted as great contrast agents for OCT.[67] PAI, by merging ultrasound detection and optical illumination, serves as a hybrid imaging modality with high sensitivity for exceptional spatial resolution.[68] Using a PAI system with a contrast agent of nanometric, near-infrared, erythrocyte-derived transducers provided suitable capability in finding the location of mimicked vulnerable plaques and stenosis/occlusion-induced MI in the left anterior descending coronary artery ligation mice models (Figure 1.3).[69]

A smart activatable MRI nanosensor, composed of iron oxide NP coated with a gadolinium layer, showed promising results in detecting thrombosis

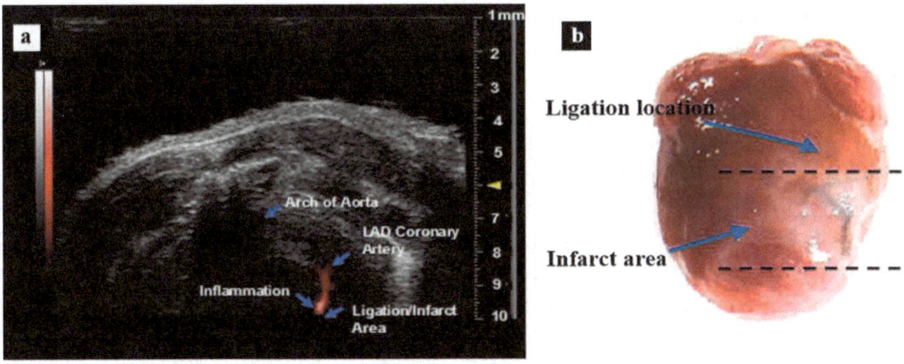

Figure 1.3 Detection of coronary artery disease/myocardial infarction in a mouse model by photoacoustic imaging and a contrast agent of near-infrared erythrocyte-derived transducer: (a) a high photoacoustic signal was detected above the site of the ligation (illustrated with the yellow arrow). Any other signal was not seen in any other area of the heart. (b) Ligation location and infarct region in the heart of the left anterior descending artery mouse model. Reproduced from ref. 69, https://doi.org/10.1038/s41598-020-75966-x, under the terms of the CC BY 4.0 license, https://creativecommons.org/licenses/by/4.0/.

and distinguishing fresh and constituted thrombi through switchability between the T1 and T2 signals. Thrombus targeting of the nanosensor was achieved by its functionalization with fibrin-binding peptide.[70] An innovative contrast agent for noninvasive magnetic resonance/fluorescence imaging constructed by covalently attaching a 16-mer peptide (LSLERFLRCWSDAPAK), PP1, to gadolinium-functionalized gold nanoclusters demonstrated superb potential for the precise diagnosis of vulnerable atherosclerotic plaques. This dual-modality imaging probe was designed based on targeting the scavenger receptor AI (one of the isoforms of scavenger receptor A) and its high affinity to PP1. Fluorescence characteristics and longitudinal relaxivity were promoted due to the co-existence of gold and gadolinium species. The images obtained from *in vivo* studies on established $ApoE^{-/-}$ mice models exhibited prolonged and robust enhancement of the contrast of the vulnerable plaques, validating the targeting efficacy of these nanosystems.[71] Paramagnetic/fluorescent micellar NPs were developed and used for *in vivo* cardiac magnetic resonance visualization and for detecting the MI onset by targeting the highly expressed extracellular matrix metalloproteinase inducer.[72] PET/CT imaging with extracellular loop 1 inverso peptide-conjugated gold nanoclusters radiolabeled with ^{64}Cu showed promising potential for clinical translation and acted as a molecular imaging probe for precisely detecting the initiation and progression of AS in the $ApoE^{-/-}$ mice. This technique exploited the chemokine receptor 2 as an AS marker.[73]

The detection of biomarkers by nanomaterial-based biosensors has been proposed as a rapid and precise diagnosis strategy for early-stage

CVDs. Furthermore, signals can be amplified through the benefits of multiple structures of hybrid nanomaterials and/or high loading capacity for recognition elements by exploiting the high surface-area-to-volume ratio and porous structures of nanomaterials. Troponin T, a cardiac biomarker, was detected by a label-free electrochemical biosensor consisting of $ZnSnO_3$ perovskite nanomaterials with sub-femtomolar detection sensitivity.[74] Detection of myoglobin, an MI biomarker, was performed using a label-free, surface-enhanced Raman scattering sensor composed of 3D silver anisotropic nano-pinetree array-modified indium tin oxide substrates, with a detection limit of 10 ng mL^{-1}.[75] An electrochemical aptasensor demonstrated a promising potential for the early diagnosis of MI in clinical applications. It was constructed using gold NP-decorated boron nitride nanosheets and developed for selectively detecting myoglobin with high sensitivity and a low detection limit.[76] A porous early detector for CVD biomarkers, composed of hydrogen-substituted graphdiyne and nanodiamonds, was developed as an impedimetric aptasensor for detecting myoglobin and cardiac troponin I with low limits of 9.04 and 6.29 fg mL^{-1}, respectively.[77] One of the point-of-care diagnosis strategies for CVDs is the use of smartphone-based methodologies and microRNAs as biomarkers.[78] A nanosystem composed of a gold NP core and a G-quadruplex DNAzyme layer was used as a catalytic nanolabel for chemiluminescence microRNA imaging by using a smartphone. This technique exhibited great performance in imaging and analyzing miRNA-133a, a biomarker for MI onset, in patients' serum.[79]

1.4 Theranostic Cardiovascular Nanomedicine

The design and development of nanosystems for simultaneously delivering therapeutic and diagnostic agents and continuously monitoring CVDs have received a great deal of attention in recent decades. Multimodal imaging-guided SDT using ramucirumab-modified PLGA NPs encapsulating hematoporphyrin monomethyl ether, manganese ferrite and perfluoropentane demonstrated an effective approach for theranostics of plaque angiogenesis. These NPs could accumulate in neovessels of the plaque by actively targeting the mitochondria of rabbit aortic endothelial cells, and MRI/PAI/ultrasound (US) imaging provided real-time observation of their distribution in plaques.[80] Hematoporphyrin monomethyl ether and perfluoropentane were used for SDT and ultrasound imaging, respectively. The manganese ferrite was encapsulated into NPs for PAI and MRI T1. NP - mediated SDT generated ROS, prevented migration and proliferation of aortic endothelial cells, induced mitochondrial-caspase apoptosis and inhibited the neovascularization in plaque (Figure 1.4).

Dual-targeted, high-density lipoprotein-mimicking NPs, incorporating the apolipoprotein A1-mimetic L-4F peptide, stearyl triphenylphosphonium, stearyl mannose and mito-magneto, exhibited the potential ability

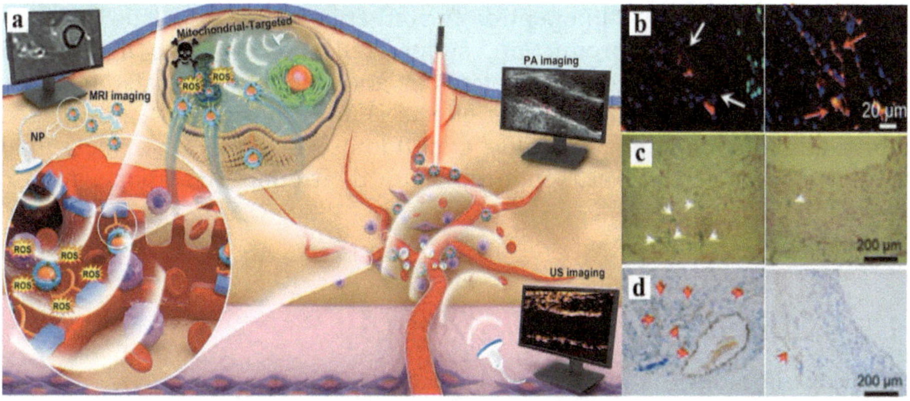

Figure 1.4 Low-intensity focused ultrasound-responsive ferrite-loaded multifunctional NPs: (a) for plaque angiogenesis theranostics and (b) for apoptosis of the neovessel endothelial cells on the third day after plaque treatment (right) in comparison with control (left). Red arrows illustrate TUNEL-positive microvessels and white arrows indicate TUNEL-negative microvessels; (c, d) for plaque stabilization and angiogenesis inhibition on the 28th day after treatment (right) in comparison with control (left). White and red arrows demonstrate intraplaque hemorrhage and abnormal neovessels, respectively. Reproduced from ref. 80, https://doi.org/10.1002/advs.202100850, under the terms of the CC BY 4.0 license, https://creativecommons.org/licenses/by/4.0/.

to act as nanotheranostic agents for CVDs. The surface functionalization conferred the NPs with the capability to remove lipids and target the M2 macrophages and mitochondria. In parallel, encapsulating the mito-magneto, a hydrophobic contrast agent for MRI, in the core of these NPs led to MRI contrast enhancement.[81] Accurate early diagnosis and site-specific treatment of MI were achieved by a theranostic nanosystem composed of PS, poly(lactide)-polycarboxybetaine and magnetic iron oxide nanocubes. The nanosystems were intravenously administered into MI rat models and accumulated at the infarcted site *via* external magnetic field induction, mimicking apoptosis. The binding of PS to its receptors on the macrophage surface induced the modulation of the macrophage phenotype from M1 to M2, which facilitated the upregulation of the anti-inflammatory regulators IL-10 and TGF-β1, downregulation of the pro-inflammatory cytokine TNF-α, resolution of the initial inflammation, revascularization, improvement of MI healing and promotion of cardiac function. Lysosomal/endosomal escape of magnetic iron oxide nanocubes was mediated by protonated polycarboxybetaine, resulting in promoted cytoplasmic accumulation of nanocubes. A high-sensitivity and precise MRI with reduced signal intensity in T2* was obtained by these nanosystems.[82]

A theranostic nanosystem for thrombosed vessels was synthesized by self-assembling the fluorescent IR820-conjugated H_2O_2-responsive boronate polymers and lipopeptide for targeting the fibrin. This nanosystem prevented H_2O_2 production, suppressed soluble CD40 ligand and TNF-α expression in activated platelets and exerted anti-inflammatory, anti-oxidant and anti-thrombotic effects. Furthermore, it demonstrated fibrin-targeted imaging capability and fluorescence/photoacoustic signal enhancement in the $FeCl_3$-induced carotid arterial thrombosis mouse model.[83]

1.5 Conclusion

In recent decades, nanomedicine has attracted increasing attention as an emerging strategy to deal with the challenges of treatment and diagnosis of cardiovascular disorders. Despite the great achievements reviewed in this chapter, cardiovascular nanomedicine has a long way to pave for successful global clinical translation and commercialization. In this regard, quality control, reproducibility, cost-effective scale-up process, safety and toxicity issues, legislation and establishment of standards are among the most important aspects that have to be considered to achieve a precise and efficient approach with improved patient outcomes. More efforts are needed for the rapid development of theranostic nanomedicine as a non-invasive simultaneous diagnostic and therapeutic strategy. Integrating these advances with those in materials engineering, chemistry, biology and physics will lead to a revolution in cardiovascular nanomedicine in the future.

References

1. G. D. Flora and M. K. Nayak, *Curr. Pharm. Des.*, 2019, **25**(38), 4063.
2. K. Wang, X. Z. Chen, Y. H. Wang, X. L. Cheng, Y. Zhao, L. Y. Zhou and K. Wang, *Cell Death Discovery*, 2022, **8**(1), 394.
3. M. Agrawal, S. Saraf, S. Saraf, S. K. Dubey, A. Puri, R. J. Patel, V. Ravichandiran, U. S. Murty and A. Alexander, *J. Controlled Release*, 2020, **321**, 372.
4. M. Cheng, Q. Liu, W. Liu, F. Yuan, J. Feng, Y. Jin and L. Tu, *J. Drug Delivery Sci. Technol.*, 2021, **63**, 102473.
5. O. Pechanova, E. Dayar and M. Cebova, *Molecules*, 2020, **25**(15), 3322.
6. D. Skourtis, D. Stavroulaki, V. Athanasiou, P. G. Fragouli and H. Iatrou, *Pharmaceutics*, 2020, **12**(12), 1160.
7. R. Pala, V. T. Anju, M. Dyavaiah, S. Busi and S. M. Nauli, *Int. J. Nanomed.*, 2020, 3741.
8. P. Sachdeva, K. Kaur, S. Fatima, F. N. Mahak, M. Noman, S. M. Siddenthi, M. A. Surksha, M. Munir, F. N. Fatima, S. S. Sultana, G. Varrassi, M. Khatri, S. Kumar, M. Elder and T. Mohamad, *Cureus*, 2023, **15**(9), e45578.

9. Q. Zhang, L. Wang, S. Wang, H. Cheng, L. Xu, G. Pei, Y. Wang, C. Fu, Y. Jiang, C. He and Q. Wei, *Signal Transduction Targeted Ther.*, 2022, **7**(1), 78.
10. C. Hu, W. Liu, L. Long, Z. Wang, W. Zhang, S. He, L. Lu, H. Fan, L. Yang and Y. Wang, *Biomaterials*, 2022, **290**, 121849.
11. S. Mao, L. Wang, P. Chen, Y. Lan, R. Guo and M. Zhang, *Artif. Cells, Nanomed., Biotechnol.*, 2018, **46**, 707.
12. J. O'Dwyer, R. Murphy, E. B. Dolan, L. Kovarova, M. Pravda, V. Velebny, A. Heise, G. P. Duffy and S. A. Cryan, *Drug Delivery Transl. Res.*, 2020, **10**, 440.
13. A. J. Rufaihah, I. C. Yasa, V. S. Ramanujam, S. C. Arularasu, T. Kofidis, M. O. Guler and A. B. Tekinay, *Acta Biomater.*, 2017, **58**, 102.
14. Y. Guan, H. Niu, J. Wen, Y. Dang, M. Zayed and J. Guan, *ACS Nano*, 2022, 11, **16**(11), 19551.
15. B. Yang, C. L. Lu, H. Zhao and R. Dong, *J. Nanosci. Nanotechnol.*, 2021, **21**(2), 1272.
16. L. F. Buckley and A. Abbate, *Eur. Heart J.*, 2018, **39**(22), 2063.
17. E. Holte, O. Kleveland, T. Ueland, G. Kunszt, M. Bratlie, K. Broch, A. E. Michelsen, B. Bendz, B. H. Amundsen, S. Aakhus and J. K. Damås, *Heart*, 2017, **103**(19), 1521.
18. L. Wang, Y. Peng, L. Song, D. Xia, C. Li, Z. Li, Q. Li, A. Yu, C. Lu and Y. Wang, *Cardiovasc. Drugs Ther.*, 2022, **36**(6), 1075.
19. H. Yang, X. Qin, H. Wang, X. Zhao, Y. Liu, H. T. Wo, C. Liu, M. Nishiga, H. Chen, J. Ge and N. Sayed, *ACS Nano*, 2019, **13**(9), 9880.
20. Y. Xue, G. Zeng, J. Cheng, J. Hu, M. Zhang and Y. Li, *Bioeng. Transl. Med.*, 2020, **6**, e10197.
21. K. El Hadri, R. Smith, E. Duplus and C. El Amri, *Int. J. Mol. Sci.*, 2021, **23**(1), 77.
22. M. Kim, A. Sahu, Y. Hwang, G. B. Kim, G. H. Nam, I. S. Kim, I. C. Kwon and G. Tae, *Biomaterials*, 2020, **226**, 119550.
23. R. Hu, C. Dai, C. Dong, L. Ding, H. Huang, Y. Chen and B. Zhang, *ACS Nano*, 2022, **16**(10), 15959.
24. M. de Castro Leão, A. R. Pohlmann, A. D. Alves, S. H. Farsky, M. K. Uchiyama, K. Araki, S. Sandri, S. S. Guterres and I. A. Castro, *Eur. J. Pharm. Biopharm.*, 2021, **159**, 99.
25. Y. Wang, K. Zhang, T. Li, A. Maruf, X. Qin, L. Luo, Y. Zhong, J. Qiu, S. McGinty, G. Pontrelli and X. Liao, *Theranostics*, 2021, **11**(1), 164.
26. C. Camaré, M. Pucelle, A. Nègre-Salvayre and R. Salvayre, *Redox Biol.*, 2017, **12**, 18.
27. J. W. Shen, C. Li, M. Y. Yang, J. F. Lin, M. D. Yin, J. J. Zou, P. Y. Wu, L. Chen, L. X. Song and J. W. Shao, *Int. J. Pharm.*, 2022, **611**, 121297.
28. M. Yin, J. Lin, M. Yang, C. Li, P. Wu, J. Zou, Y. Jiang and J. Shao, *Colloids Surf., B*, 2022, **214**, 112464.

29. A. Darwitan, Y. S. Wong, L. T. Nguyen, B. Czarny, A. Vincent, A. M. Nedumaran, Y. F. Tan, A. Muktabar, J. K. Tang, K. W. Ng and S. Venkatraman, *Adv. Healthcare Mater.*, 2020, **9**(14), 2000465.
30. N. Kiaie, A. M. Gorabi, P. E. Penson, G. Watts, T. P. Johnston, M. Banach and A. Sahebkar, *Drug Discovery Today*, 2020, **25**(1), 58.
31. F. Gong, Z. Wang, R. Mo, Y. Wang, J. Su, X. Li, C. T. Omonova, A. M. Khamis, Q. Zhang, M. Dong and Z. Su, *J. Controlled Release*, 2022, **349**, 940.
32. S. Gadde and K. J. Rayner, *Arterioscler., Thromb., Vasc. Biol.*, 2016, **36**(9), e73.
33. Q. Bai, Y. Xiao, H. Hong, X. Cao, L. Zhang, R. Han, L. K. Lee, E. Y. Xue, X. Y. Tian and C. H. Choi, *Proc. Natl. Acad. Sci.*, 2022, **119**(39), e2201443119.
34. A. E. Nadimi, S. Y. Ebrahimipour, E. G. Afshar, S. K. Falahati-Pour, Z. Ahmadi, R. Mohammadinejad and M. Mohamadi, *Eur. J. Med. Chem.*, 2018, **157**, 1153.
35. F. Xiong, K. Liu, S. Liu, J. Chen, J. Liu, H. Wang and N. Gu, *Drug Delivery Transl. Res.*, 2018, **8**, 1471.
36. D. Sergeevichev, V. Fomenko, A. Strelnikov, A. Dokuchaeva, M. Vasilieva, E. Chepeleva, Y. Rusakova, S. Artemenko, A. Romanov, N. Salakhutdinov and A. Chernyavskiy, *Mar. Drugs*, 2020, **18**(8), 410.
37. Z. Wang, Q. Tong, T. Li and Y. Qian, *Front. Cardiovasc. Med.*, 2022, **9**, 906350.
38. E. Robinson, S. Kaushal, J. Alaboson, S. Sharma, A. Belagodu, C. Watkins, B. Walker, G. Webster, P. McCarthy and D. Ho, *Nanoscale*, 2016, **8**(7), 4267.
39. J. M. Flack and B. Adekola, *Trends Cardiovasc. Med.*, 2020, **30**(3), 160.
40. S. K. Masenga and A. Kirabo, *Front. Cardiovasc. Med.*, 2023, **10**, 1205475.
41. S. M. Farag, H. M. Rabea, M. E. Abdelrahim and H. B. Mahmoud, *Curr. Hypertens. Rev.*, 2022, **18**(2), 138.
42. M. Rezvani, J. Hesari, S. H. Peighambardoust, M. Manconi and H. Hamishehkar, *Adv. Pharm. Bull.*, 2018, **8**(4), 609.
43. M. Rezvani, J. Hesari, S. H. Peighambardoust, M. Manconi, H. Hamishehkar and E. Escribano-Ferrer, *Food Chem.*, 2019, **293**, 368.
44. P. R. Amarachinta, G. Sharma, N. Samed, A. K. Chettupalli, M. Alle and J. C. Kim, *J. Nanobiotechnol.*, 2021, **19**, 1.
45. A. Ahad, M. Aqil, K. Kohli, Y. Sultana and M. Mujeeb, *Artif. Cells, Nanomed., Biotechnol.*, 2016, **44**(3), 1002.
46. P. Rathee, A. Kamboj and S. Sidhu, *Nanotechnol. Rev.*, 2017, **6**(6), 517.
47. N. Dudhipala, K. Y. Janga and T. Gorre, *Artif. Cells, Nanomed., Biotechnol.*, 2018, **46**, 616.
48. V. Segura-Ibarra, S. Wu, N. Hassan, J. A. Moran-Guerrero, M. Ferrari, A. Guha, H. Karmouty-Quintana and E. Blanco, *Front. Physiol.*, 2018, **9**, 890.
49. N. Nafee, S. Makled and N. Boraie, *Eur. J. Pharm. Sci.*, 2018, **125**, 151.

50. N. A. Mohamed, H. Abou-Saleh, Y. Kameno, I. Marei, G. de Nucci, B. Ahmetaj-Shala, F. Shala, N. S. Kirkby, L. Jennings, D. E. Al-Ansari and R. P. Davies, *Sci. Rep.*, 2021, **11**(1), 4336.
51. T. Kanaya, S. Miyagawa, T. Kawamura, Y. Sakai, K. Masada, N. Nawa, H. Ishida, J. Narita, K. Toda, T. Kuratani and Y. Sawa, *Sci. Rep.*, 2021, **11**(1), 7292.
52. S. Akagi, K. Nakamura, H. Matsubara, M. Kondo, D. Miura, T. Matoba, K. Egashira and H. Ito, *J. Cardiovasc. Pharmacol.*, 2016, **67**(4), 290.
53. N. A. Mohamed, B. Ahmetaj-Shala, L. Duluc, L. S. Mackenzie, N. S. Kirkby, D. M. Reed, P. D. Lickiss, R. P. Davies, G. R. Freeman, B. Wojciak-Stothard and A. H. Chester, *J. Cardiovasc. Transl. Res.*, 2016, **9**, 162.
54. Q. Zhao, C. Hou, Y. Lu, Q. Qiu, L. Xie, S. Xu, T. Xiao and M. Zhong, *J. Drug Delivery Sci. Technol.*, 2022, **68**, 102970.
55. Y. Lee, S. B. Pai, R. V. Bellamkonda, D. H. Thompson and J. Singh, *J. Pharmacol. Exp. Ther.*, 2018, **366**(1), 66.
56. T. Wu, X. Chen, Y. Wang, H. Xiao, Y. Peng, L. Lin, W. Xia, M. Long, J. Tao and X. Shuai, *Nanomedicine*, 2018, **14**(7), 2215.
57. C. Hu, W. Liu, L. Long, Z. Wang, W. Zhang, S. He, L. Lu, H. Fan, L. Yang and Y. Wang, *Biomaterials*, 2022, **290**, 121849.
58. K. Wu, W. Hua, X. Li and J. Lin, *Chem. Eng. J.*, 2022, **446**, 136835.
59. M. Rezvani, *BioChem.*, 2024, **4**(3), 189.
60. K. X. Vazquez-Prada, S. S. Moonshi, Z. P. Xu and H. T. Ta, *Appl. Mater. Today*, 2023, **35**, 101967.
61. A. Refaat, B. del Rosal, V. Bongcaron, A. P. Walsh, G. Pietersz, K. Peter, S. E. Moulton and X. Wang, *Adv. Funct. Mater.*, 2023, **33**(4), 2209019.
62. N. Düzgüneş, J. Piskorz, P. Skupin-Mrugalska, T. Goslinski, J. Mielcarek and K. Konopka, *Ther. Delivery*, 2018, **9**(11), 823.
63. F. Zhang, Y. Liu, J. Lei, S. Wang, X. Ji, H. Liu and Q. Yang, *Adv. Sci.*, 2019, **6**(17), 1901378.
64. Y. Xu, Z. Dong, R. Zhang, Z. Wang, Y. Shi, M. Liu, J. Yang, T. Yang, R. Zhang, T. Wang and J. Zhang, *Free Radical Biol. Med.*, 2023, **195**, 36.
65. L. Jiang, J. Wang, J. Jiang, C. Zhang, M. Zhao, Z. Chen, N. Wang, D. Hu, X. Liu, H. Peng and M. Lian, *Eur. J. Pharm. Biopharm.*, 2020, **146**, 101.
66. Y. Tu, Y. Sun, Y. Fan, Z. Cheng and B. Yu, *Cell. Physiol. Biochem.*, 2018, **48**(4), 1401.
67. T. Muñoz-Ortiz, J. Hu, F. Sanz-Rodríguez, D. H. Ortgies, D. Jaque, D. Méndez-González, R. Aguilar, F. Alfonso, F. Rivero, E. M. Rodríguez and J. G. Solé, *Nanomedicine*, 2022, **43**, 102556.
68. A. B. Attia, G. Balasundaram, M. Moothanchery, U. S. Dinish, R. Bi, V. Ntziachristos and M. Olivo, *Photoacoustics*, 2019, **16**, 100144.
69. Y. Liu, T. Hanley, H. Chen, S. R. Long, S. S. Gambhir, Z. Cheng, J. C. Wu, G. E. Fakhri, B. Anvari and R. T. Zaman, *Sci. Rep.*, 2020, **10**(1), 5983.
70. H. T. Ta, N. Arndt, Y. Wu, H. J. Lim, S. Landeen, R. Zhang, D. Kamato, P. J. Little, A. K. Whittaker and Z. P. Xu, *Nanoscale*, 2018, **10**(31), 15103.

71. J. Wang, M. Wu, J. Chang, L. Li, Q. Guo, J. Hao, Q. Peng, B. Zhang, X. Zhang and X. Li, *Nanomedicine*, 2019, **19**, 81.
72. I. Cuadrado, M. J. Piedras, I. Herruzo, M. del Carmen Turpin, B. Castejón, P. Reventun, A. Martin, M. Saura, J. L. Zamorano and C. Zaragoza, *Theranostics*, 2016, **6**(4), 545.
73. D. Sultan, W. Li, L. Detering, G. S. Heo, H. P. Luehmann, D. Kreisel and Y. Liu, *Nanomedicine*, 2021, **36**, 102416.
74. P. Supraja, V. Sudarshan, S. Tripathy, A. Agrawal and S. G. Singh, *Anal. Methods*, 2019, **11**(6), 744–751.
75. W. A. El-Said, D. M. Fouad and S. A. El-Safty, *Sens. Actuators, B*, 2016, **228**, 401.
76. M. Adeel, M. M. Rahman and J. J. Lee, *Biosens. Bioelectron.*, 2019, **126**, 143.
77. C. Wang, J. Li, M. Kang, X. Huang, Y. Liu, N. Zhou and Z. Zhang, *Anal. Chim. Acta*, 2021, **1141**, 110–119.
78. L. Mi, Y. Sun, L. Shi and T. Li, *ACS Appl. Mater. Interfaces*, 2020, **12**(7), 7879.
79. Y. Sun, L. Shi, Q. Wang, L. Mi and T. Li, *Anal. Chem.*, 2019, **91**(5), 3652.
80. J. Yao, Z. Yang, L. Huang, C. Yang, J. Wang, Y. Cao, L. Hao, L. Zhang, J. Zhang, P. Li and Z. Wang, *Adv. Sci.*, 2021, **8**(19), 2100850.
81. B. Banik, B. Surnar, B. W. Askins, M. Banerjee and S. Dhar, *ACS Appl. Mater. Interfaces*, 2020, **12**(6), 6852.
82. J. Chen, J. Yang, R. Liu, C. Qiao, Z. Lu, Y. Shi, Z. Fan, Z. Zhang and X. Zhang, *Theranostics*, 2017, **7**(17), 4149.
83. C. Kang, S. Gwon, C. Song, P. M. Kang, S. C. Park, J. Jeon, D. W. Hwang and D. Lee, *ACS Nano*, 2017, **11**(6), 6194.

CHAPTER 2

Photoluminescent Nanomaterials for Cardiovascular Diagnosis and Therapeutics

SÍLVIA RODRIGUES GAVINHO[a], SURESH KUMAR JAKKA[a], PAVANI KRISHNAPURAM[a], UJJAWAL SHARMA[b], VALLAMKONDU JAYALAKSHMI[c] AND MANUEL PEDRO FERNANDES GRAÇA*[a]

[a]i3N and Physics Department, Aveiro University, 3800-193 Aveiro, Portugal; [b]School of Health Science, Central University of Punjab, India; [c]National Institute of Technology Warangal, India
*E-mail: mpfg@ua.pt

2.1 Introduction

Cardiovascular disease (CVD; cardio means heart and vascular means blood vessels) is the designation given to the set of pathological changes that affect the heart and/or blood vessels, including arteries, veins, and capillaries. The report from the World Health Organization (WHO) revealed that, in 2019, CVDs were the main cause of death in the world. A study reported by Our World in Data showed that CVDs have been, with great prominence relative to other causes, the main reason for mortality worldwide in the last three decades (Figure 2.1).[1] The risk factors associated with this disease

Nanoscience & Nanotechnology Series No. 64
Cardiovascular Nanomedicine
Edited by Roger Narayan and Tanveer Tabish
© The Royal Society of Chemistry 2024
Published by the Royal Society of Chemistry, www.rsc.org

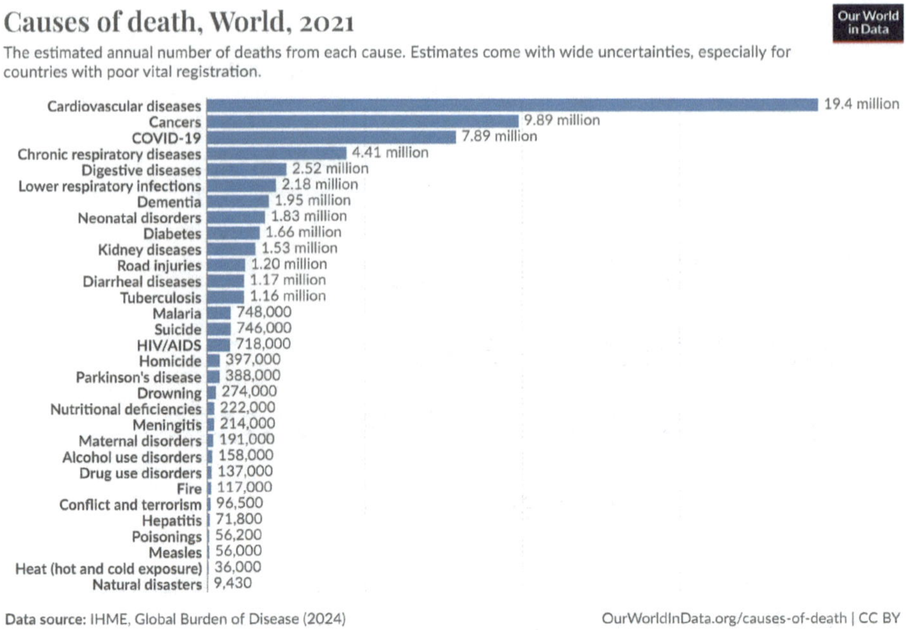

Figure 2.1 Graph of the main causes of death worldwide by Our World in Data report (2024). Reproduced from ref. 1, https://ourworldindata.org/causes-of-death, under the terms of the CC BY 4.0 license, https://creativecommons.org/licenses/by/4.0/legalcode.

include behavioral factors, such as alcohol consumption, smoking, an unhealthy diet, and lack of exercise. Studies have reported that, in 2019, 32% of global deaths were due to CVDs, and 85% of these cases were associated with heart attacks and strokes (WHO) (Figure 2.1).[2]

Heart attacks and strokes are mainly caused by the inability of the blood flow to reach the heart or brain. The reason for the blocked flow is the accumulation of fat in the inner walls of the blood vessels supplying these two organs (atherosclerosis). Atherosclerosis is a chronic, progressive inflammatory condition and is initiated by the retention of apolipoprotein B (Apo B), which is the primary apolipoprotein of low-density lipoprotein (LDL) responsible for carrying cholesterol to the tissues.[3,4] The presence of these lipoproteins of cholesterol and oxidized phospholipids activates the endothelial cells (ECs), which subsequently recruit monocytes. The recruitment of monocytes to the arterial wall is one of the first events in atherosclerosis (Figure 2.2b). Specifically, monocytes differentiate into macrophages, mediating the process of inflammation and innate immune response in atherosclerosis. Macrophages are the most abundant immune cells in atherosclerotic plaques and may adopt a pro-inflammatory phenotype (M1-like macrophages) or an anti-inflammatory phenotype (M2-like macrophages).[6–9] The M1-like macrophages amplify the inflammatory

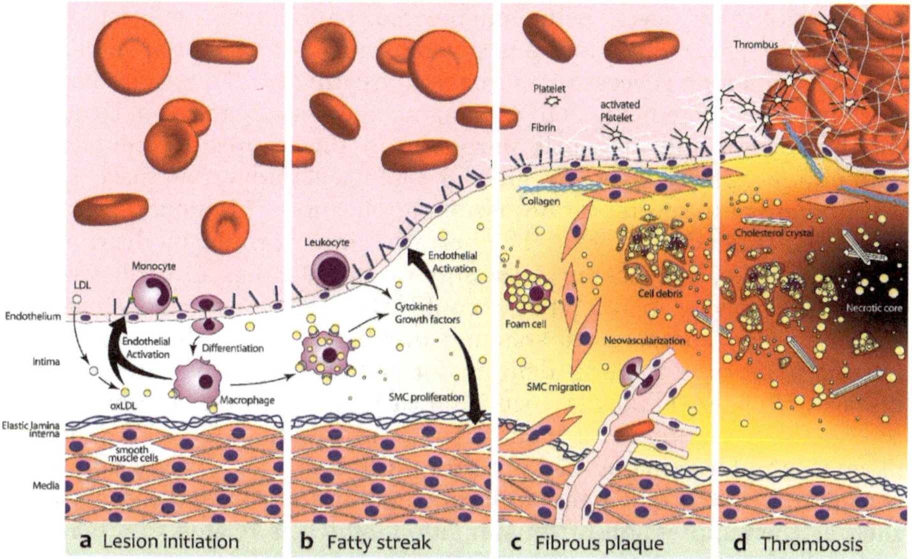

Figure 2.2 Disease stages and biomolecules involved in atherosclerosis. (a) Formation of OxLDL by the oxidation of low-density lipoprotein-cholesterol, stimulating the expression of adhesion molecules on endothelial cells and the monocyte recruitment. These monocytes in the intima are differentiated into macrophages; (b) the internalization of OxLDL by macrophages activates endothelial cells and recruits leukocytes; (c) the plaque increases in volume and vascularization, and there is agglomeration of platelets and fibrin in dysfunctional endothelium cells, forming a pre-thrombotic state.[5] Reproduced from ref. 5, https://doi.org/10.3390/ijms16059749, under the terms of the CC BY 4.0 license, https://creativecommons.org/licenses/by/4.0/.

response locally and increase the production of pro-inflammatory cytokines, reactive nitrogen species, and reactive oxygen species (ROS). The M2-like macrophages are able to moderate inflammation and stimulate tissue repair. In addition, macrophages also engage lipoproteins in the intima to form foam cells, promoting early atherosclerotic lesions (Figure 2.2b and c). The atherosclerotic lesions, in advanced stages, are characterized by increased apoptosis of macrophages and their consequent elimination. This sequence of events promotes necrosis of the atherosclerotic plaques, typical of vulnerable plaques. Vulnerable plaques are characterized by a thin fibrous cap and a large necrotic core and are therefore more prone to rupture. Plaque rupture exposes pro-thrombotic substances in the plaque, which cause platelet aggregation, leading to thrombus formation and arterial occlusion (Figure 2.2d).[10,11]

Cardiovascular diseases are non-transmissible diseases, so early detection, targeted therapy, and subsequent monitoring are quite important. Nanomedicine, which is a subset of nanotechnology, defined as the "monitoring, repair, construction, and control of human biological systems at the molecular level, using engineered nanodevices and nanostructures",[12] has presented numerous advantages to address these demanding requirements for a successful approach to treating cardiovascular diseases. The aim of nanomedicine is to use engineered devices and nanostructures for therapeutic and diagnosis purposes. In this context, nanomaterials, with unique physical, chemical, and biological properties, have shown great potential in several cardiovascular applications.[13] Nanoparticles (NPs) have been studied to improve the pharmacokinetic profile, biodistribution, cellular uptake, and chemical stability of stored drugs in cardiovascular diseases. Applying nanoparticles could also reduce adverse effects on healthy tissues compared to systemic-level drug release, thus overcoming the main barriers in drug delivery for the treatment of cardiovascular diseases.[3] Therefore, for therapeutic efficacy, nanoparticles involved in the process of circulation in the bloodstream; extravasation into atherosclerotic plaques; and targeting of macrophages, ROS, lipids, or proteins should be considered. The use of nanoparticles could address several mechanisms, such as the suppression of inflammation, which is related to atherosclerosis progression, and overexpression of some target receptors in macrophages for drug delivery or laser exposure therapeutics. In addition to the advantages of using nanoparticles to treat cardiovascular diseases, they can improve the delivery of imaging agents for the on-site diagnosis and monitoring of atherosclerotic vascular diseases.[6,14,15]

This bivalent application of nanoparticles in therapy and diagnosis on a single platform led John Funkhouser, in 2002, to define the concept as theranostics. For this purpose, scientists have searched for new nanomaterials, such as carbon-based, silica-based, metallic, and polymeric materials, with specific active properties, such as magnetic, electric, or optic. In this context, the nanomaterials that are reactive to light and capable of emitting it have been used, such as luminescent materials.[16]

The phenomenon of luminescence is characterized by the spontaneous absorption and emission of an electrically or vibrationally excited species. When the absorption/emission is through photons, it is called photoluminescence. The photoluminescence mechanism has been widely used in bioimaging applications, detection of biomolecules, drug delivery, and gene engineering, among others, due to the high sensitivity, the possibility of developing multi-systems with various colors or at various wavelengths that can measure various parameters, and non-invasive nature.[17]

Luminescent particles can be made from organic materials, such as lipids and carbon-based materials like graphene, and inorganic materials, such as lanthanides and quantum dots (QDs). Besides these naturally available materials, there is the possibility of using synthetic organic fluorescent dyes and artificial fluorescent proteins (FPs) as analytical tools.[18,19] The optical response of these materials exponentiate, and can be also modulated, with the decrease in the material size into the nanoscale region, resulting in a nanomaterial. A nanoparticle refers to a natural,

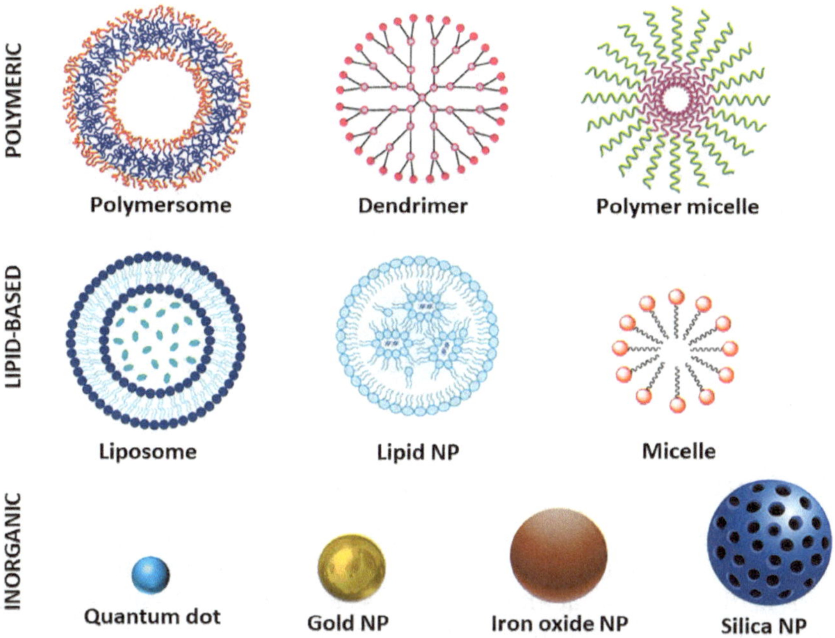

Figure 2.3 Types of nanoparticles.

incidental, or manufactured material comprising particles, either in an unbound state or as an aggregate, wherein one or more external dimensions are in the size range of 1–100 nm for ≥50% of the particles according to the number size distribution. In cases of health, environment, safety, or competitiveness concerns, the number size distribution threshold of 50% may be substituted by a threshold between 1% and 50%. Structures with one or more external dimensions below 1 nm should be considered as nanomaterials, and materials with a surface-area-to-volume ratio exceeding 60 m^2 cm^{-3} are also included according to the European Commission Recommendation 696/2011. Figure 2.3 shows the types of nanoparticles based on the structure types: organic (polymeric- and lipid-based) and inorganic.[20,21]

The nanomaterials present sizes in the order of magnitude of 10^{-9} m and can be divided into size/dimension categories. In ascending order, there are zero-dimensional (0D) particles, which include the quantum dots, upconversion nanoparticles, and fullerenes; one-dimensional (1D) particles, which include nanowires, nanorods, and nanotubes; two-dimensional (2D) particles, such as the nanoflakes or nanosheets, like graphene; and three-dimensional (3D) nanoparticles constituted by spatial arrays of 1D and/or 2D nanostructures or 3D nanoporous structures.[22] Several types of nanomaterials have already been intensively studied, such as nanocarriers of therapeutic agents and imaging contrast agents used in cancer treatment and diagnosis. Nanocarriers have several advantages over systemic administration, such as increased circulation time of the agents, good distribution in tumors, control over the kinetics of their distribution/release, and targeted delivery focusing on cancer cells. Furthermore, the physical properties and chemical compositions can be easily customized for a specific function. Drawing an analogy to the requirements for the treatment and diagnosis of atherosclerosis, the use of nanomaterials in this pathology is both possible and beneficial.[6]

The next section will discuss the several targets in CVDs and the use of luminescent nanoparticles for the early detection and monitoring of disease progression by real-time imaging. Furthermore, the use of luminescent nanoparticles (LNPs) in CVD treatment as drug delivery nanocarriers will be shown, which have improved drug efficacy, reduced side effects, and application in photodynamic or thermal therapies.

2.2 Cardiovascular Disease Targets

Due to the complexity of cardiovascular diseases, their diagnosis and treatment have been a challenge. Thus, it is important to define the targets of this disease at the level of tissues, cells, cell receptors, metabolic pathways, and their products, which may be useful for the administration and recognition of nanomaterials. This section will address the targets of cardiovascular diseases, specifically atherosclerosis, including its products resulting from inflammatory conditions (ROS), the enzymes present specifically in the immune response system, the accumulated lipids typical of this disease, and the receptors overexpressed in specific cells.

2.2.1 Reactive Oxygen Species

Oxidative stress is triggered by the imbalance of the oxidation–reduction processes or the uptake and accumulation of ROS, such as hydrogen peroxide (H_2O_2), O_2•, HClO, and $ONOO^-$. This biological condition is interconnected with the occurrence and progression of atherosclerosis by several mechanisms. One of the mechanisms is the oxidative modification of LDLs in the subendothelial space, resulting in the development of a type of cytotoxic LDL (OxLDL), which promotes endothelial dysfunction and the expression of adhesion molecules. The expression of these specific molecules allows monocytes to attach and migrate to the intima, where they differentiate into macrophages. Macrophages will express some scavenger receptors capable of internalizing OxLDL, forming foam cells. The control of monocytes in these early lesions is regulated by their exit and apoptosis. At a later stage of the inflammatory process, there is the production of chemokines, cytokines, proteases, and ROS by infiltrating immune system cells. The accumulation of these inflammatory mediators generates an inflammatory environment that promotes the recruitment and accumulation of more inflammatory cells and smooth muscle cells, promoting plaque expansion. At this stage, foam cell death also occurs, resulting in the formation of necrotic nuclei that further foment the inflammatory response. These processes can lead to plaque rupture and, consequently, thrombus formation.[23–25] The monitoring of oxidative stress, and hence atherosclerosis, can be regulated through the imbalance between H_2O_2 formation and its elimination mediated by glutathione (GSH). Thus, it is possible to control the spatial distribution and fluctuation of H_2O_2/GSH levels or specifically recognize GSH and H_2O_2 for the dynamic monitoring of the redox/reduction state to assess plaque vulnerability. With this known dynamic, it has already been possible, for example, to develop a multifunctional system for visual colorimetric detection of H_2O_2 *in vitro*, Surface-Enhanced Raman spectroscopy (SERS) imaging of H_2O_2 in living cells, and H_2O_2-activated magnetic resonance imaging (MRI) *in vivo*.[26]

Another reactive oxygen species that could be used as a biomarker in the diagnosis of atherosclerosis is O_2•. This ROS precursor, which mediates the expression of other active species, is quite abundant in vulnerable plaques generated by macrophages, endothelial cells, vascular smooth muscle cell (VSMC), *etc.*, which enhance the inflammatory response. It was already developed as an O_2-activated system for dynamic photoacoustic imaging detection, indicating oxidative stress levels in atherosclerotic plaques.[27] In addition to these, the overexpression of $ONOO^-$ in foam cells is also an important factor that is related to the ROS-controlled inflammatory process. In the case of the detection of $ONOO^-$ species, highly sensitive and selective fluorescence probes have been developed for the dynamic detection of $ONOO^-$ levels in foam cells. Involved in many mechanisms related to atherosclerosis, ROS shows potential as a biomarker for the diagnosis and monitoring of disease progression.[28]

2.2.2 Enzymes

Several enzymes are associated with pathogenic processes, as in the case of atherosclerosis, and may be an interesting target for applying therapies and detecting diseases. Myeloperoxidase (MPO) is an influential enzyme in the innate immune system, secreted by neutrophils and macrophages, and can influence the oxidation process of LDL.[29] MPO-modified LDL is unable to bind to the native LDL receptor and is identified by scavenger receptors on ECs and macrophages, promoting endothelial dysfunction and foam cell formation, respectively. The use of this enzyme as a biomarker for cardiovascular disease is justified by its role in the production of HClO, being the only enzyme capable of doing so. The presence of this enzyme and the detection of HClO levels at the site of the inflammatory condition encourage the development of fluorescent probes activated by the presence of HClO.[30,31]

In addition to their role in ROS production, these enzymes are also connected with macrophage metabolism. Macrophages show high levels of secretion of many proteinases, such as cathepsin and matrix metalloproteinases, in plaque regions, which leads to the degradation of the collagen-rich cap and increases the risk of plaque rupture. Near-infrared (NIR) fluorescent probes of cathepsin B were already developed to detect atherosclerotic plaques by using overexpressed receptors on the surface of macrophages.[32]

2.2.3 Lipoproteins

Lipoproteins are distinguished based on their density and size. LDLs and high-density lipoproteins (HDLs) bind to different proportions of triglycerides and cholesterol and are related to distinct apolipoprotein types. In a healthy organism, there is a process of lipid transport, metabolism, and storage by several types of lipoproteins. These lipoproteins are activated when glyceride/cholesterol levels increase in the blood, rebalancing lipid metabolism to normal circulating levels of these molecules. However, the action of these lipoproteins to keep lipid metabolism regulated may fail due to prolonged exposure to high levels of cholesterol and triglycerides, potentiating lipid accumulation in the atherosclerotic plaque and, thus, aggravating atherosclerosis.[10,33,34]

Targeting a large amount of lipid infiltration in vulnerable plaques is another path to combat the pathology by developing lipid-targeted imaging probes. On the other hand, by positively using lipid metabolism, it is possible to develop HDL-based systems due to their antioxidant and anti-inflammatory functions, allowing the removal of cholesterol from macrophages and thus inhibiting the formation of foam cells.[34–36]

2.2.4 Activated Cells

Considering all the stages associated with macrophages, namely proliferation, agglomeration, aging, and death, in the growth and regression of atherosclerotic lesions, these activated cells have been a critical strategy in the treatment of atherosclerotic cardiovascular diseases. In atherosclerotic plaques, macrophages overexpress on their surface signal, such as receptors transferrin receptor 1 (TfR1), folate beta receptor (FR-β), mannose receptor (MMR), transmembrane proteins (CD47, CD36, CD44, and CD68), and a collagen structure [macrophage receptor with collagenous structure (MARCO)]. Thus, by taking advantage of the overexpression of these biological targets, it is possible to use them for imaging detection of macrophages and for regression and stabilization in atherosclerotic plaques.[14,28]

Endothelial cells can control the substance exchange between blood and tissues, the blood flow, and the vascular tension by secreting various active substances. Dysfunctional ECs allow biomacromolecules, such as LDL, to move and accumulate in the vascular intima; secrete growth

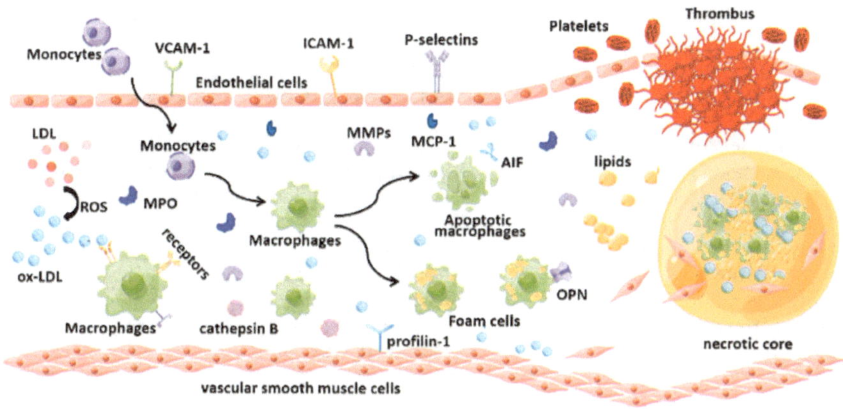

Figure 2.4 Biomarkers for the diagnosis and therapy of atherosclerosis.[28] Reproduced from ref. 28 with permission from American Chemical Society, Copyright 2022.

factors, adhesion molecules, and chemokines; and promote the occurrence and development of atherosclerosis. As potential targets for atherosclerosis detection by imaging, adhesion molecules overexpressed on the surface of these activated cells [VCAM-1, intercellular adhesion molecule-1 (ICAM-1), E-selectins, and P-selectins] are studied.[3,37]

VSMCs are another type of cell capable of acting as biomarkers. These cells are constituents of the arteries and vessels capable of regulating vasoconstriction and relaxation and maintaining blood pressure. These cells have an important role in the production of extracellular matrix molecules (interstitial collagen and elastin) and together develop a fibrous cap that lines the plaques. Overexpression of receptors on the surface of VSMCs, associated with cholesterol uptake, promotes abundant OxLDL uptake, forming foam cells. Overexpression of profilin-1 (Pfn1) present in VSMCs and ECs of atherosclerotic plaques indicates the dysfunction of this cell type and is a target for the early diagnosis of atherosclerosis.[38]

The macrophages and VSMCs phagocytize OxLDL and form foam cells, accelerating necrosis and vulnerable plaque formation. The glycoprotein, osteopontin (OPN), present on the surface of foam cells in atherosclerotic lesions is overexpressed and plays an important role in plaque formation, vascular remodeling, and luminal stenosis, presenting advantages in the highly specific recognition of these lesions.[39]

All these mechanisms and their biomarkers are depicted in Figure 2.4.

2.3 Photoluminescent – Physical Principles

Photoluminescence is a type of luminescence characterized by the emission of light from a substance that has not been heated to a temperature

Table 2.1 Different types of luminescence phenomena based on the mode of excitation.

Phenomenon	Mode of excitation
Photoluminescence	Photons in UV, visible, and NIR regions
Radioluminescence	Ionizing radiation (X-rays and α, β, and Y)
Cathodoluminescence	Cathode rays (electron beam)
Electroluminescence	Electric field
Thermoluminescence	Heating after radioactive irradiation
Chemiluminescence	Chemical reaction
Bioluminescence	Biochemical reaction
Triboluminescence	Friction and electrostatic forces
Sonoluminescence	Ultrasonic waves

that would cause incandescence. This emission can result from various processes, such as chemical reactions, electrical activation, radiation, and mechanical stress. When a material absorbs energy, some of that energy can be released as light, which is the basis of luminescence. Luminescence occurs in many different forms, including photoluminescence, bioluminescence, chemiluminescence, or electroluminescence, based on the mode of excitation of the luminescent material. It is a fundamental process in nature and has important applications in areas such as lighting, energy production, medicine, and materials science.[40,41] Table 2.1 classifies different types of luminescence based on the mode of excitation.

Photoluminescence, often referred to as PL, spectroscopy occurs when the light energy, or photons, stimulates the emission of a photon from any matter. The photons from the light source directed onto the sample are absorbed and undergo a process called photoexcitation. The photoexcitation causes the material to jump to a higher electronic state, and then, it de-excites by releasing energy (photons) back to a lower energy level. The emission of light or luminescence through this process is called PL.[17]

Photoluminescence is also categorized according to the different processes involved, namely fluorescence, phosphorescence, delayed fluorescence, and upconversion. These are well explained by the Jablonski diagram (Figure 2.5), proposed by Alexander Jablonski in 1935, through the representation of excitation and de-excitation of materials *via* simplified electronic energy-level diagrams.[42]

Jablonski diagrams are graphical representations of the energy transitions that occur within a molecule, or material, when it absorbs and emits light. They are useful for understanding the different kinds of photoluminescence, including fluorescence, phosphorescence, delayed fluorescence, upconversion, and persistent luminescence, which can be briefly described as follows:

- Fluorescence: In fluorescence, a molecule or material absorbs a photon of light (reaching an excited singlet state) and quickly returns to its ground singlet state by emitting a photon of light of lower energy. The emission occurs almost immediately, typically within nanoseconds of the absorption event. This process is shown in the Jablonski diagram (Figure 2.5a) as a rapid downward transition from the excited singlet state to the ground state. Fluorescence is a very common phenomenon and is widely used in biological imaging, chemical analysis, and materials science.[43]

- Phosphorescence: In phosphorescence, a molecule or material absorbs a photon of light and transitions to a long-lived excited state, where it can remain for microseconds to seconds before emitting a photon of light and returning to the ground state. This process is shown in the Jablonski diagram (Figure 2.5b) as a slower downward transition from the excited triplet state to the ground state. Phosphorescence is less common than fluorescence, but it has important applications in materials science, such as in organic light-emitting diodes (OLEDs) and sensors.[43]
- Delayed fluorescence: This is a type of photoluminescence that occurs when a molecule or material has a long-lived excited state but emits light after a delay. This delay can range from microseconds to milliseconds and is due to a specific type of energy transfer process. In the Jablonski diagram (Figure 2.5c), delayed fluorescence is shown as a slow upward transition from a long-lived triplet state to a higher-energy singlet state, followed by a rapid downward transition to the ground state.[44]

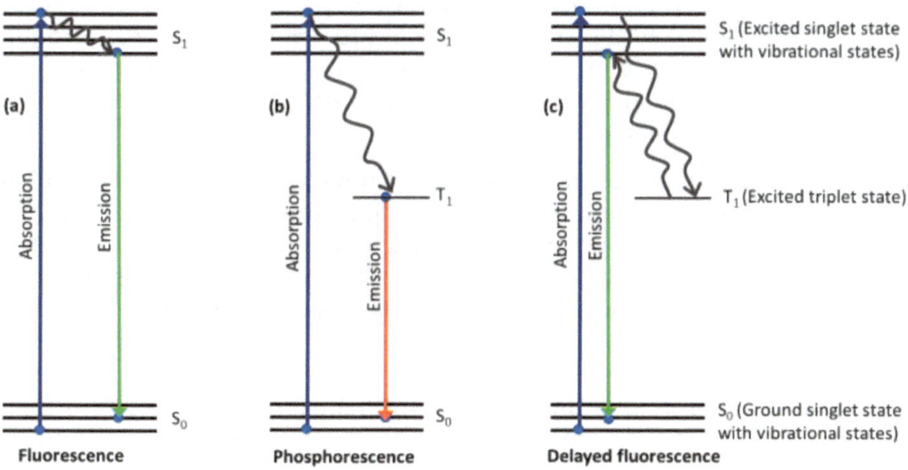

Figure 2.5 Jablonski diagrams depicting (a) fluorescence, (b) phosphorescence, and (c) delayed fluorescence, respectively.

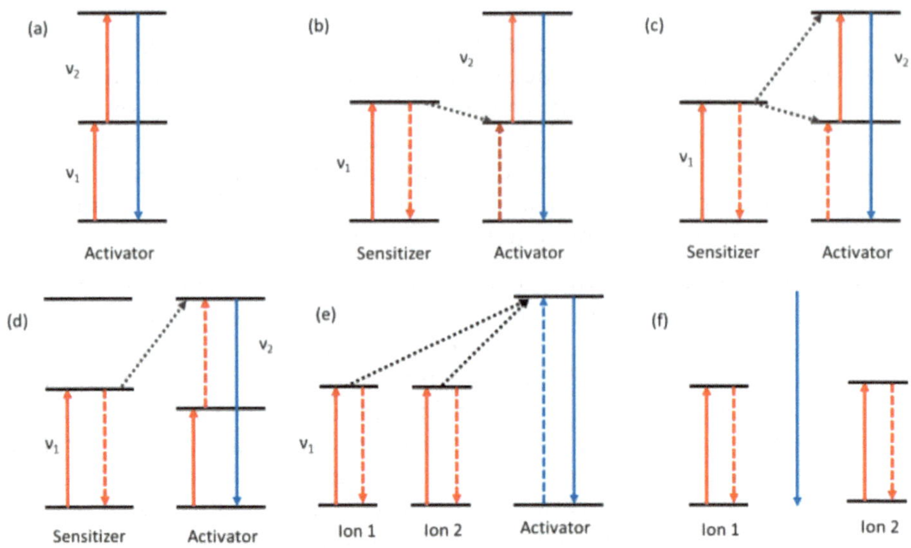

Figure 2.6 Illustrative representation of simplified upconversion processes, namely (a) excited state absorption, (b) energy transfer followed by excited state absorption, (c) successive energy transfer, (d) cross-relaxation upconversion, (e) cooperative sensitization, and (f) cooperative luminescence, respectively.

- Upconversion: This is a process that converts low-energy photons, such as infrared or near-infrared light, into higher-energy photons, such as visible or ultraviolet light. This process is useful for a variety of applications, including bioimaging, solar cells, and data storage. In the Jablonski diagram, upconversion is shown as a series of upward transitions from lower-energy states to higher-energy states, followed by a rapid downward transition to the ground state, emitting a higher-energy photon. There are different types of upconversion mechanisms that could be generalized in the categories depicted in Figure 2.6.[45]
- Persistent luminescence: This is another type of process in which the emission/luminescence is maintained/trapped in the material for minutes to hours without de-excitation after the removal of the excitation source. The luminescence is trapped by electron-trapping centers around the active ion or defects such as oxygen vacancies in the material.[46]

All the different photoluminescence mechanisms, along with their time correlations, *i.e.* the decrease in their intensities with time, are actively tested and applied to different cases, including imaging and sensing in nanomedicine for the diagnosis and therapy of CVD. Nowadays, one focus of nanomedicine has been on the development and characterization of nanomaterials capable of delivering agents to prevent, diagnose, monitor, and/or treat some diseases in a non-invasive way and with fewer associated risks. The next section is dedicated to the application of these optical phenomena in the treatment and diagnosis of CVDs.

2.4 CVD Diagnosis and Therapy Based on Photoluminescence Phenomenon

2.4.1 Upconversion

Nowadays, upconversion nanoparticles (UCNPs) conjugated with rare earth ions have been a good option for the treatment and therapy of atherosclerotic diseases due to their abilities to absorb NIR light and emit ultraviolet, visible light, and NIR light, depending on the amount and type of activator used. UCNPs show a narrow emission peak, trace autofluorescence, detection sensitivity, low noise relative to signal, large depth range *in vivo* tissues, light stability, long fluorescence times, adjustable color, and biocompatibility. These advantages make this particle type ideal for bioimaging and photodynamic therapy (PDT) applications. Moreover, because of their easy surface functionalization, they are quite useful in targeted drug delivery.[47]

As already mentioned in Section 1.4, macrophages are entirely related to atherosclerotic plaque formation, and their detection is essential for the prevention of plaque progression. Wang *et al.* developed NaGdF4:Yb,Er@NaGdF4 UCNPs by conjugating polyclonal MARCO antibody to its surface. This type of system is capable of dual-modality upconversion fluorescence imaging and weighted MRI. Thus, this research allowed obtaining *in vivo* upconversion fluorescence images, namely in atherosclerotic plaques, reaching the maximum fluorescence intensity at 1 hour, which decreased to the baseline at 24 hours. These UCNP probes can be further improved, distinguishing stable from vulnerable plaques, with the insertion of covalently binding OPN antibody, being a biomarker present in vulnerable plaques.[48–50]

Continuing to take advantage of the overexpression on the surface of foam cells of OPN and CD36 and their relationship with atherosclerotic plaque rupture, UCNPs of Yb and Tm as dopant ions and cerium fluoride as a matrix of luminescent materials (CeF_3:Yb,Tm@SiO_2-CD36/OPN) were developed. These UCNPs can be excited using near-infrared radiation with deep tissue penetration, while emitting radiation in the blue spectral region forms images, identifying and detecting vulnerable atherosclerotic plaques.[51]

UCNPs also have been studied to mediate PDT. Researchers have developed UCNPs by inserting Chlorin e6 (Ce6) as an ROS generator, inhibiting the inflammatory process by inducing macrophage apoptosis and cholesterol efflux. Photodynamic therapy has presented a new approach to the treatment of non-invasive cardiovascular diseases with minimal side effects. This therapy requires the presence of a specific light source to provide energy for a specific type of photodynamic reaction, photosensitizers that can be activated by this light to induce photodynamic reactions, and oxygen-containing substrates that produce ROS after electron transfer from the excited photosensitizer.[52] In contrast to other types of photosensitizers, Ce6 has advantages due to its activation by NIR light (980 nm), which penetrates deeper than visible light (660 nm).[53,54]

The results of these studies suggest that these nanoagents have a giant potential for use in selective imaging and photodynamic treatment in cardiovascular diseases due to their ability to target specific cells, high solubility, lack of toxicity to non-targeted cells, and strong phototoxicity. Furthermore, the ability to generate ROS selectively in targeted cells makes it an attractive option for the treatment of chronic inflammatory diseases, such as atherosclerosis.

2.4.2 Fluorescence

Unmodified polymer-based NPs, lipid-based NPs, micelles, and dendrimers do not induce any response to treatment and imaging. However, due to their easy modification by encapsulation of therapeutic agents, their functionalization to specific targets, the ability to release therapeutic agents at the precise site, activated by external mechanisms, and the addition of photoluminescent nanoparticles, they have been extensively studied in the therapy and diagnosis of cardiovascular diseases.

Biodegradable photoluminescent polymers (BPLPs) based on citrates have been developed for bioimaging and drug delivery due to their intense photoluminescence. The use of these polymers allows controlled degradation, unlike other contrast agents. Kuriakose *et al.* developed nanocarriers formed by polylactone, conjugated with another type of polymer [*e.g.* polylactic-*co*-glycolic acid (PLGA)], avoiding agglomeration and preserving the photoluminescent properties of BPLPs. One advantage is that the fluorescence emission region can be tuned by the amino acid composition. These drug delivery polymers have already been investigated for use in endothelial cells, both as imaging agents to track endothelial cell delivery and as vascular drug carriers to promote *in situ* endothelialization post-arterial damage.[55]

The targeted therapy can be performed using HDL NPs. As already mentioned, natural HDL consists of phospholipids and apolipoprotein A-I, which are capable of acting as anti-inflammatories, antioxidants, and anti-atherosclerotics due to their ability to remove cholesterol from foam cells. Considerable emphasis has been placed on synthesizing reconstituted HDL (rHDL), which mimics endogenous HDL, consists of apoA-I from human plasma and phospholipids, and is used for delivering drugs, antagonists, or genes for the treatment of atherosclerosis. This delivery is achieved by modifying these NPs with plaque-directed ligands, which carry therapeutic agents capable of interfering with specific pathways related to atherosclerosis. Nanocarriers based on those rHDL NPs were developed to deliver statins (a drug to reduce cholesterol levels) conjugated with a green-fluorescent dye (DiO) and were detected in macrophages.[56]

In the case of micelles, besides their advantages such as ease of synthesis, they are effective in encapsulating drugs, which increases bioavailability and reduces adverse effects, and they are biocompatible and allow controlled and prolonged release kinetics. In the presence of an aqueous medium, the micelle forms a sphere with a hydrophobic center, where it is possible to load drugs or other therapeutic agents through physical encapsulation, chemical bonding, and electrostatic interaction. In addition, this type of device, depending on its composition, can be controlled through its sensitivity to ROS, pH, and temperature.[7,57,58]

In a groundbreaking study, Hou *et al.* explored the potential of utilizing a unique method to harness the power of curcumin – a natural antioxidant – for the treatment of oxidative stress in macrophages. Through a process of self-assembly, the team incorporated curcumin into a specialized compound, oligomeric hyaluronic acid-2′-[propane-2,2-diyllbls (thio)] diacetic acid-hydroxymethylferrocene (oHATKL-Fc), to form nanomicelles known as HASF@Cur. The results of the study were highly promising, as the team observed that HASF@Cur effectively aggregated within macrophages and significantly reduced the level of ROS compared to encapsulated curcumin. *In vivo* studies, on atherosclerotic rat models, revealed that HASF@Cur led to fewer lesions than free curcumin. This highlights the potential of using these nanoparticles as both a therapeutic agent and an imaging substance. The curcumin within the nanoparticles could be easily detected by fluorescence imaging, making it an ideal candidate for future research. However, the precise mechanism of curcumin's action on macrophages remains to be fully understood and requires further studies to fully understand its potential implications in the treatment of chronic inflammatory diseases, such as atherosclerosis.[59]

Beyond the micelles, polyamidoamine dendrimers (PAMAM) have been studied due to their ease of modification, biocompatibility, lack of immune response triggering, and ability to encapsulate or attach molecules in the internal apolar cavity or on the surface of the dendrimer, respectively, thereby stabilizing drugs (Figure 2.7). Studies using PAMAM have been developed to prevent thrombosis. It has been mentioned that in more advanced stages of atherosclerosis, thrombus formation can obstruct blood vessels or release and enter the blood circulation. Durán-Lara *et al.* conjugated arginine, which increases nitric acid production and regulates platelet aggregation, endothelial function, and intravascular thrombosis, and showed that PAMAM generation 4 (G4)-arginine-Tos presents hemocompatibility and antiplatelet/antithrombotic effect, increasing the

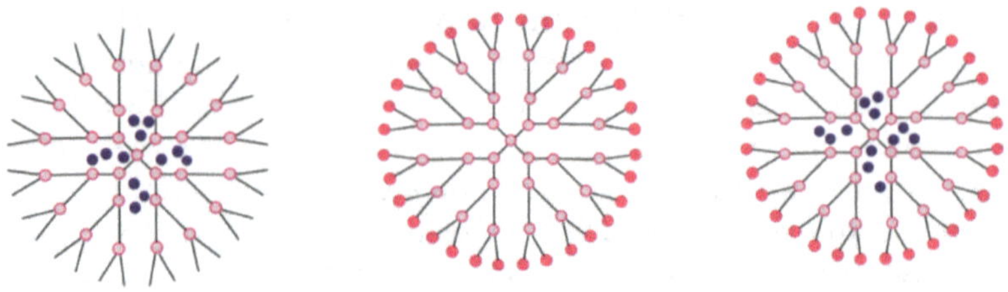

Figure 2.7 Representative scheme of complexing drug molecules or therapeutic agents with dendrimers.[58] Reproduced from ref. 58, https://doi.org/10.3390/molecules25173982, under the terms of the CC BY 4.0 license, https://creativecommons.org/licenses/by/4.0/.

concentration of the platelet response inhibitor cyclic adenosine monophosphate.[60]

Dendrimers, modified with polyethylene glycol (PEG), as imaging agents have shown advantages over traditional contrast agents. The encapsulation in dendrimers decreases renal toxicity and increases specific delivery and half-life. Furthermore, as mentioned earlier, the possibility of functionalizing the surface of dendrimers with multifunctional elements allows these devices to be more adaptable for more targeted molecular imaging. Functionalized dendrimers have assisted as contrast agents in magnetic resonance imaging and fluorescence imaging, and they can be applied to computed tomography if functionalized with Mn(II) or Gd(III) chelators, dyes, and Au nanoparticles or with iodinated compounds, respectively.[61,62]

Carbon nanotubes (CNTs) have shown very promising results in the prevention, diagnosis, and therapy of atherosclerosis through their modification and functionalization. CNTs have been used as highly sensitive biosensors to detect biological molecules and as nanocarriers to deliver drugs and bioactive genes. Researchers have demonstrated the application of single-walled carbon nanotubes (SWNTs) in the photothermal ablation of macrophages, prioritizing the ability of these structures to present strong optical absorbance on near-infrared fluorescence (NIRF).[63] Subsequently, this treatment was optimized with the functionalization of SWNTs with phenoxylate dextran to avoid the aggregation of SWNTs, targeting macrophages and increasing their anti-inflammatory effect (Figure 2.8).[65,66]

SWNTs have also been used in systems that control transmembrane protein overexpression, thereby preventing efferocytosis. Efferocytosis is a process that protects tissues from exposure to inflammatory content by inducing phagocytosis of macrophages or endothelial cells with compromised function through biomarkers. Conversely, these cells can overexpress these ligands, preventing their removal/phagocytosis and allowing

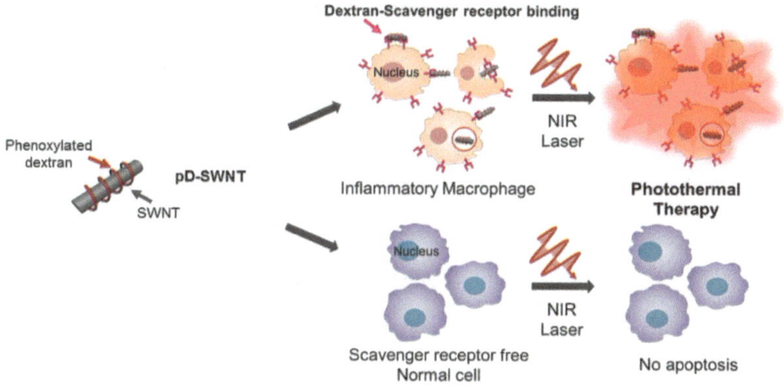

Figure 2.8 Schematic illustration of photothermal ablation by phenoxylated dextran-coated single-walled carbon nanotubes (pD-SWNT). Adapted from ref. 64 with permission from John Wiley & Sons, Copyright © 2016 WILEY-VCH Verlag GmbH & Co. KGaA, Weinheim.

the progression of the inflammatory process. One of these molecules involved in the progression of atherosclerosis is CD47. Therefore, to restore efferocytic activity, researchers developed a system called SWNT-SHP1i, conjugated with PEG-functionalized SWNTs, loaded with Cy5.5 (far-red/NIR fluorescent dye) for fluorescence and intrinsic NIR imaging, and incorporating a small-molecule inhibitor of CD47 in monocytes and macrophages. Thus, this treatment allows an unprecedented high degree of specificity (<3% uptake on any other circulating immune cell type due to the uptake of SWNTs and local excitation by laser light).[67]

The QDs, as fluorophores, have shown potential in the imaging of cardiovascular diseases, such as atherosclerosis, myocardial infarction, and thrombosis. These semiconductor nanoparticles with unique electronic and optical properties exhibit broad absorption spectra and a narrow emission band, which are useful for their application as a fluorescent probe. The sizes of the quantum dots range from 2 to 10 nm, and the control of their size and composition allows an easy adjustment in the emission spectra, ranging from ultraviolet to infrared (Figure 2.9a). Depending on their structure, quantum dots can be divided into three groups: core-type QDs, core–shell QDs, and alloying QDs (homogeneous and gradient) (Figure 2.9b). Core-type QDs are composed of a single heavy metal. Some of them are cytotoxic, and therefore, the development of biocompatible QDs is needed, such as carbon-based QDs and silicon or germanium QDs. To increase the photostability and photoluminescence efficiency of core-type QDs, core–shell QD systems were designed with a core coated with types II–VI IV–VI, and III–V semiconductors, such as ZnS. The alloying QDs consist of various core-type QDs of a uniform shape, presenting the same composition at all locations of the QD, or a gradient shape, in which the composition varies radially.[68,69]

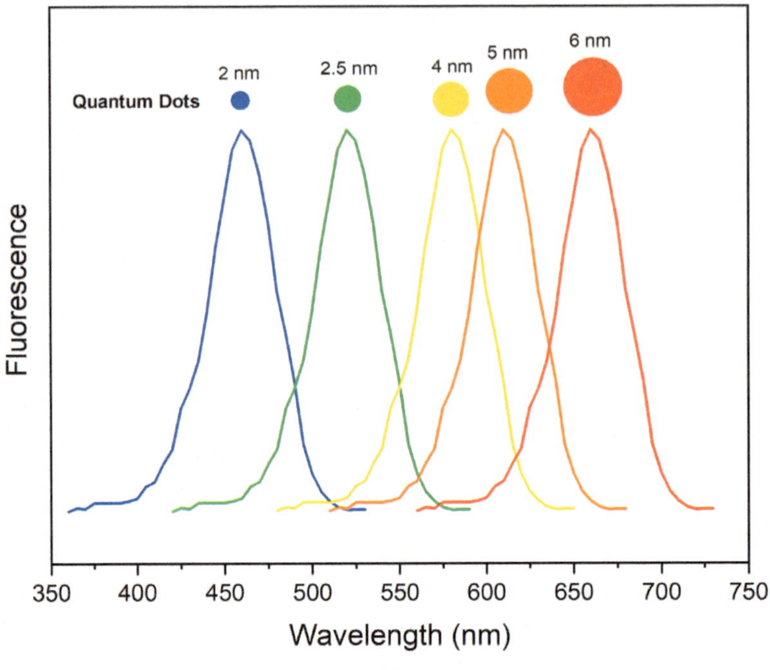

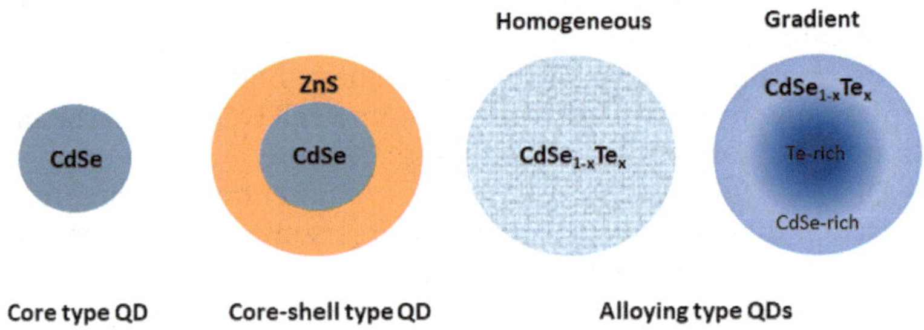

Figure 2.9 (a) Particle size-dependent emission spectra of fluorescent QDs and (b) different structures of QDs.

Researchers studied the fabrication of doped QDs (ZAISe/ZnS) with core–shell structures exploited for dual emission (visible excitonic emission and surface defect emission in the near-infrared). The ZAISe core ($Zn(Ac)_2$, Ag_2NO_3, $In(Ac)_3$, and Se) enhances the photoluminescence intensity of the QDs and ZnS enhances the surface defect emission. This system was encapsulated in bovine serum albumin and functionalized with macrophage-derived microvesicles to improve biocompatibility and further enhance the targeted imaging capability of the QDs, respectively. Furthermore, the biomimetic ZAISe/ZnS@BSA@MMV was successfully used for multilevel imaging in foam cells and atherosclerotic plaque, showing potential for an effective and safe treatment of atherosclerosis.[70]

Lanthanides have been used together with metal organics (Ln-MOFs) as nanoprobes due to the intrinsic photoluminescence of lanthanides and the unique properties of MOF materials, such as biocompatibility. Thus, these nanoprobes exhibit a sharp PL emission, long PL lifetime, large Stokes shift, high color purity, and high photostability.[71]

Nanoprobes based on these structures were developed by inserting Eu^{3+} into H4QPTCA ([1,1′:4′,1″:4′,1‴:4‴,1⁗-quinquephenyl]-2,2″,2⁗,5″-tetracarboxylic acid). The resulting system exhibited an intense red emission upon UV excitation, which was capable of detecting the activity of the cardiac biomarker creatine kinase (CK) through its interaction with ATP/ADP. The different luminescent responses to ATP and ADP make these systems promising for detecting CK, which acts as a switch for the transformation from ATP to ADP.[71,72]

Similar to organic particles, inorganic nanoparticles can also be functionalized with multiple molecules to target specific sites and produce highly specific multimodal images of atherosclerosis.[37] Gold nanoparticles (Au NPs) may be involved in several mechanisms that allow imaging and therapy through thermal ablation, energy absorption, and subsequent emission. Au NPs applied to imaging use localized surface plasmon resonance absorption and scattering, generating a powerful electromagnetic field (dark field microscopy, optical coherence tomography, and photoacoustic tomography), luminescence (photoluminescence imaging), and high electron density (X-ray imaging and computed tomography). In the thermal ablation treatment of cells (*e.g.* macrophages), Au NPs are able to absorb light energy and then convert it into thermal energy.

Lu *et al.*[10] developed gold nanorods for diagnostics using NIR and CT and for photothermal therapy. The *in vitro* and *in vivo* experimental results showed that, even at low concentrations of those nanorods and under low power, it was possible to kill macrophages and obtain quite satisfactory results in imaging.[11]

A study was also conducted to compare conventional detection by fluorophores with the use of Au NPs on silica systems functionalized with antibodies to ICAM-1, and Au NPs exhibited better detection compared with fluorophore–antibody systems. In addition, they also allowed an enhanced spectral definition and a higher resolution when applied to in-depth *in vitro*, *ex vivo*, and *in vivo* models.[3,68,73]

In nanomedicine, iron oxide nanoparticles (IONPs) are typically complexed with organic fluorochromes, such as Cy5.5 dye, rhodamine, and fluorescein isothiocyanate, where the core can be coated with silica, lipids, and polymers. These coatings will improve the properties of IONPs by preventing their aggregation, increasing hydrophilicity, reducing binding to plasma proteins, and improving binding to target ligands. Researchers developed magnetofluorescent IONPs, coated with dextran and conjugated with a photosensitizer capable of manifesting phototoxicity and fluorescence in the presence of light, which were detected by magnetic resonance imaging and near-infrared fluorescence, killing macrophages *in vitro*.[68,74]

In a study by Lu *et al.*, a new type of selenium nanoparticles (CAT/2nd-SeNPs) was developed as a potential treatment for pro-inflammatory stimulated macrophages. These nanoparticles were designed to suppress inflammation-induced H_2O_2 and NO production, which are both known to contribute to the progression of these diseases. One of the key innovations of these CAT/2nd-SeNPs is their incorporation of catalase, an enzyme that converts H_2O_2 into oxygen. This enhances the photodynamic efficiency of the nanoparticles, as the oxygen generated by the catalase can be used to inhibit the activation and proliferation of macrophages. The SeNPs are also effectively taken up by pro-inflammatory activated macrophages and release catalase upon intracellular reduction of disulfide bonds, further depleting H_2O_2 and promoting O_2 evolution. The use of these SeNPs for fluorescence imaging and photodynamic therapy in chronic inflammatory diseases is considered to be safe and effective by researchers. PDT is a treatment modality that uses a photosensitizer, which is preferentially accumulated in the targeted cells, and light to generate ROS that can kill the targeted cells. The SeNPs have been found to have good biocompatibility and low toxicity, making them suitable for *in vivo* applications. Overall, the development of these CAT/2nd-SeNPs represents a promising new approach for the treatment of chronic inflammatory diseases, such as rheumatoid arthritis and atherosclerosis. Further research will be necessary to fully understand the mechanisms of action of these nanoparticles and to optimize their therapeutic efficacy.[75]

2.4.3 Phosphorescence

Among the metal complexes used as activatable phosphorescence probes, cyclometalated iridium(III) complexes are the most widely used in luminescence assays in biological systems. In addition to their stability and long emission duration, among other advantages, the Ir(III) complex is responsive to the presence of ROS or other characteristic biomarkers in inflammatory processes. Scientific researchers have developed a unique Ir(III) complex phosphorescence probe for the highly sensitive detection of thiol-containing cysteine amino acid (Cys), where its abnormal intracellular levels are related to several diseases, namely cardiovascular diseases. This probe is composed of Ir(III)–phenylpyridine complex derivative, [Ir(ppy)2(NTY-bpy)](PF6) [ppy: 2-phenylpyridine; NTYbpy: 4-methyl-4′-(2-nitrovinyl)-2,2′-bipyridine], which attenuates its phosphorescence emission due to an intramolecular charge transfer process. In the presence of Cys, a reaction occurs forming the product [Ir(ppy)2(Cys-bpy)](PF6). With this specific reaction, a remarkable "off–on" phosphorescence response can be obtained for the detection of Cys. This reaction was visualized in inflammatory macrophage-mimicking cells, suggesting that [Ir(ppy)2(NTYbpy)](PF6) may be a promising probe for early diagnosis and treatment monitoring of inflammatory diseases.[76]

Other studies have developed an Ir(III) complex to sequentially recognize Cu^{2+} and Cys. In addition to Cys being associated with cardiovascular disease, Cu^{2+} deficiency can also indicate the presence of this type of disease. It is worth noting that Cys is a strong Cu^{2+} ligand and can remove Cu^{2+} from sensors. Ir(III) responded rapidly, was highly selective, showed an excellent sensitivity for Cu^{2+} detection, and served as an on–off-type probe. The generated Ir–Cu array was useful as an "off–on" phosphorescence probe for the specific detection of Cys.[77]

In addition to Cu^{2+} and Cys, there are other types of targets, such as intracellular O_2, associated with cardiovascular diseases, which can provide information at an early stage of the disease. Liu *et al.* have developed an O_2 probe based on 2-ureido-4-[1H] pyrimidinone (UPy) (as a quadruple hydrogen-bonded unit) functionalized with bromine-substituted naphthalimide derivative (BrNpA-UPy) (as an organic chromophore) and a bis-UPy-functionalized benzyl (Ph-bisUPy) (as a skeleton molecule). When present in an organic solvent, BrNpA-UPy emits fluorescence at room temperature, whereas they show phosphorescence and fluorescence simultaneously when dispersed in water. The room temperature phosphorescence and fluorescence of BrNpA-UPyA are due to the synergistic interaction of quadruple hydrogen bonds and the heavy atom effect of bromine. This dual emission from a single chromophore is obviously advantageous over conventional sensor designs with two chromophores, presenting advantages in terms of their stability, accuracy, reproducibility, and homogeneity.[78]

2.4.4 Persistent Luminescence

Persistent luminescent nanoparticles (PLNPs) are capable of storing energy under several excitation light sources and monitoring a long afterglow lifetime of hours or even days. Researchers have shown interest in persistent luminescent nanoparticles as promising optical materials in biomedical imaging and guided therapy due to their deep tissue penetration and lack of need for on-site excitation.

The presence of heavy metals, such as arsenic (As(III)), is commonly associated with several diseases, such as cardiovascular diseases. The World Health Organization defines the maximum levels for these types of materials, in the case of arsenic as 10 μg L^{-1}. Chen *et al.* developed a system based on persistent luminescence to detect As(III). This complex is formed by persistent luminescent nanoparticles $Zn_2GeO_4 : Mn$ (energy donor) linked to graphene oxide (efficient acceptor) (Figure 2.10). When these two elements are linked, the persistent green luminescence of those nanoparticles is completely quenched by the graphene oxide through a process of energy transfer by luminescence resonance. The introduction of

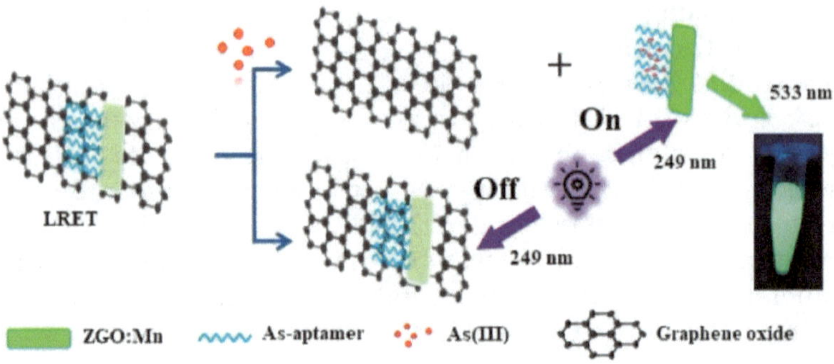

Figure 2.10 The detection mechanism of As(III) by persistent luminescence.[79] Reproduced from ref. 79, https://doi.org/10.3390/nano10030551, under the terms of the CC BY 4.0 license, https://creativecommons.org/licenses/by/4.0/.

As(III) interfered with the system and interrupted the energy transfer by recovering the persistent luminescence. Thus, it is possible to quantify the concentration of As(III) through the level of luminescence recovery.[79,80]

Another study found that it was possible to track macrophages using persistent luminescent nanoparticles with near-infrared luminescence emission for a long period and thereafter with red light-emitting diode (LED) excitation. The PLNPs were composed of $Zn_{1.1}Ga_{1.8}Ge_{0.1}O_4 : Cr^{3+}$, carboxyl-signaled, and conjugated with a transactivator of transcription (TAT) peptide (YGRKKRRRQRRRR-NH_2) capable of promoting the entry of the complex into macrophages while reducing cytotoxicity. These nanosystems can be used as a tool for *in vitro* and *in vivo* bioimaging without autofluorescence to track inflamed tissues.[81]

References

1. H. Ritchie, F. Spooner and M. Roser, Causes of death, Available from: https://ourworldindata.org/causes-of-death.
2. Cardiovascular diseases (CVDs)-Word Heatlh Organization, Available from: https://www.who.int/news-room/fact-sheets/detail/cardiovascular-diseases-(cvds).
3. K. A. Choi, J. H. Kim, K. Ryu and N. Kaushik, Current Nanomedicine for Targeted Vascular Disease Treatment: Trends and Perspectives, *Int. J. Mol. Sci.*, 2022, 12397.
4. N. A. Mohamed, I. Marei, S. Crovella and H. Abou-Saleh, Recent Developments in Nanomaterials-Based Drug Delivery and Upgrading Treatment of Cardiovascular Diseases, *Int. J. Mol. Sci.*, 2022, 1404.
5. D. C. Steinl and B. A. Kaufmann, Ultrasound Imaging for Risk Assessment in Atherosclerosis, *Int. J. Mol. Sci.*, 2015, **16**, 9749–9769.

6. J. Shi and I. Tabas, Macrophage-Targeted Nanomedicine for the Diagnosis and Treatment of Atherosclerosis, *Nat. Rev. Cardiol.*, 2022, **19**, DOI: 10.1038/s41569-021-00629-x.
7. J. Chen, X. Zhang, R. Millican, J. Sherwood, S. Martin, H. Jo, Y. S. Yoon, B. C. Brott and H. Jun, Recent Advances in Nanomaterials for Therapy and Diagnosis for Atherosclerosis, *Adv. Drug Delivery Rev.*, 2021, **170**, 142–199.
8. M. Batty, M. R. Bennett and E. Yu, The Role of Oxidative Stress in Atherosclerosis, *Cells*, 2022, 3843.
9. K. Malekmohammad, E. E. Bezsonov and M. Rafieian-Kopaei, Role of Lipid Accumulation and Inflammation in Atherosclerosis: Focus on Molecular and Cellular Mechanisms, *Front. Cardiovasc. Med.*, 2021, **8**, 1–16.
10. Y. Lu, X. Cui, L. Zhang, X. Wang, Y. Xu and Z. Qin, The Functional Role of Lipoproteins in Atherosclerosis: Novel Directions for Diagnosis and Targeting Therapy, *Aging Dis.*, 2022, **13**(2), 491–520.
11. P. P. Hu, S. X. Luo, X. Q. Fan and D. Li, Macrophage-Targeted Nanomedicine for the Diagnosis and Management of Atherosclerosis, *Front. Pharmacol.*, 2022, 1–10.
12. R. A. Freitas, The Future of Nanofabrication and Molecular Sscale Devices in Nanomedicine, *Stud. Health Technol. Inform.*, 2002, **80**, 45–59.
13. S. H. Nasr and X. Huang, Nanotechnology for Targeted Therapy of Atherosclerosis, *Front. Pharmacol.*, 2021, **12**, 1–12.
14. X. Zhang, F. Centurion, A. Misra, S. Patel and Z. Gu, Molecularly Targeted Nanomedicine Enabled by Inorganic Nanoparticles for Atherosclerosis Diagnosis and Treatment, *Adv. Drug Delivery Rev.*, 2023, **194**, 114709.
15. L. J. F. Peters, A. Jans and M. Bartneck, Immunomodulatory Nanomedicine for the Treatment of Atherosclerosis, *J. Clin. Med.*, 2021, **10**, 1–15.
16. R. Pala, S. Pattnaik, S. Busi and S. M. Nauli, Nanomaterials as Novel Cardiovascular Theranostics, *Pharmaceutics*, 2021, 348.
17. W. R. Algar, M. Massey, K. Rees, R. Higgins, K. D. Krause, G. H. Darwish, W. J. Peveler, Z. Xiao, H. Tsai, R. Gupta, K. Lix, M. V. Tran and H. Kim, Photoluminescent Nanoparticles for Chemical and Biological Analysis and Imaging, *Chem. Rev.*, 2021, **121**, 9243–9358.
18. S. N. Iyer, N. Behary, V. Nierstrasz, J. Guan and G. Chen, Study of Photoluminescence Property on Cellulosic Fabric Using Multifunctional Biomaterials Riboflavin and Its Derivative Flavin Mononucleotide, *Sci. Rep.*, 2019, 1–16.
19. R. Dong, Y. Li, W. Li, H. Zhang, Y. Liu, L. Ma, X. Wang and B. Lei, Recent developments in luminescent nanoparticles for plant imaging, *J. Rare Earths*, 2019, **37**, 903–915.

20. C. Ferraris, C. Rimicci, S. Garelli, E. Ugazio and L. Battaglia, Nanosystems in Cosmetic Products: A Brief Overview of Functional, Market, Regulatory and Safety Concerns, *Pharmaceutics*, 2021, **13**, 1–30.
21. M. J. Mitchell, M. M. Billingsley, R. M. Haley, M. E. Wechsler, N. A. Peppas and R. Langer, Engineering precision nanoparticles, *Nat. Rev. Drug Discovery.*, 2021, DOI: 10.1038/s41573-020-0090-8.
22. K. Yadav, P. Kumar, D. R. Teja, S. Chakraborty, M. Chakraborty, S. S. Mohapatra, A. Sahoo, M. M. C. Chou, C. Liang and D. Hang, A Review on Low-Dimensional Nanomaterials: Nanofabrication, Characterization and Applications, *Nanomaterials*, 2023, **13**, 1–44.
23. E. M. Espinoza, J. J. Røise, I. C. Li, R. Das and N. Murthy, Advances in Imaging Reactive Oxygen Species, *J. Nucl. Med.*, 2021, **62**, 457–461.
24. K. Gwozdzinski, A. Pieniazek and L. Gwozdzinski, Reactive Oxygen Species and Their Involvement in Red Blood Cell Damage in Chronic Kidney Disease, *Oxid. Med. Cell. Longevity*, 2021, **2021**, 1–19.
25. D. Moris, M. Spartalis, E. Spartalis, G. Karachaliou, I. Georgios, G. Tsourouflis, D. I. Tsilimigras, E. Tzatzaki and S. Theocharis, The Role of Reactive Oxygen Species in the Pathophysiology of Cardiovascular Diseases and the Clinical Significance of Myocardial Redox, *Ann. Transl. Med.*, 2017, **5**(2), 1–11.
26. R. J. Mailloux, C. Grayson and O. Koufos, Regulation of Mitochondrial Hydrogen Peroxide Availability by Protein S-Glutathionylation, *Cells*, 2023, 107.
27. L. Xu, B. Yin, J. Shang, F. Chen, J. Xu, Z. Song, B. Nan, G. Song and X. Zhang, Ratiometric Semiconducting Polymer Nanoparticle for Reliable Photoacoustic Imaging of Pneumonia-Induced Vulnerable Atherosclerotic Plaque in Vivo, *Nano Lett.*, 2021, **21**, 4484–4493.
28. J. Li, K. Wang, W. Pan, N. Li and B. Tang, Targeted Imaging in Atherosclerosis, *Anal. Chem.*, 2022, **94**, 12263–12273.
29. C. Frangie and J. Daher, Role of Myeloperoxidase in Inflammation and Atherosclerosis (Review), *Biomed. Rep.*, 2022, **16**, 1–11.
30. J. Zhou, R. Yang, Y. Sun, F. Luo, J. Zhang, H. Ma and M. Guan, HClO-Triggered Interventional Probe Enabled Early Detection and Intervention of Atherosclerosis, *Analyst*, 2022, **148**(1), 163–174.
31. B. Wang, F. Zhang, S. Wang, R. Yang, C. Chen and W. Zhao, Imaging Endogenous HClO in Atherosclerosis Using a Novel Fast-Response Fluorescence Probe, *Chem. Commun.*, 2020, **56**(17), 2598–2601.
32. E. R. H. Walter, S. M. Cooper, J. J. Boyle and N. J. Long, Enzyme-Activated Probes in Optical Imaging: A Focus on Atherosclerosis, *Dalton Trans.*, 2021, **50**(41), 14486–14497.
33. M. Chandarana, A. Curtis and C. Hoskins, The Use of Nanotechnology in Cardiovascular Disease, *Appl. Nanosci.*, 2018, **8**(7), 1607–1619.
34. L. J. Hassen, S. R. Scarfone and M. W. Milks, Lipid-Targeted Atherosclerotic Risk Reduction in Older Adults: A Review, *Geriatrics*, 2022, **7**(2), 1–16.

35. S. Bhargava, S. de la Puente-Secades, L. Schurgers and J. Jankowski, Lipids and Lipoproteins in Cardiovascular Diseases: A Classification, *Trends Endocrinol. Metab.*, 2022, **33**(6), 409–423.
36. E. Dayar and O. Pechanova, Targeted Strategy in Lipid-Lowering Therapy, *Biomedicines*, 2022, 1090.
37. T. Dai, W. He, C. Yao, X. Ma, W. Ren, Y. Mai and A. Wu, Applications of Inorganic Nanoparticles in the Diagnosis and Therapy of Atherosclerosis, *Biomater. Sci.*, 2020, **8**, 3784.
38. A. Allen, D. Gau and P. Roy, The Role of Profilin-1 in Cardiovascular Diseases,*J. Cell Sci.*, 2021, **134**, 1–11.
39. S. Li, T. Gou, Q. Wang, M. Chen, Z. Chen, M. Xu, Y. Wang, D. Han, R. Cao, J. Liu, P. Liang, Z. Dai and F. Cao, Ultrasound/Optical Dual-Modality Imaging for Evaluation of Vulnerable Atherosclerotic Plaques with Osteopontin Targeted Nanoparticles, *Macromol. Biosci.*, 2020, **20**(2), DOI: 10.1002/mabi.201900279.
40. K. E. Sapsford, L. Berti and I. L. Medintz, Materials for Fluorescence Resonance Energy Transfer Analysis: Beyond Traditional Donor–Acceptor Combinations, *Angew. Chem.*, 2006, **45**, 4562–4588.
41. D. Kong, K. Zhang, J. Tian, L. Yin and X. Sheng, Biocompatible and Biodegradable Light-Emitting Materials and Devices, *Adv. Mater. Technol.*, 2021, **2100006**, 1–17.
42. A. L. Miller, The Fates of Electronic Excitation Energy, *J. Chem. Educ.*, 1966, **43**(9), 469–473.
43. N. Siraj, B. El-Zahab, S. Hamdan, T. E. Karam, L. H. Haber, M. Li, S. O. Fakayode, S. Das, B. Valle, R. M. Strongin, G. Patonay, H. O. Sintim, G. A. Baker, A. Powe, M. Lowry, J. O. Karolin, C. D. Geddes and I. M. Warner, Fluorescence, Phosphorescence, and Chemiluminescence, *Anal. Chem.*, 2016, **88**, 170–202.
44. Y. Duan, L. Wen, Y. Gao, Y. Wu, L. Zhao, Y. Geng, G. Shan, M. Zhang and Z. Su, Fluorescence, Phosphorescence, or Delayed Fluorescence?—A Theoretical Exploration on the Reason Why a Series of Similar Organic Molecules Exhibit Different Luminescence Types, *J. Phys. Chem. C*, 2018, DOI: 10.1021/acs.jpcc.8b06533.
45. J. Zhou, Q. Liu, W. Feng, Y. Sun and F. Li, Upconversion Luminescent Materials: Advances and Applications, *Chem. Rev.*, 2015, **115**(1), 395–465.
46. L. L. Noto, H. C. Swart, B. M. Mothudi, P. S. Mbule and M. S. Dhlamini, The Dynamics of Luminescence, in *Luminescence - An Outlook on the Phenomena and their Applications*, InTech, 2016, vol. **11**, p. 13, https://doi.org/10.5772/65050.
47. G. Liang, H. Wang, H. Shi, H. Wang, M. Zhu, A. Jing and J. Li, Recent Progress in the Development of Upconversion Nanomaterials in Bioimaging and Disease Treatment,*J. Nanobiotechnol.*, 2020, **18**, 1–22.
48. R. Qiao, H. Qiao, Y. Zhang, Y. Wang, C. Chi, J. Tian, L. Zhang, F. Cao and M. Gao, Molecular Imaging of Vulnerable Atherosclerotic Plaques

in Vivo with Osteopontin-Specific Upconversion Nanoprobes, *ACS Nano*, 2017, **11**(2), 1816–1825.
49. Y. Wang, Y. Zhang, Z. Wang, J. Zhang, R. R. Qiao, M. Xu, N. Yang, L. Gao, H. Qiao, M. Gao and F. Cao, Optical/MRI Dual-Modality Imaging of M1 Macrophage Polarization in Atherosclerotic Plaque with MARCO-Targeted Upconversion Luminescence probe, *Biomaterials*, 2019, **219**, 119378.
50. Z. H. O. U. Bing-Shuai, X. U. Shi-Han, H. U. Song-Tao, S. U. N. Li-Heng, L. Y. U. Jie-Kai, S. U. N. Rui, L. I. U. Wei, B. A. I. Xue, X. U. Lin, W. A. N. G. Lin, H. A. N. Bing and D. O. N. G. Biao, Recent Progress of Upconversion Nanoparticles in the Treatment and Detection of Various Diseases, *Chin. J. Anal. Chem.*, 2022, **50**(2), 19–32.
51. Y. Cheng, H. Xu, W. Cao, W. Gao and B. Tang, Cerium Fluoride Nanoparticles as a Theranostic Material for Optical Imaging of Vulnerable Atherosclerosis Plaques,*J. Am. Ceram. Soc.*, 2022, 2375–2383.
52. Z. Zhou, J. Song, L. Nie and X. Chen, Reactive Oxygen Species Generating Systems Meeting Challenges of Photodynamic Cancer Therapy, *Chem. Soc. Rev.*, 2016, **45**(23), 6597–6626.
53. X. Han, Z. Zhong, J. Kou, Y. Zheng, Z. Liu, Y. Jiang, Z. Zhang, Z. Gao, L. Cong, Y. Tian and L. Yang, ROS Generated by Upconversion Nanoparticle-Mediated Photodynamic Therapy Induces Autophagy via PI3K/AKT/MTOR Signaling Pathway in M1 Peritoneal Macrophage, *Cell. Physiol. Biochem.*, 2018, **48**(4), 1616–1627.
54. X. Zhu, H. Wang, L. Zheng, Z. Zhong, X. Li, J. Zhao, J. Kou, Y. Jiang, X. Zheng, Z. Liu, H. Li, W. Cao, Y. Tian, Y. Wang and L. Yang, Upconversion Nanoparticle-Mediated Photodynamic Therapy Induces THP-1 Macrophage Apoptosis via ROS Bursts and Activation of the Mitochondrial Caspase Pathway, *Int. J. Nanomed.*, 2015, **10**, 3719–3736.
55. A. E. Kuriakose, N. Pandey, D. Shan, S. Banerjee and J. Yang, Characterization of Photoluminescent Polylactone-Based Nanoparticles for Their Applications in Cardiovascular Diseases, *Front. Bioeng. Biotechnol.*, 2019, **7**, 1–12.
56. J. Chen, X. Zhang, R. Millican, J. E. Creutzmann, S. Martin and H. W. Jun, High Density Lipoprotein Mimicking Nanoparticles for Atherosclerosis, *Nano Convergence*, 2020, **7**(1), 1–14.
57. M. Cheng, Q. Liu, W. Liu, F. Yuan, J. Feng and Y. Jin, Micelles for the Treatment and Diagnosis of Atherosclerosis, *J. Drug Delivery Sci. Technol.*, 2021, **63**, 102473.
58. A. A. Chis, C. Dobrea, C. Morgovan, A. M. Arseniu, L. L. Rus, A. Butuca, A. M. Juncan, M. Totan, A. L. Vonica-tincu, G. Cormos, A. C. Muntean, M. L. Muresan, F. G. Gligor and A. Frum, Applications and Limitations of Dendrimers in Biomedicine, *Molecules*, 2020, **25**, 3982.
59. X. Hou, H. Lin, X. Zhou, Z. Cheng, Y. Li, X. Liu, F. Zhao, Y. Zhu, P. Zhang and D. Chen, Novel Dual ROS-Sensitive and CD44 Receptor Targeting

Nanomicelles Based on Oligomeric Hyaluronic Acid for the Efficient Therapy of Atherosclerosis, *Carbohydr. Polym.*, 2020, **232**, 115787.
60. E. Durán-Lara, L. Guzmán, A. John, E. Fuentes, M. Alarcón, I. Palomo and L. S. Santos, PAMAM Dendrimer Derivatives as a Potential Drug for Antithrombotic Therapy, *Eur. J. Med. Chem.*, 2013, **69**, 601–608.
61. G. Anwaier, C. Chen, Y. Cao and R. Qi, A Review of Molecular Imaging of Atherosclerosis and the Potential Application of Dendrimer in Imaging of Plaque, *Int. J. Nanomed.*, 2017, **12**, 7681–7693.
62. L. Aranda-Lara, B. E. O. García, K. Isaac-Olivé, G. Ferro-Flores, L. Meléndez-Alafort and E. Morales-Avila, Drug Delivery Systems-Based Dendrimers and Polymer Micelles for Nuclear Diagnosis and Therapy, *Macromol. Biosci.*, 2021, **2000362**, 1–13.
63. H. Kosuge, S. P. Sherlock, T. Kitagawa, R. Dash and J. T. Robinson, Near Infrared Imaging and Photothermal Ablation of Vascular Inflammation Using Single-Walled Carbon Nanotubes, *J. Am. Heart Assoc.*, 2012, **1**, 1–10.
64. S. Han, T. Kwon, J. E. Um, S. Haam and W. J. Kim, Highly Selective Photothermal Therapy by a Phenoxylated-Dextran-Functionalized Smart Carbon Nanotube Platform, *Adv. Healthcare Mater.*, 2016, **5**(10), 1147–1156.
65. S. Han, T. Kwon, J. Um, S. Haam and W. Kim, Highly Selective Photothermal Therapy by a Phenoxylated-Dextran-Functionalized Smart Carbon Nanotube Platform, *Adv. Healthcare Mater.*, 2016, 1147–1156.
66. Y. Cao and Y. Luo, Pharmacological and Toxicological Aspects of Carbon Nanotubes (CNTs) to Vascular System: A Review, *Toxicol. Appl. Pharmacol.*, 2019, **385**, 114801.
67. A. M. Flores, N. Hosseini-Nassab, K. Jarr, J. Ye, X. Zhu, R. Wirka, A. L. Koh, P. Tsantilas, Y. Wang, V. Nanda, Y. Kojima, Y. Zeng, M. Lotfi, R. Sinclair, I. L. Weissman, E. Ingelsson, B. R. Smith and N. J. Leeper, Pro-Efferocytic Nanoparticles Are Specifically Taken up by Lesional Macrophages and Prevent Atherosclerosis, *Nat. Nanotechnol.*, 2020, **15**, 154–161.
68. J. Li, F. Centurion, R. Chen and Z. Gu, Intravascular Imaging of Atherosclerosis by Using Engineered Nanoparticles, *Biosensors*, 2023, **13**, 1–17.
69. Z. Liang, M. B. Khawar, J. Liang and H. Sun, Bio-Conjugated Quantum Dots for Cancer Research: Detection and Imaging, *Front. Oncol.*, 2021, **11**, 1–14.
70. Q. Yu, J. Li, X. Zhang, S. Yang, P. Zhou, J. Xia and T. Deng, Dual-Emission ZAISe/ZnS Quantum Dots for Multi-Level Bio-Imaging: Foam Cells and Atherosclerotic Plaque Imaging, *J. Colloid Interface Sci.*, 2023, **629**, 399–408.
71. X. Li, S. Zhou, S. Lu, D. Tu, W. Zheng, Y. Liu, R. Li and X. Chen, Lanthanide Metal–Organic Framework Nanoprobes for the In Vitro Detection of Cardiac Disease Markers, *ACS Appl. Mater. Interfaces*, 2019, **11**, 43989–43995.

72. Y. Cui, Y. Yue, G. Qian and B. Chen, Luminescent Functional Metal-Organic Frameworks, *Chem. Rev.*, 2012, 1126–1162.
73. R. McQueenie, R. Stevenson, R. Benson, N. MacRitchie, I. McInnes, P. Maffia, K. Faulds, D. Graham, J. Brewer and P. Garside, Detection of Inflammation in Vivo by Surface-Enhanced Raman Scattering Provides Higher Sensitivity Than Conventional Fluorescence Imaging, *Anal. Chem.*, 2012, 5968–5975.
74. E. Alphandéry, Iron Oxide Nanoparticles as Multimodal Imaging Tools, *RSC Adv.*, 2019, **9**, 40577–40587.
75. K.-Y. Lu, P.-Y. Lin, E.-Y. Chuang, C.-M. Shih, T.-M. Cheng, T.-Y. Lin, H.-W. Sung and F.-L. Mi, H2O2-Depleting and O2-Generating Selenium Nanoparticles for Fluorescence Imaging and Photodynamic Treatment of Proinflammatory-Activated Macrophages, *ACS Appl. Mater. Interfaces*, 2017, **9**(6), 5158–5172.
76. Z. Du, R. Zhang, B. Song, W. Zhang and Y. Wang, Iridium(III) Complex-Based Activatable Probe for Phosphorescent/Time-Gated Luminescent Sensing and Imaging of Cysteine in Mitochondria of Live Cells and Animals, *Chem. - Eur. J.*, 2019, **25**, 1498–1506.
77. P. Deng, Y. Pei, M. Liu, W. Song, M. Wang, F. Wang, C. Wu and L. Xu, A Rapid "on–off–on" Mitochondria-Targeted Phosphorescent Probe for Selective and Consecutive Detection of Cu^{2+} and Cysteine in Live Cells and Zebrafish, *RSC Adv.*, 2021, **11**, 7610–7620.
78. X. Liu, K. Zhang, J. Gao, Y. Chen, C. Tung and L. Wu, Monochromophore-Based Phosphorescence and Fluorescence from Pure Organic Assemblies for Ratiometric Hypoxia Detection, *Angew. Chem.*, 2020, **59**, 23662–23666.
79. T. Chen, H. Wang, Z. Wang and M. Tan, Construction of Time-Resolved Luminescence Nanoprobe and Its Application in As(III) Detection, *Nanomaterials*, 2020, **10**, 1–12.
80. Y. Feng, Y. Su, R. Liu and Y. Lv, Engineering Activatable Nanoprobes Based on Time-Resolved Luminescence for Chemo/Biosensing, *TrAC, Trends Anal. Chem.*, 2021, **140**, 1–12.
81. L. J. Chen, X. Zhao and X. Yan, Cell-Penetrating Peptide-Functionalized Persistent Luminescence Nanoparticles for Tracking J774A.1 Macrophages Homing to Inflamed Tissues, *ACS Appl. Mater. Interfaces*, 2019, **11**, 19894–19901.

CHAPTER 3

Nanoparticles for the Treatment of Ischemic Heart Diseases

NURA A. MOHAMED*[a], FATIHA BENSLIMANE[a] AND ZAIN ZAKI SALIM ZAKARIA[a,b]

[a]Biomedical Research Center, Qatar University, 2713 Doha, Qatar;
[b]Department of Biomedical Sciences, College of Health Sciences, QU-Health, Qatar University, 2713 Doha, Qatar
*E-mail: nura.adam@qu.edu.qa

3.1 Introduction to Ischemic Heart Diseases (IHDs)

Cardiovascular disease (CVD) is a term used to describe illnesses affecting the heart or blood vessels. The major subtypes of cardiovascular disease are stroke, aortic disease, coronary artery disease (CAD), including myocardial infarction and angina pectoris (AP), peripheral artery disease, and cardiomyopathies.[1] The most prevalent type of cardiovascular disease, known as IHD or coronary heart disease (CHD), is a leading global cause of death and disability.[2,3] IHD reportedly afflicted 126 million persons worldwide in 2017 (1655 per 100 000) or 1.72% of the world's population.[4] Worldwide, IHD was responsible for nine million fatalities. More males than females were affected, and the onset of symptoms often occurred in the fourth decade. The prevalence of IHD is increasing worldwide. By 2030, the current prevalence rate of 1655 cases for every 100 000 people is predicted

Nanoscience & Nanotechnology Series No. 64
Cardiovascular Nanomedicine
Edited by Roger Narayan and Tanveer Tabish
© The Royal Society of Chemistry 2024
Published by the Royal Society of Chemistry, www.rsc.org

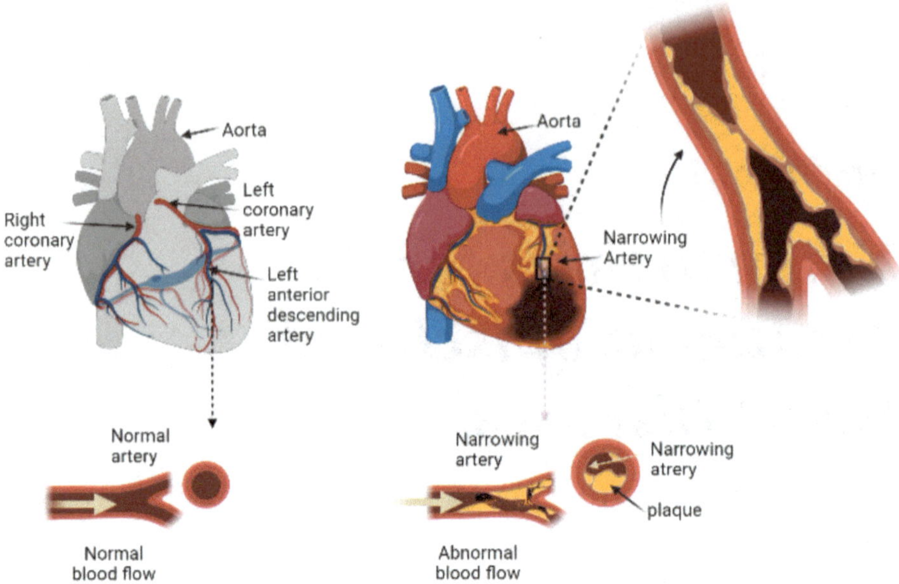

Figure 3.1 A simple diagram depicting the location of the coronary arteries within the heart and a cross-section demonstrating atherosclerosis within the artery wall.

to rise to 1845. IHD is caused by atherosclerosis (AS) of the coronary arteries. The gradual narrowing of coronary artery blood vessel walls by lipid-laden plaques eventually leads to symptoms of ischemia, a condition in which blood vessel narrowing, or atherothrombotic occlusion, causes a lack of oxygen and nutrient supply to the myocardium (heart muscle) and can eventually cause a heart attack[5] (Figure 3.1). IHD develops silently and may remain asymptomatic for many years or a person's entire lifetime. IHD typically manifests clinically as one of four syndromes ranging from moderate to severe: stable angina, unstable angina, myocardial infarction, and sudden cardiac death.[6]

3.1.1 Clinical Manifestations of IHD

Atherosclerosis is a progressive, chronic, immune-inflammatory, and fibroproliferative disease characterized by lipid accumulation, active cellular contact, inflammation, and matrix remodeling in the major arteries.[7] Complex lesions in coronary, cerebrovascular, and peripheral vascular disorders are caused by atherosclerosis.[8] The earliest stage of atherosclerosis is characterized by a fatty streak composed of cholesterol and macrophage. In the first decade of a person's life, they are only found in the aorta, but later, they spread to the coronary and peripheral arteries.

The lesion leads to acute blockage of the artery by forming a thrombus, typically due to intimal rupture or erosion; clinical manifestations of the lesion include unstable angina, myocardial infarction (MI), and sudden cardiac mortality.[9] Atherosclerosis lesion formation in stable coronary artery disease (SCAD) is a slow process, allowing for the emergence of collateral circulation.[9] Traditional cardiovascular risk factors, such as diabetes, hypertension, obesity, dyslipidemia, smoking, and heredity, can speed up atherosclerosis. Stroke symptoms include weakness or paralysis on one side of the body, difficulty in speaking or understanding speech, dizziness, and trouble walking.

Even though percutaneous coronary intervention (PCI) has significantly decreased MI-related mortality since its inception in 1977, MI is still the world's top cause of death. Revascularization with PCI restores blood flow to the ischemic myocardium, greatly extending the life of the heart muscle. Contrarily, PCI treatment frequently results in unanticipated muscle damage that is accompanied by fatal or shocking arrhythmias in the heart, a condition known as myocardial ischemia–reperfusion (I/R) injury.[10] Recent research suggests that myocardial I/R injury is mainly caused by the large generation of reactive oxygen species (ROS) that occurs when blood flow is restored to the ischemic myocardium.[11] It has been demonstrated that ROS play crucial roles in the advancement of certain clinical diseases, such as diabetes, hypertension, and hypercholesterolemia, that contribute to IHD.[10] However, ROS created during cellular metabolism have an important role as second messengers in cells and, at normal amounts, are required for cell growth, survival, and signaling. Moreover, ROS control normal cellular operations, such as gene transcription, signal transmission, and homeostasis,[12] and cardiovascular problems, such as endothelial dysfunction, atherosclerosis, and I/R injury, which occur from high ROS overpowering cellular antioxidant capacity.

The prevalence of risk factors, such as obesity, diabetes, and metabolic syndrome, as well as an aging population, points to a future of rising rates of IHD.[13] Global population aging has increased dramatically during the last two decades.[14] In fact, the United Nations predicts that by 2050, one in six people will be over the age of 65, up from one in 11 in 2019.[15] IHD in the current generation is also influenced by new social relationship problems, psychological distress, and sleeping for fewer than six hours each night.[16,17]

Clinical diagnosis is based on clinical signs of stroke and examinations with computed tomography and magnetic resonance imaging (MRI). The discovery of novel clinically relevant biomarkers could lead to earlier disease diagnosis, facilitating earlier treatment and improved prognosis, improved risk stratification and prediction, and personalized novel treatments (stratification of patients based on disease subgroups, leading to treatments with fewer side effects and improved prognosis).

3.1.2 Stages of IHD

3.1.2.1 Angina Pectoris

AP literally means "a strangling of the chest" in Greek. Patients usually describe it as a tight, heavy, or gripping chest pain that gets worse with activity and can spread to the jaw, neck, shoulder, or arms but goes away after a few minutes of rest. Angina usually happens when the narrowing caused by AS is severe enough (usually 75% stenosis of one or more coronary arteries) to cut off blood flow to the myocardium. Therefore, it usually happens when you work out. For the same reason, it can get worse when you are angry, sad, and cold. There may also be nausea, sweating, and trouble breathing.[18]

3.1.2.2 Myocardial Infarction

MI translates literally to "heart muscle death" in Greek. It is produced by an ischemic episode resulting from the full obstruction of one or more coronary arteries, lasting long enough to cause cardiomyocyte necrosis. The symptoms of myocardial infarction are identical to those of angina, except that they are more severe and occur at rest, *i.e.* abrupt, intense, and crushing chest pain (sometimes with referred pain in the neck, jaw, or arm), dyspnea, nausea, weakness, anxiety, and sweating.[19]

3.1.2.3 Congestive Heart Failure

Congestive heart failure (CHF) is a common disease that worsens over time and has a high death rate. CHF is a condition in which the heart muscle cannot pump enough blood to meet the needs of the body, or it can only do so at a higher pressure.[20] Chronic work overload (such as hypertension or valve disease) or chronic IHD after an MI and a significant level of ischemic damage can cause cardiac hypertrophy of the non-infarcted myocardium as a compensatory response to increased mechanical work, leading to CHF in the late stages of the disease. However, when IHD is severe enough, CHF can develop without a previous infarction or any obvious ischemic symptoms. In this case, heart failure is the initial symptom of IHD.[20]

3.1.3 Risk Factors in the Development and Progression of IHD

IHD is multifactorial, involving genetic, environmental, and lifestyle risk factors; clinicians and public health physicians need to have a thorough understanding of these risk factors because even a small lifestyle change can lower an individual's and the population's overall risk of IHD and help to lower cardiovascular morbidity and mortality. The age of onset and severity of IHD depend on these risk factors. Modifiable risk factors for IHD include blood pressure, cholesterol levels, and weight, whereas

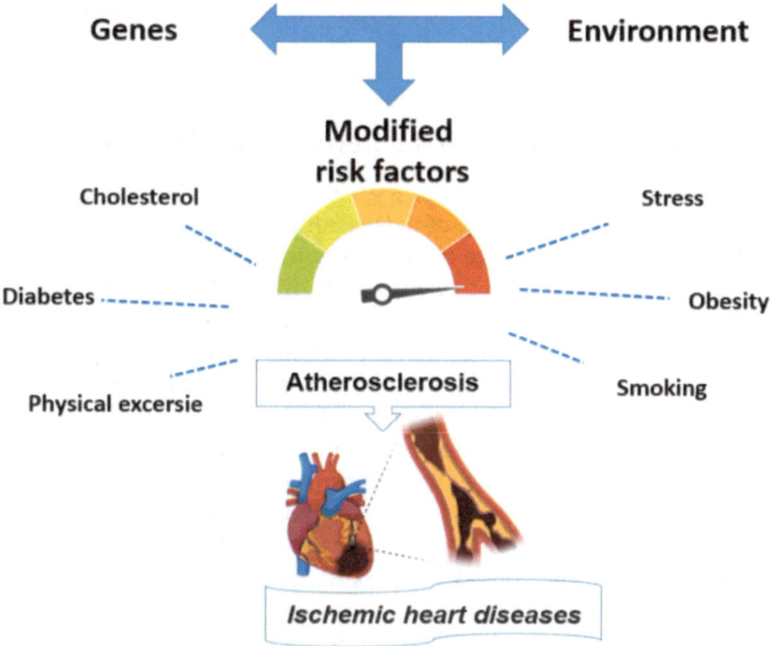

Figure 3.2 Diagram illustrating how hereditary, environmental, and modifiable risk factors interact to cause IHD.

non-modifiable risk factors include age, gender, and a family history of heart disease (Figure 3.2). Numerous randomized placebo-controlled clinical trials have supported the epidemiological data for many modifiable risk factors for IHD, including smoking, hypertension, and hypercholesterolemia. These risk factors have also been linked to a 30% to 40% lower incidence of clinical events related to IHD.[21] Individuals with many main risk factors for IHD are inclined to a cumulative effect. In the Framingham investigation, for instance, the risk of myocardial infarction was shown to be seven times higher for those with all three risk factors (hyperlipidemia, smoking, and hypertension) compared to those with none of these risk factors.[22]

The prevalence of hypertension, hypercholesterolemia, and diabetes among people globally is 32%, 39%, and 9.3%, respectively.[10] They have been recognized as harmful variables for developing IHD and are among the most frequent and independent risk factors that can enhance oxidative stress. Oxidative stress, in turn, plays a significant role in causing both microvascular and macrovascular problems.

Such vascular phenotypes are significantly tied to genetic background, despite being frequently associated with behavioral risk factors, such as smoking, a sedentary lifestyle, and poor food habits. Population and sibling studies have led researchers to conclude that genetics account for 40–60% of the IHD risk.[2,3]

3.1.4 Biomarkers in Stable Ischemic Heart Disease

Non-invasive methods for measuring genetic markers have made it possible to screen for a family history of IHD. Furthermore, genetic markers are stable throughout time, unlike circulating indicators, such as cholesterol or triglycerides. Better preventative measures might be possible with earlier screening (drug and lifestyle modifications), given that modifying risk variables before the onset of the disease can delay or prevent it.[23]

Many circulating biomarkers have been studied and proposed as possible strategies for risk classification in patients with stable IHD throughout the past two decades. Despite the fact that several of these markers also provide predictive information in general population cohorts, end-organ damage may play a more significant role in outcomes in the stable IHD context; this suggests that myocardial stress and damage measures are incredibly robust prognostic markers in patients with stable IHD.

Combined efforts from the scientific and clinical communities have led to the discovery of potential biomarkers of ischemia and the development of easy-to-use diagnostics. Similar analytical (limit of detection, accuracy, and reference intervals) and clinical performance (sensitivity, specificity, risk stratification, and predictive value) indicators of necrosis, such as high-sensitivity cardiac troponin assays, are required. Several potential biomarkers have been found. However, only a tiny percentage of assays leap a research grade to a fully licensed automated test for clinical use.

Improvement in MI and acute coronary syndrome (ACS) diagnosis was the primary focus of the initial success reports of blood-based biomarkers, giving clinical value for cardiovascular illness (Figure 3.3). In the 1950s, the first biomarkers for MI were discovered. Aspartate aminotransferase (AST) was known as glutamic oxaloacetic transaminase, which was the first biomarker, followed by lactate dehydrogenase (LDH).[25] These markers were found to be higher in acute myocardial infarction (AMI),[26] but they lacked cardiac specificity.[27] Therefore, the search was on for more accurate markers of myocardial damage. With the development of MI treatments, such as lytic therapy, it has become crucial to target the proper treatment to the right patient. Kott *et al.* thoroughly discuss the many new indicators that have evolved throughout time to represent heart failure and vascular inflammation.[24]

Blood-based biomarkers, and diagnostic tests in general, have played an essential role in cardiovascular disease for over 60 years, particularly in the diagnosis of AMI, where it was first shown that the enzymatic activity of serum aspartate aminotransferase was elevated in patients with AMI. Cardiac troponin has risen to prominence since the turn of 2000 and is

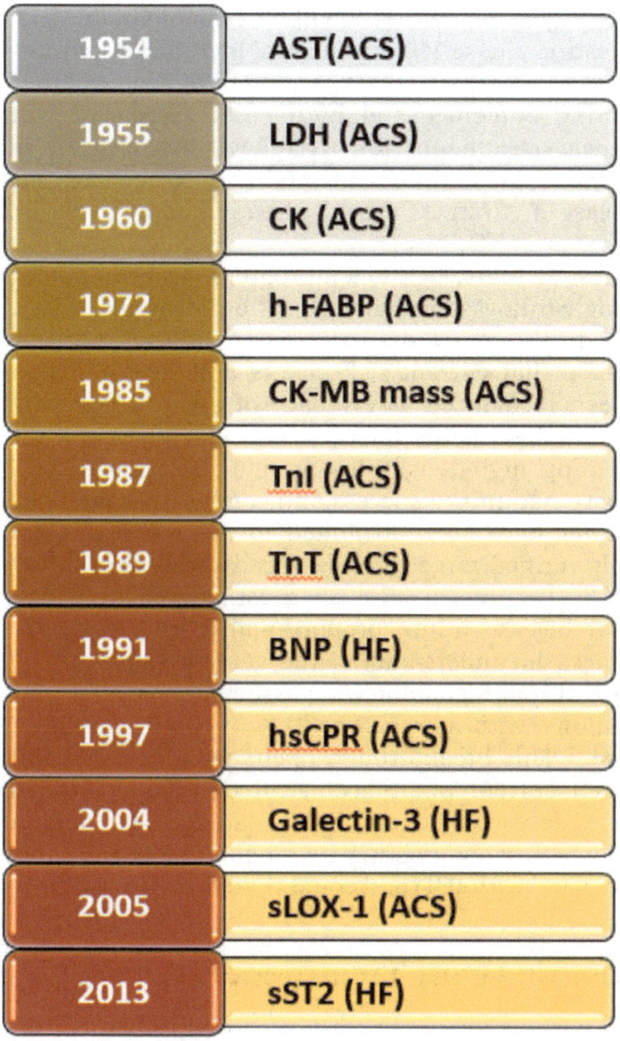

Figure 3.3 The evolution of cardiac biomarkers through time. AST = aspartate aminotransferase; ACS = acute coronary syndromes; LDH = lactate dehydrogenase; CK = creatine kinase; CK-MB = myocardial creatine kinase isoenzyme; h-FABP = heart-type fatty acid-binding protein; TnI = troponin I; TnT = troponin T; BNP = brain natriuretic peptide; hsCRP = high-sensitivity C-reactive protein; sLOX-1 = soluble lectin-like oxidized low-density lipoprotein receptor-1; sST2 = soluble suppression of tumorigenesis-2. Adapted from ref. 24, https://doi.org/10.3390/cells11030588, under the terms of the CC BY 4.0 license, https://creativecommons.org/licenses/by/4.0/.

now widely regarded as the best blood-based biomarker for detecting AMI. Troponin, creatine kinase-MB, myoglobin, BNP, and N-terminal-proBNP are just a few biomarkers used to diagnose and treat heart failure.[28]

Disappointing outcomes and high costs associated with developing pharmacological treatments for IHD have discouraged investment in innovative prevention and control methods. A new treatment's odds of success increase if it targets a gene already known to be involved in the disease.

Over the past 50 years, significant progress has been made in preventing and managing cardiovascular disease.[29] Cardiovascular disease is still the top cause of death worldwide. Patterns of disease and trends in cardiovascular disease that current risk factors and treatments cannot explain have long been recognized as evidence of an insufficient understanding of the disease.[29] Treatments for cardiovascular disease with targeted risk factors, including high-density lipoprotein-cholesterol and inflammatory indicators, have had mixed success and costly late-stage failures.[30–34] Some of these failures have been attributed to the medicines not functioning on genetically verified targets, as has been the case with varespladib and darapladib.[34,35] The present level of investment in drug development for cardiovascular disease is not proportional to the disease's prevalence.[36] Recent advances in understanding the genetic causes of IHD have been made possible through genome-wide association studies (GWAS) of single genetic variations, with as much as 21% of the heritability of IHD possibly explicable.[37] This breakthrough opens up a new possibility for evaluating how well current pharmacological treatments for IHD prevention and control make use of genetically validated targets and, conversely, for determining whether any existing treatments target the products of genes strongly associated with IHD and could therefore be repurposed.

3.2 Ischemic Heart Treatments

The target of current treatment for myocardial ischemia (MI) is to restore blood flow to the heart muscle. Depending on the patient's condition, doctors may prescribe medication, suggest surgery, or suggest a combination of the two. Overall, ischemic heart disease treatment involves medications that are antithrombotic and/or antianginal, reduction of cardiovascular risk factors, and, when necessary, revascularization.

3.2.1 Anticoagulant Therapy

People at a moderate to high risk of bleeding and vascular disease can benefit from antiplatelet medication, as it lowers their risk of major vascular events, such as myocardial infarction, stroke, and vascular death.[38,39] The risk of anticoagulants, however, is bleeding. That being said, the benefits of 75–100 mg per day of aspirin, which is considered as a low dose, far outweigh that risk in patients who have had a myocardial infarction or revascularization.[40,41] Primary prevention has weaker supporting evidence. Patients with imaging evidence of CAD who lack a history of myocardial infarction or PCI and those at a moderate risk of CAD should consider taking aspirin.[42] For patients with aspirin allergies, clopidogrel is indicated instead. Those who have had a stroke or who suffer from peripheral artery disease should take clopidogrel.[43] Patients with chronic coronary syndromes undergoing PCI should receive dual antiplatelet therapy with aspirin and clopidogrel 75 mg once daily. The length of the treatment varies depending on the symptoms. Several studies have shown that twelve months of dual antiplatelet therapy after PCI is not more beneficial than six months.[44–52] Therefore, it is recommended that patients at a high risk of bleeding should have a shorted dual antiplatelet therapy course.[53,54] A twelve-month dual antiplatelet therapy treatment is recommended post percutaneous coronary intervention in patients with ACS. For these patients, treatment with ticagrelor and prasugrel has been shown to be more effective compared to clopidogrel.[53,54] It is possible that those at a high risk of cardiovascular events could benefit from long-term, over one year, dual antiplatelet therapy or aspirin in combination with low-dose rivaroxaban, but this benefit comes at the cost of an increased risk of bleeding.[55–57] Patients at a high risk of cardiovascular events, but with a low risk of bleeding, may benefit from taking these combinations, but they are not standard treatment recommendations.

It is important to note that patients who have an indication for oral anticoagulants, such as those with atrial fibrillation or a mechanical heart valve, should avoid receiving antiplatelet medication after PCI because of the increased risk of bleeding. Because of the lack of data on the effects of combining anticoagulants with ticagrelor, prasugrel, or clopidogrel is recommended instead. The patient's risk of hemorrhage and cardiovascular disease must be carefully considered when determining the duration of combination therapy, whether double or triple. Although there was no significant difference in thrombolysis in myocardial infarction-related major bleeding, the ISAR-TRIPLE trial showed differences in the primary outcome between patients treated with six weeks of triple therapy compared to those treated for six months.[58]

3.2.2 Antianginal Therapy

Acute angina is typically treated with sublingual nitroglycerin, which has a rapid onset of action and rapid resolution of symptoms. Antianginal drugs include beta-blockers, calcium channel blockers, and long-acting nitrates, which are all used to treat or prevent angina.

Patients with symptoms and a reduced ejection fraction should take beta-blockers because they decrease mortality and morbidity.[59,60] For patients with chronic coronary syndromes in the absence of heart failure or preceding myocardial infarction, beta-blockers and calcium channel blockers did not show an improvement in patient outcome.[61,62] Despite the fact that no randomized trial has compared these treatments to other antianginal drugs,[63] the European and American standards recommend single or combination therapy of beta-blockers and calcium channel blocks as the first-line treatment for angina.[64] When angina persists despite treatment with beta-blockers and/or calcium channel blockers, current guidelines recommend adding several antianginal medications.[40,41,65]

Some examples of these are long-acting nitrates, such as ivabradine, ranolazine, trimetazidine, and nicorandil. No data suggest that these chemicals affect mortality or morbidity; at best, they could provide relief symptoms. Long-acting nitrates, trimetazidine and ranolazine, have been shown to not only reduce symptoms but also lengthen the time it is possible to be physically active before experiencing chest pain or angina.[66] Even though ivabradine has not been shown to reduce angina attacks, it has a similar effect.[67] Therapeutic options for angina include beta-blockers, calcium channel blockers, and long-acting nitrates for microvascular angina and calcium channel blockers and long-acting nitrates for vasospastic angina.[68,69]

3.2.3 Modification of Cardiovascular Risk Factors

Another treatment for ischemic heart disease is to illuminate, modify, or reduce the causes or symptoms that lead to the development of cardiovascular complications, including dyslipidemia and lifestyle changes.

The ability of statins to lower low-density lipoprotein (LDL) levels has a major bearing on the frequency and severity of cardiovascular problems and deaths. For every 1.0 mmol L^{-1} drop in LDL, regardless of the starting point, there is a 10% drop in all-cause mortality.[70] It has been recommended for many years that when the LDL level is between 1.8 and 3.5 mmol L^{-1} as a starting point, the LDL level should be lowered to 1.8 mmol L^{-1} or at least reduced by 50%. This recommendation comes from both European and American guidelines.[40,41,65] The LDL target for secondary prevention in people with ischemic heart disease was raised to 1.4 mmol L^{-1} in the European Society of Cardiology (ESC) guidelines for the management of dyslipidemias.[71] It is recommended to use high-dose statin as the initial treatment. Ezetimibe is indicated for use when the LDL target is not met, as it reduces the risk of cardiovascular death, coronary revascularization, nonfatal myocardial infarction, unstable angina that requires re-hospitalization, and nonfatal stroke. Additional statin treatment with a proprotein convertase subtilizing/kexin type 9 (PCSK9) inhibitor was associated with a significant reduction in LDL cholesterol (by 29–68%) and the primary composite endpoint risk (cardiovascular death, myocardial infarction, hospitalization for unstable angina pectoris, coronary revascularization, and stroke) by 15%.[72,73] Individuals at a very high risk for cardiovascular disease who fail to achieve LDL goals despite maximally tolerated statin and ezetimibe are still only recommended to take PCSK9 inhibitors.[74] Avoiding relapse by giving up cigarettes, for instance, is crucial. According to prospective cohort studies, patients with ischemic heart disease who quit smoking had a 36% lower risk of dying from their condition.[75]

3.2.4 Revascularization

Adjuvant to medical therapy, myocardial revascularization is an important component of chronic coronary syndrome care. When it comes to alleviating symptoms, slowing the development of atherosclerosis, and avoiding atherothrombotic events, the best medical treatment is essential. An improved prognosis and/or relief from angina symptoms are the two main motivations for revascularization in patients with angina.

Angina can be relieved, and the threat of significant acute cardiovascular complications, such as myocardial infarction and cardiovascular mortality, can be decreased by undergoing revascularization procedures such as PCI, which is minimally invasive, or more complicated procedure such as coronary artery bypass grafting (CABG). Thus, if symptoms and evidence indicate ischemia from functional testing or if diameter stenosis is above 90%, and fractional flow reserve (FFR) or instantaneous wave-free ratio (iFR) in the main vessel and left ventricular ejection fraction are below or equal to 0.8, 0.89 and 35%, respectively, due to CAD, revascularization should be considered.[76,77] Revascularization may be considered even in asymptomatic individuals if the same criteria are met, and functional testing demonstrates substantial areas of reversible ischemia.[78,79] Coronary pathophysiology based on coronary anatomy decides which revascularization technique will be used. Therefore, according to the 2019 ESC guidelines, guided revascularization based on FFR or iFR is considered a Class I recommendation in chronic coronary syndromes patients; however, if a diameter stenosis of over 90% is present, immediate revascularization should be considered. As both iFR and FFR are closely linked,[80] the risk of major cardiac complications following iFR-guided PCI is the same as that following FFR-guided PCI.[41]

Considering that both the FFR ratio and iFR are continuous,[81,82] there is no clear cutoff for when revascularization should be attempted. Which of the two thresholds, an FFR ratio of 0.75 or 0.80, is better is a matter of debate. For patients with an FFR ratio of 0.80, PCI reduced primary endpoints in randomized clinical trials and meta-analyses, most notably the need for urgent revascularization but not mortality.[83–86] Positive outcomes and a low event rate are associated with optimal medical care alone in patients with an FFR ratio of >0.80.[86–88] Delaying PCI in patients with an FFR ratio of 0.75 is also safe, according to other observational studies[89,90] and randomized trials,[91] although this finding is less conclusive.[89] It is also important to remember that the FFR and iFR readings lose precision as you get closer to the FFR or iFR cutoff values. The measurement certainty of an FFR value is greater than 95% outside of the range of an FFR ratio from 0.75 to 0.85, but it drops to less than 80% close to this cutoff.[92] If you are getting close to the edge, FFR and iFR can actually complement each other.[93] However, a thorough clinical evaluation is always necessary when deciding on a treatment strategy. Revascularization *via* CABG is preferable to PCI in patients with multi-vessel disease or left main CAD. However, contrary to the findings of some studies, which found no difference between PCI and CABG in terms of mortality and stroke, PCI appears to be inferior to CABG in patients with multi-vessel disease[94] or CAD of the left major coronary artery.[95] Revascularization is more likely to be necessary after PCI, according to studies.[96,97] The severity of coronary lesion complexity, as quantified by the Synergy Between PCI With Taxus and Cardiac Surgery (SYNTAX) score, which is an anatomical scoring system used to quantify the complexity of angiographic lesions, determines the outcome.

When comparing PCI and CABG in patients with low anatomical complexity, there was no statistically significant difference in the primary or secondary endpoints. The primary endpoints are major cardiac complications and cerebrovascular event (MACCE) with a SYNTAX score of 22, a complex endpoint that is defined as all mortality causes, stroke, MI, and repeated revascularization, while the secondary endpoints are death, MI, or revascularization. Patients with a SYNTAX score between 23 and 32 had an increased risk of MI and repeat revascularization in the PCI group but not of death or stroke. Higher rates of MACCE, cardiac mortality, and myocardial infarction after PCI in patients with a high SYNTAX score above 33 were seen compared to patients who underwent CABG.[98–100] Therefore, patients with multi-vessel disease or left main CAD and a low SYNTAX score below or equal to 22 may undergo revascularization *via* PCI or CABG, depending on the patient's other clinical parameters.[40,101] Selecting patients with moderate-to-high SYNTAX complexity for CABG is recommended.

Patients with reduced ejection fraction, below 35, due to obstructive CAD, benefit prognostically from revascularization.[77] Myocardial viability can be assessed with a fluorodeoxyglucose (FDG)-positron emission tomography (PET), cardiac magnetic resonance (CMR), and stress echocardiography. However, research has been unable to show that revascularization helps when myocardium is still viable.[102,103] The degree of myocardial viability is a critical factor in choosing patients who will benefit most from revascularization in terms of enhanced left ventricular function. In order to assess the severity of dysfunction, the extent of transmural affection, and the likelihood of restoring function in reversibly damaged areas, a battery of tests, including PET, CMR, and stress echocardiography, is often necessary.

Those suffering from ACS who have had ST-elevation myocardial infarction (STEMI) symptoms for more than 12 hours should undergo immediate revascularization of the offending lesion. Within 48 hours of a non-ST elevation myocardial infarction diagnosis, revascularization should be performed. No offending lesions must be treated during FFR-guided revascularization.[83,85,104–108]

3.2.5 Non-conversional Therapy

3.2.5.1 Protein Therapy

Several therapies have been proposed for the treatment of ischemic heart disease. However, the treatments are yet to be approved and validated for clinical treatment. For instance, animal models of ischemic heart disease demonstrated substantial improvements in collateral growth, blood perfusion, and cardiac function when treated with a protein-based angiotherapy strategy based on simultaneous administration of arteriogenic and angiogenic agents. Protein therapies based on heat shock protein 27 (HSP27) have shown promise in protecting cardiomyocytes from ischemic assaults. Recent research using cardiomyocytes and a rat model of the left anterior descending coronary artery ligation found that increased intracellular delivery of HSP27 *via* a transactivating transcriptional activator (TAT)-protein transduction domain (PTD) could retain cytoprotective function following ischemia insult that triggers cardiac cell death.[109] The goal of angiogenic growth factor infusion for the treatment of ischemic heart disease, which includes fibroblast growth factor, vascular endothelial growth factor (VEGF), and hepatocyte growth factor, is to induce neovascularization in the ischemic heart. Endothelial dysfunction and endothelial function deficiencies are well-known phenomena in ischemic heart disorders, and they share a similar risk factor: nitric oxide (NO)-associated oxidative stress. Recent investigations have shown that VEGF and the stimulation of NO generation are key factors in the development of coronary collaterals.[110] Studies in animals also demonstrated an increase in collateral blood flow in the ischemic myocardium after the delivery of VEGF and FGF proteins to the ischemic heart. Anti-apoptotic fusion proteins, such as TAT-FNK fusion protein, have been shown to have a protective effect in hepatocytes that have undergone programmed cell death and in hearts that have been damaged by ischemia and reperfusion.[111] Left ventricular function was temporarily restored, procaspase-3 cleavage was inhibited, and the number of terminal deoxynucleotidyl transferase dUTP nick end labeling (TUNEL)-stained cells was decreased after treatment with TAT-FNK, all of which point to a decrease in infarct size caused by the prevention of myocardial apoptosis in response to ischemia and reperfusion.[112]

3.2.5.2 Cell Therapy

Cell therapy is the practice of using donor cells to treat disease or restore function in a patient. Transdifferentiation of implanted cells into cardiomyocytes offers promising new avenues for improving heart function as part of cell therapy. Although preliminary research suggested that cell therapy could be able to restore cardiomyocytes, further research has been unable to build upon these findings. It appears that the mechanistic foundations of cardiac cell treatment are significantly more sophisticated. It has been hypothesized that the injected cells in cell therapy might stimulate vascularization by fusing with cells in the perivascular matrix or developing into new capillaries. In ischemic heart disease and heart failure, cell transplantation is seen as a novel and promising treatment approach to boost endogenous regenerative pathways. In the past decade, many different types of cells have been tried for cardiac repair. These include cells taken from bone marrow, skeletal myoblasts, embryonic stem cells, adipose tissue, and umbilical cords.[113]

3.2.5.3 Gene Therapy

Both viral and non-viral gene treatments have shown promise in cardiovascular medicine recently. Animal and human clinical trials of various catheter or surgical methods for *in vivo* cardiac gene transfer showed encouraging results. Transgene expression would be necessary only during times of high risk, such as cardiac remodeling following a myocardial infarction. Angiogenic gene therapy, such as angiogenic protein therapy, is garnering a lot of attention as a potential treatment for ischemic heart failure. Increased collateral blood flow and improved cardiac function following an injection of plasmid VEGF or FGF were seen in an animal model of chronic myocardial ischemia.[114,115] Clinical trials have focused heavily on angiogenic gene therapy. Human myocardial ischemia was successfully treated by injecting plasmid VEGF into the heart by a minor left anterior thoracotomy.[116] Different VEGF isoforms, administered via different routes, have been studied in clinical trials to enhance its expression level as it greatly varies.[117–119] Overall, transfection efficiency, stability, and long-term expression of therapeutic genes need to be enhanced before gene therapy may be used to treat ischemic heart disorders.

3.2.5.4 Polymeric Hydrogels for Delivery of Therapeutic Proteins

Although there have been significant advancements in the manufacture of diverse therapeutic proteins, it remains difficult to control enzymatic susceptibility, storage stability, and efficacy in the body.[120] Many studies have been described to overcome these restrictions and to prolong the biological activity of proteins in the body. Using delivery mechanisms to produce the regulated protein release, either systemically or locally, over time is one possible strategy. These delivery systems can keep protein medications in the body for a longer time by shielding them from breakdown by enzymes and neutralization by antibodies. Particular attention has been paid to polymeric hydrogels for their potential as a vehicle for the transport of peptides and proteins.[121,122] When it comes to water content, hydrogels are unparalleled. Chemical or physical crosslinking of polymers yields hydrogels, which are structurally analogous to the body's macromolecular-based components. Patients may experience less discomfort due to the minimally invasive nature of hydrogel injections.[123,124]

3.2.5.5 Nanomaterial-based Drug-targeted Therapy

Rapid progress has been made in recent years in the field of nanomedicine, which uses tailored biomaterials and advanced nanotechnologies as a supplement to traditional pharmaceuticals. It has been proven effective in improving early diagnostics, resulting in better patient prognosis as well as drug delivery.[125] Targeted drug delivery is set to eliminate or reduce the side effects and bypass the limitations, as summarized in Table 3.1. In this chapter, we will be covering in detail the use of nanoparticles in the treatment of ischemic heart disease.

3.3 Nanomedicine in the Treatment of Ischemic Heart Disease

The concept of targeted drug delivery after systemic administration implies the drug's selective accumulation within the affected tissues/cells only with minimum contact with irrelevant cells/tissues to avoid any possible side effects.[126] As an idea, targeted drug delivery was first proposed by Paul Ehrlich in 1906, who described it as finding the "magic bullet" that is aimed at a chosen cell alone, avoiding any damage to the other healthy cells and bringing solutions to the drug toxicity, solubility/stability, and insufficient bioavailability problems.[127] This magic bullet can be achieved through the use of the advantages nanomedicine field provides. Nanomedicine links advanced nanotechnologies and engineered biomaterials, generating carriers that can serve in delivering conventional drugs. Moreover, nanomedicine can be designed to have unique properties and perform certain functions that serve the needs of the desired disease. In doing so, nanomedicine uses tiny devices called nanoparticles that act as drug carriers; these nanoparticles can often be modified to (1) present adjustable shapes and sizes, (2) have modifiable electrical charges, (3) have a high surface-to-volume ratio, (4) have other physicochemical properties, and (5) present different drug pharmacokinetic and pharmacodynamic profiles[128] (Figure 3.4).

Additionally, nanoparticles can be decorated with multifunctional linkers or ligands to precisely deliver the loaded cargo to the desired cells and/or tissues, achieving targeted drug delivery. In addition, they can be coated with different materials to prolong their blood circulatory half-life.[129] A wide range of materials are used in synthesizing nanoparticles, including inorganic and organic materials, such as polymers, liposomes, micelles, carbon nanotubes, mesoporous silica, and metal–organic frameworks, with

Table 3.1 Ischemic heart disease treatment and its side effects and limitations.

Treatment	Drugs	Side effects and limitations
Anticoagulant therapy	Aspirin Rivaroxaban	• Allergies • Bleeding • Need for dual therapy for the effective relieving of anti-ischemic symptoms
Antianginal therapy	Nitrate	• Headache • Flushing • Palpitations • Hypotension • Reflex tachycardia • Syncope in rare cases associated with bradycardia • Nitrates are contraindicated in patients diagnosed with constrictive pericarditis, or severe aortic or mitral stenosis, or hypertrophic obstructive cardiomyopathy • Bioconversion of nitrates into nitric oxide results in endothelium-independent vasodilation • Small arterioles are nitrate resistant
	β-Blockers	• Bradycardia • Hypotension • Bronchospasms

(continued)

Table 3.1 (*continued*)

Treatment	Drugs	Side effects and limitations
		• Blunting of the tachycardia response to hypoglycemia • Need to be admitted as a dual therapy to enhance the anti-ischemic effect • Contradiction with blood pressure medications • Nonselective, and a few selective, β-blockers can cause coronary constriction as it shifts β-adrenoreceptors to α-adrenoreceptors • Pregnant ladies, those with low systolic blood pressure, Raynaud's syndrome, and diabetes all need to use with caution • β-Blockers are contraindicated in vasospastic angina and not effective in microvascular angina
	Calcium channel blockers	• Systemic hypotension • Headache • Dizziness • Flushing • Palpitations • Leg edema • Some calcium channel blockers might reduce left ventricular contractility

(*continued*)

Table 3.1 (*continued*)

Treatment	Drugs	Side effects and limitations
		• Decreases intestinal motility, leading to constipation • Patients with chronic stable angina, whether or not they have had a myocardial infarction or left ventricular dysfunction, may not benefit from its use, and those with heart failure may actually be harmed by it • Contradicted with drugs that are metabolized by cytochrome P450 3A4 to avoid the risk of severe bradycardia
Cardiovascular risk factor modification	Dyslipidemia drugs	• Regulates causative agents not necessary for cardiovascular symptoms • Needs to be administered with other ischemic heart medication for effectiveness
	Lifestyle changes	• Cannot be monitored • Patient-dependent; several patients do not adhere to the necessary life changes required to reduce the risk of cardiovascular event
Revascularization	Surgical intervention	• Surgical complications • Revascularization is needed on some occasions

(*continued*)

Table 3.1 (*continued*)

Treatment	Drugs	Side effects and limitations
Other therapies	Protein therapy	• Short half-life • For targeted effect, proteins need to be delivered directly to the affected area • Risk of diabetic retinopathy and cancer • Not enough data from the clinics
	Cell therapy	• Source of the cells • Integration of transplanted cells with host cardiomyocytes • Most available data are laboratory- and animal model-based
	Gene therapy	• Off-target effects • Achieving sufficient alteration of gene expression levels • Risk of cancer
	Polymeric hydrogels	• For targeted effect, polymeric hydrogels need to be delivered directly to the affected area • Allergies

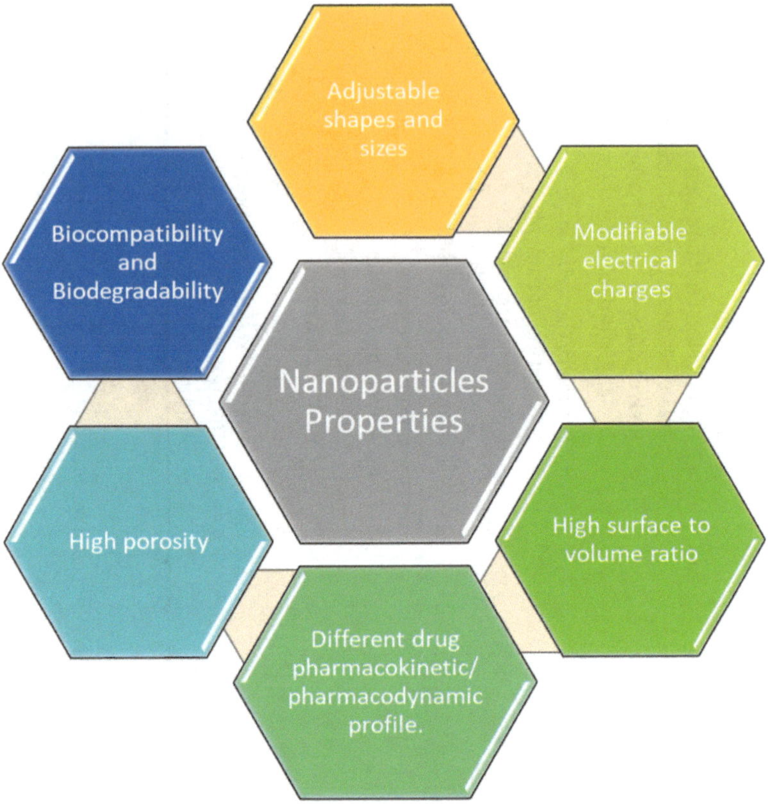

Figure 3.4 Summary of different attractive properties of nanoparticles.

each type having different attractive properties that could be used to improve the IHD treatment and detection strategies.[130,131]

In general, a site-targeted nanoparticle will have several functional elements, with the major part being the drug-loaded nanoparticle complex. Other elements include the functionalization of drug-loaded nanoparticles, which insinuates the binding and decoration of the nanoparticle's surface with a targeting ligand. Accordingly, the right ligand must be used and the ligand/nanoparticle interaction should be studied precisely. Furthermore, nanoparticles can also be tagged with radioactive isotopes or fluorescent dyes, enabling their visualization in the tissue of interest. Effective targeted drug delivery with the use of nanoparticulate carriers can bring solutions to problems, such as drug toxicity, insufficient bioavailability, poor solubility, and stability.[128] Apart from annexin V, there are no other means to target the ischemic-reperfused myocardium; therefore, we cannot rule out the benefits passive targeting can bring to this field.[129] Specifically, many nanoparticles have the ability to accumulate in the ischemic-reperfused area due to their increased microvascular permeability in these areas.[132]

3.3.1 History of Nanomedicine in IHD

Implementing nano-mediated drug delivery in IHD is important as it can help treat this highly vicious disease. Looking back, ischemic heart disease became a leading cause of death worldwide, with over 7.2 million deaths per year in 2004.[133] Since then, microscale instruments have been developed by biomedical engineers to open occluded arteries, alleviate the cardiac pump failure, and/or retain the normal cardiac rhythm. Unfortunately, tools developed back then were bulky and infection-prone, along with other limitations. These limitations inspired nanotechnologists to seek new tools, leading them to design cardiac devices that are 10^3–10^5 smaller than current ones.[134] Looking through PubMed, we can find that the studies on the use of nanomedicine in ischemic heart disease were not published until 2007. Research in this area then peaked in 2008, with a total of 249 results for the word ischemic heart disease and nanomedicine in PubMed as of 6 November 2022. Torchilin *et al.*[135] were among the first research groups that developed an IHD nanomedicine prototype, which was the Q10-loaded liposomes. CoQ10 is known to have quantifiable positive effects on the ischemic heart inflammatory markers, which could be enhanced and better unlocked by improving its delivery mechanism. From this perspective, Torchilin *et al.*[135] investigated the effect of coenzyme Q10-loaded liposomes *in vivo* using rabbits as an animal model, with results showing that CoQ10 liposomes provided effective protection for the ischemic heart as it enhanced the CoQ10 intracellular delivery, particularly in rabbits with myocardial infarction. Devices developed then were very primitive; however, this did not exclude the possibility that fully nanomechanical entities would someday see the light. Such hope was the key motivator for improving the previous primitive nanotechnology tools to the developed state they are in now. For sure, we have not yet fully benefited from the sophisticated nanotechnology platform, and there will always be room at the bottom for development, but we cannot deny how far this technology has moved. Another study explored the possibility of using gene therapy for the treatment of myocardial ischemia, which is still a novel and promising approach. This study suggested the possibility of enhancing gene delivery to target cells by using a double-targeted delivery system that has the ability to both accumulate extracellularly and penetrate targeted cells.[136] With this in mind, Ko *et al.* designed low-cationic dual-targeted liposomes–plasmid DNA complexes (lipoplexes) that have a high cell-penetration ability through the use of either the anti-myosin monoclonal antibody 2G4 (mAb 2G4) specific toward cardiac myosin or the TAT (GRKKRRQRRRPQ) peptide (TATp). Both mAb 2G4 and TATp cab provided the targeted gene delivery to the ischemic myocardium, with results showing that the genetic transformation of the normoxic/hypoxic cardiomyocytes can be strengthened by the TATp- and/or mAb 2G4-modified lipoplexes. In addition, this peptide is capable of recognizing and binding to the ischemic cells, which explains the enhanced transfection rate of the dual-modified lipoplexes. Consequently, *in vivo* experiments showed that TATp- and/or mAb

2G4-loaded liposomes effectively targeted the hypoxic cardiomyocytes as they accumulated in the ischemic myocardium (*i.e.* ischemic part of the heart), where enhanced gene expression levels were detected.[136]

3.3.2 Use of Reactive Oxygen Species in Developing IHD Nanotherapies

Myocardial ischemia is the end result of the IHD, and the rapid restoration of blood flow in the defective site often increases the ROS generation. Such a surge in ROS endangers the structure and function of the myocardium, resulting in myocardial ischemia.[137] The increase in ROS production is faced by the antioxidant defense system, which initiates a cascade of antioxidants, leading to a complex response to oxidative stress.[138] Despite the benefits antioxidants have, the surge in both ROS and antioxidants can have catastrophic effects on the myocardium, leading to the progression of the IHD to myocardial ischemia.[138] Furthermore, ROS were shown to play key roles in the pathological progression of various conditions, including hypertension and hypercholesterolemia, which lead to IHD. Physiological levels of ROS are important for regulating normal cellular functions[127]; however, excessive levels that can exceed the antioxidant capacity can overwhelm the cells, resulting in I/R injury, a phenomenon caused by unexpected muscle damage accompanied by myocardial stunning or lethal arrhythmias, as well as other cardiovascular diseases. Despite this fact, since 1977, PCI has significantly reduced the mortality caused by MI. However, revascularization by PCI is paradoxically known to lead to I/R injury.[139] Such an event is attributable to the massive surge in ROS production upon restoration of blood flow to the ischemic myocardium.[140] Moreover, evidences suggest that oxidative stress is the deleterious factor for the development of IHD, particularly with the current antioxidants having a short half-life, which is where nanomedicine turns out to be useful. Therefore, developing an efficient drug carrier system that can achieve a sustainable release of IHD drugs and targeted delivery options can help in better managing and preventing the disease progression into myocardial ischemia.[141] In this aspect, Gao *et al.* developed a matrine-loaded covalent organic framework (COF), with findings demonstrating the cardioprotective effect of matrine-loaded COFs, unleashing the immense potential of using nanoparticles as smart drug carriers for the treatment of IHD.[141]

Another study investigated the effect of loading superoxide dismutases (SODs) and catalase, a tetrameric enzyme in polyethylene glycol (PEG) nanoparticle, where drug–nanoparticle conjugates showed higher affinity to endothelium and exhibited protective effects in MI.[142] Other studies suggested conjugating SOD and catalase with specific antibodies (*e.g.* anti-platelet endothelial cell adhesion molecule (PECAM) antibody) for improved endothelial cell targeting.[143] Besides polymers, other nanocarriers were also investigated to serve as vascular delivery carriers for antioxidants, such as liposomes, magnetic nanoparticles, and ferritin nanocages.[144] For instance, Muzykantov *et al.* reported the successful synthesis of an SOD/catalase-mimetic drug (EUK-134)-PEG–liposome-based conjugate.[145] The conjugate was further coated with PECAM-1 antibody to target the endothelium; the modified nanoparticle was then shown to significantly alleviate the excessive ROS-related pathological condition by targeting the endothelium and inhibiting cytokine-induced inflammation.[145] Another group synthesized a biomimetic porous manganese-substituted Prussian blue nanocubes (PMPB-NC) to carry the antioxidative drug simvastatin and reduce the inflammatory response caused by the ischemic myocardium.[146] Furthermore, Zhou *et al.* synthesized a dual-shell polymeric nanoplatform to target the ischemic myocardium by decorating the nanoparticle with the peptide STSMLKA and the mitochondria targeting ligand SS-31. The nanoparticle was then loaded with the ROS scavenger drug resveratrol, with results revealing that STSMLKA enhanced the nanoparticle's accumulation in the ischemic myocardium; in addition, SS-31 further increased the accumulation in the mitochondria, which finally led to increasing the resveratrol's ROS elimination efficacy.[147]

Wu *et al.* generated a smart drug delivery system that focused on ROS depletion, where an ROS-consuming block copolymer PEG and poly(propylene sulfide) (PEG–PPS) was loaded with andrographolide, which yielded a prototype that exhibited a combinational antioxidant effect.[148]

3.3.3 Use of Inflammation in Developing IHD Nanotherapies

Inflammation is another factor that is involved in the development of IHD. Eckmann *et al.* designed a lysozyme dextran nanogel (NG) to carry dexamethasone, which is a classical long-lasting glucocorticoid for anti-inflammatory action. The nanoparticle was further decorated with intercellular adhesion molecule-1 (ICAM-1), which resulted in having a nanomedicine formulation that specifically accumulated in lesion sites and minimized the inflammatory response.[149] Another study showed that fusing thrombomodulin (TM), a potent antioxidant, with single-chain variable fragment generating (scFv/TM) and decorating the nanoparticles (*e.g.* liposome) with antibodies for PECAM and ICAM enhanced the endothelium binding, which significantly improved the therapeutic effect of the TM.[150,151] Besides these two molecules, there are various targetable endothelial adhesion molecules expressed in the ischemic myocardium[152], including (1) ICAM-1 and platelet-EC adhesion molecule 1 (PECAM-1),[153] (2) TM[154], (3) VCAM-1 and selectins,[155] (4) angiotensin-converting enzyme (ACE),[156] (5) Caveolae[157], and (6) adiponectin (APN) and integrins $\alpha_v\beta_3$, $\alpha_v\beta_5$, and $\alpha_5\beta_1$.[158] Subsequently, magnetically responsive nanomedicines were developed, such as calcium oleate, which was loaded with the antioxidant enzymes SOD and catalase. The nanoformulation could then be guided by a high-gradient magnetic field, targeting the therapeutic enzymes to the endothelium and protecting the enzyme cargo from proteolysis and early leakage in plasma.[159] Furthermore, Kartha *et al.*[160] designed a nanoformulation that consisted of ligands polybutadiene (PBD) and poly(propylene oxide) (PPO), which was then coated by PEG polymer. The newly synthesized copolymers PEG–PBD and PEG–PPO–PEG were then loaded with SOD, which alternatively prolonged the half-life of SOD. Similarly, Muzykantov *et al.*[161] synthesized another antioxidant catalase decorated with PEG, which was further encapsulated by a PEG–PLA polymer carrier (PNC). Such modification produced excellent protease degradation resistance ability, which in turn prolonged the SOD half-life.[161]

3.3.4 Use of Gene Therapy in Developing IHD Nanotherapies

Ischemic heart disease is a major health concern in both developing and industrialized countries, as it induces serious damage to cardiomyocytes, which can sometimes be irreversible; thus, the use of the gene therapy strategy involving mesenchymal stem cells (MSCs) and human adipose-derived stem cells (hASCs) is widely studied. MSCs obtained from elderly, diabetic, and IHD patients were shown to have significantly reduced proliferation, differentiation, and neovascularization potentials.[162] However, in ischemic tissues, hASCs are known to overexpress VEGF.[163] Incorporating VEGF along with the angiogenic growth factor, Ang1, was shown to promote a significant increase in neovascularization. Moreover, their combined action was shown to stimulate the formation of mature and functional vasculature.[164–166] Hence, Prakash *et al.*[167] suggested that creating genetically modified hASCs that can secret Ang1, along with their natural ability to secret VEGF, and trans-differentiate to functioning cardiomyocytes, can generate superior myocardial regeneration therapy with synergistic therapeutic effects. For this purpose, they used peptide nanoparticles to carry the TAT peptide/deoxyribonucleic acid, generating the nanoparticles (Bac-NP) that were further functionalized with Ang1 ($\text{Bac-NP}_{\text{Ang1}}$). Results showed that hASC activated with $\text{Bac-NP}_{\text{Ang1}}$ had higher cell retention and cardiac performance, along with a reduction in infarct sizes, which demonstrates the potential $\text{Bac-NP}_{\text{Ang1}}$ has in the treatment of IHD using nanomedicine and a stem cell–gene therapy strategy.[167] Similarly, cell-penetrating peptide (CPP)-modified drug delivery systems were also proven to facilitate the delivery of gene therapy. In addition to significantly promoting drug delivery, CPPs can also be used in stem cell therapy. Qi *et al.*[168] generated porous silicon nanoparticles that were then decorated with the cell-penetrating peptide human apolipoprotein (apoA-I) mutants fused with cationic CPPs (TAT) to fulfill the cellular internalization of the Wnt3a protein, the so-called TAT-porous silicon (TAT-PSi) NPs. The formulated nanocomplex displayed a prolonged Wnt3a protein release, and it protected the labeled MSCs in myocardial infarction. In addition, the intracellular aggregation of thermoresponsive polysiloxane (TPSi) NPs enabled precise stem cell transplantation into the myocardium, demonstrating great potential for the clinical translation of this approach.[169]

3.3.5 Use of Geometry of Nanoparticles in Developing IHD Nanotherapies

The geometry of nanoparticles, such as shape, size, and charge, determines the circulation time, cellular interactions, tissue penetration, and pharmacokinetics.[170] Nanoparticles smaller than 10 nm can rapidly undergo renal clearance, reducing the circulation time.[171] Spherical nanomedicine is eligible for vascular, mucosal, and dermal administrations. On the other hand, nanoparticles with 100–200 nm size have often prolonged the blood circulation time and high blood concentration.[172]However, nanoparticles >200 nm often require airway or vascular injection as well as have reduced access to targets due to the enhanced blood dragging force.[145,173] Moreover, endothelial cells show more desirability toward elongated nanoparticles than spherical ones.[174] This was evidenced by Mitragotri *et al.*, who showed that rod-shaped nanoparticles showed higher accumulation levels at the ischemic myocardium compared to their spherical counterparts.[175] Similarly, Muzykantov *et al.* reported long and flexible polymeric filomicelle nanoparticles that were more desirable to endothelial cells.[176]

Delivering drugs to the ischemic area of the heart through inhalation is a unique non-invasive approach that Miragoli *et al.*[177] explored. In their study, they formulated negatively charged biocompatible calcium phosphate nanoparticles named CaPs, which were then loaded with a cell-penetrating mimetic peptide (R7W-MP and DQRPDREAPRS). Results illustrated the accumulation of the CaPs-R7W nanoparticles in the myocardium, which recovered the heart contractility and function.[169] Another nanomedicine approach is using nanoparticle-mediated gene delivery to improve IHD treatment. In this area, researchers constructed a polyethylene glycol–peptide-optimized DNA (PEG–POD)/DNA by linking the PEG with a peptide, inducing a twenty-one-fold increase in the transgene expression and inducing a 50% decline in the cellular neovascularization when tested *in vivo*. Furthermore, mannosylated chitosan nanoparticles were shown to be excellent carriers for DNA and siRNA delivery, readily taken up by macrophages.[178] Moreover, chitosan–alginate nanoparticles were synthesized to effectively deliver placental growth factor (PIGF), which is a useful angiogenesis stimulator and can improve cardiac function at the ischemic myocardium, as these nanoparticles function by protecting the growth factor.[179] Furthermore, the injection of the irbesartan (antagonist of angiotensin II type 1 receptor)-loaded bioabsorbable nanoparticles (PIGA) at the time of reperfusion was shown to inhibit the recruitment of the monocytes, subsequently reducing the myocardium infarct through its anti-inflammatory activity.[180] Drug delivery through injectable hydrogels formed by nanofibers is considered to be a non-invasive method that can reduce the IHD recovery time. For instance, treating myocardial infarction animal models with RAD16-II hydrogels loaded with vascular endothelial growth factors showed improvement in heart function and vascular density.[181] Moreover, mitochondrial dysfunction is involved in the development of the IHD, making the mitochondria an important therapeutic target. Despite having several mitoprotective drugs, these drugs are associated with some drawbacks, such as poor permeability through the cellular and mitochondrial membranes. Hence, nanoparticles loaded with mitoprotective drugs can be used to solve this hurdle. An example is the mdivi-1 (a small-molecule inhibitor of mitochondrial fission protein)-loaded PLGA nanoparticles, which showed great protection against the H_2O_2-mediated death of rat cardiomyocytes.[182] Another drug encapsulated by PLGA is CsA, which when tested in a perfusion mice model showed reduced myocardial infarct size, protected against mitochondrial permeability transition pore opening, and prevented ventricular remodeling.[183] Of the metal nanoparticles, gold nanoparticles of size below 200 nm were shown to accumulate in the ischemic muscle through their enhanced permeability and retention effect in the ischemic myocardium. In addition, gold nanoparticles were shown to promote ischemic tissue recovery and enhance angiogenesis by delivering the exogenous growth factors.[184]

3.3.6 Nano-immunotherapy in IHD

Nano-immunotherapies are new emerging tools that have recently been introduced and are considered to clinically target the inflammatory component of IHD. An example is the CANTOS trial,[185] which showed that chronic use of the interleukin (IL)-1β-blocking antibody therapy had the ability to lower the cardiovascular events in patients. Such discovery opened new therapeutic horizons with the concept of manipulating the immune response being suggested to prevent IHD. In this aspect, Santamaria *et al.*[186,187] reported that coating major histocompatibility complex (pMHC) nanoparticles with autoimmune disease-relevant peptides blunted autoimmune responses. This action was due to the fact that the pMHC–peptide conjugate triggered the cellular differentiation and expansion of antigen-specific regulatory T cells *in vivo*.[186,187]

3.3.7 Nanotheranostics in IHD

In addition to using nanomedicine in IHD treatment, it can also be used in developing theranostic and detection tools. Having said that and knowing that IHD is the earliest stage of MI, Feiner *et al.*[188] developed a novel hybrid cardiac patch by combining the heart cells with flexible stand-alone electronics and 3D nanocomposite scaffolds to be used for the early detection of IHD. Another study conducted by Cheng *et al.*[189] used stem cell transplantation technology as a therapeutic strategy for cardiac regeneration after MI. Used iron oxide nanoparticles were modified with anti-CD34 and anti-CD45 as a cell ligation mediator. Interestingly, nanoparticles acted as magnetic bifunctional cell engagers (MagBICE), connecting the endogenous injured cardiomyocytes recognized by anti-CD34 with the exogenous bone marrow-derived stem cells recognized by anti-CD45. The used iron oxide nanoparticles were localized at the damaged ischemic myocardium, which were then visualized using MRI. It is noteworthy that using the iron oxide nanoparticles promoted the efficient delivery of the stem cells to the target location by magnetic attraction. In another study, Cowan *et al.*[190] designed labeled mitochondria to be used as a cardioprotection tool. In this approach, fluorine-18-rhodamine 6G and iron oxide nanoparticles were used to label the mitochondria, allowing the non-invasive detection using multiple detection tools, including MRI, fluorescence imaging, transmission electron microscopy (TEM), PET, and microcomputed tomography (μCT). Results indicate that designed nanoformulations successfully attached to the mitochondrial surface as intended. Finally, Lee *et al.*[191] synthesized hydrogen peroxide-responsive copolyoxalate nanoparticles to be used for the detection of ischemic injury. In this study, rubrene (Rb) was loaded into copolyoxalate nanoparticles (HPOX)-conjugated hydroxybenzyl alcohol (HBA), generating HPOX/RB. Results showed that the generated nanoformulation had an excellent anti-inflammatory

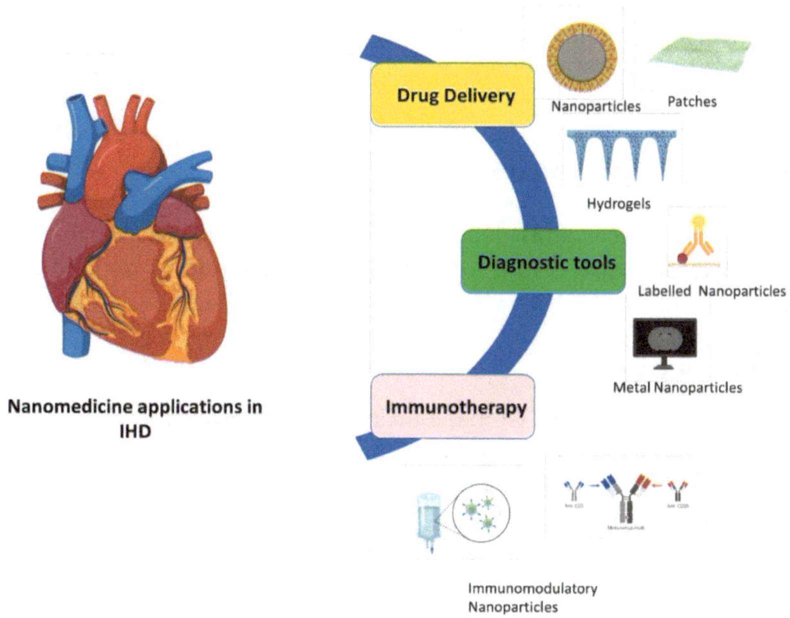

Figure 3.5 Summary of the nanomedicine applications in IHD.

activity. *In vivo* chemiluminescent images demonstrate that Rb-labeled nanoparticles, HPOX/RB, reduced the H_2O_2 levels in the injured areas of mice, and the images are taken from the hind limb of the I/R mice model, having a strong fluorescence intensity 2 minutes after injection.[191]

Despite the substantial progress that has been made in the nanomedicine field and the great potential it has in better treating and diagnosing IHD (Figure 3.5), there are still several challenges that face the future development of this field, and they need to be addressed before it can be fully translated into the clinic. Such challenges include (1) nanoformulation tolerance[144]; (2) nanoformulation toxicity, biosafety, and side effects[192]; (3) the ability to use the nanoformulation for disease prevention; and (4) immunogenicity.[193] In addition to these, there are other challenges associated with the translation, cost of production, quality control, tedious methods for characterization and purity analysis, and scaling up of the nanoformulation production while maintaining its properties and efficacy in delivering the loaded cargo.[193]

3.4 Conclusion

Nanomedicine is a rapidly developing field, with many therapeutic tools that could be further improved to have a greater and more substantial impact on the treatment of IHD. By benefiting from the current state of the nanomedicine field and its advances, we can forever change the concepts and approaches used in diagnosing and treating IHD patients. However, to do so and to fully benefit from this fascinating field, we will need to thoroughly investigate the changes and issues facing the nanomedicine field to facilitate a smoother and faster translation to clinical use.

Acknowledgements

We would like to acknowledge our work colleagues, department, and the biomedical research center for the continuous support provided to us.

References

1. G. A. Roth, *et al.*, Global burden of cardiovascular diseases and risk factors, 1990-2019: update from the GBD 2019 study, *J. Am. Coll. Cardiol*, 2020, **76**, 2982–3021.
2. M. F. Yuyun, K. Sliwa, A. P. Kengne, A. O. Mocumbi and G. Bukhman, Cardiovascular Diseases in Sub-Saharan Africa Compared to High-Income Countries: An Epidemiological Perspective, *Glob. Heart.*, 2020, **15**(1), 15.

3. G. A. Roth, *et al.*, Global, Regional, and National Burden of Cardiovascular Diseases for 10 causes, 1990 to 2015, *J. Am. Coll. Cardiol.*, 2017, **70**, 1–25.
4. M. A. Khan, M. J. Hashim, H. Mustafa, M. Y. Baniyas, S. K. B. M. Al Suwaidi, R. AlKatheeri, F. M. K. Alblooshi, M. E. A. H. Almatrooshi, M. E. H. Alzaabi, R. S. Al Darmaki and S. N. A. H. Lootah, Global Epidemiology of Ischemic Heart Disease: Results from the Global Burden of Disease Study, *Cureus*, 2020, **12**(7), e9349.
5. H. A. H. Alfarisi, Z. B. H. Mohamed and M. B. Ibrahim, Basic pathogenic mechanisms of atherosclerosis, *Egypt. J. Basic Appl. Sci.*, 2020, **7**, 116–125.
6. J. J. Regieli, *et al.*, Coronary collaterals—insights in molecular determinants and prognostic relevance, *Int. J. Cardiol.*, 2007, **116**, 139–143.
7. R. A. Brown, E. Shantsila, C. Varma and G. Y. Lip, Current understanding of atherogenesis, *Am. J. Med.*, 2017, **130**, 268–282.
8. I. Tabas, G. García-Cardeña and G. K. Owens, Recent insights into the cellular biology of atherosclerosis, *J. Cell Biol.*, 2015, **209**, 13–22.
9. T. Wang and J. Butany, Pathogenesis of atherosclerosis, *Diagn. Histopathol.*, 2017, **23**, 473–478.
10. Z. Zhang, *et al.*, Reactive Oxygen Species Scavenging Nanomedicine for the Treatment of Ischemic Heart Disease, *Adv. Mater.*, 2022, 2202169.
11. D. N. Granger and P. R. Kvietys, Reperfusion injury and reactive oxygen species: The evolution of a concept, *Redox Biol.*, 2015, **6**, 524–551.
12. L. Milkovic, A. Cipak Gasparovic, M. Cindric, P.-A. Mouthuy and N. Zarkovic, Short overview of ROS as cell function gegulators and their implications in therapy concepts, *Cells*, 2019, **8**, 793.
13. UN, *World population prospects 2019: highlights*, 2019, https://www.un.org/en/desa/world-population-prospects-2019-highlights. Accessed 12 Mar 2020.
14. S. Barquera, *et al.*, Global overview of the epidemiology of atherosclerotic cardiovascular disease, *Arch. Med. Res.*, 2015, **46**, 328–338.
15. *Burden of Disease, Our World Data*, 2016, https://ourworldindata.org/burden-of-disease. Accessed 12 Mar 2020.
16. C. J. Pepine, in , American College of Cardiology Foundation, Washington, DC, 2004, vol. **43**, pp. 1727–1730.
17. B. Wolff, K. Machill, I. Schulzki, D. Schumacher and D. Werner, Acute reversible cardiomyopathy with cardiogenic shock in a patient with Addisonian crisis: a case report, *Int. J. Cardiol*, 2007, **116**, e71–e73.
18. C. Hermiz and Y. R. Sedhai, Angina, in *StatPearls*, StatPearls Publishing LLC, Treasure Island, FL, 2024.
19. N. Ojha and A. S. Dhamoon, Myocardial Infarction, in *StatPearls*, StatPearls Publishing LLC, Treasure Island, FL, 2024.
20. A. Malik, D. Brito, S. Vaqar and L. Chhabra, Congestive Heart Failure, in *StatPearls*, StatPearls Publishing LLC, Treasure Island, FL, 2024.

21. N. J. Wald and M. R. Law, A strategy to reduce cardiovascular disease by more than 80%, *BMJ*, 2003, **326**, 1419.
22. R. Hajar, Risk factors for coronary artery disease: historical perspectives, *Heart Views*, 2017, **18**, 109.
23. A. L. Catapano, *et al.*, Prevalence of potential familial hypercholesteremia (FH) in 54,811 statin-treated patients in clinical practice, *Atherosclerosis*, 2016, **252**, 1–8.
24. K. A. Kott, *et al.*, Biomarker Development in Cardiology: Reviewing the Past to Inform the Future, *Cells*, 2022, **11**, 588.
25. E. Danese and M. Montagnana, An historical approach to the diagnostic biomarkers of acute coronary syndrome, *Ann. Transl. Med.*, 2016, **4**, 194.
26. B. E. Sobel and W. E. Shell, Serum enzyme determinations in the diagnosis and assessment of myocardial infarction, *Circulation*, 1972, **45**, 471–482.
27. C. C. Johnston and E. C. Bolton, Cardiac enzymes, *Ann. Emerg. Med.*, 1982, **11**, 27–35.
28. P. A. Kavsak and A. H. Wu, in *Contemporary Practice in Clinical Chemistry*, Elsevier, 2020, pp. 519–543.
29. M. Ezzati, *et al.*, Contributions of risk factors and medical care to cardiovascular mortality trends, *Nat. Rev. Cardiol.*, 2015, **12**, 508–530.
30. N. Jackson, *et al.*, Improving clinical trials for cardiovascular diseases: a position paper from the Cardiovascular Round Table of the European Society of Cardiology, *Eur. Heart J.*, 2016, **37**, 747–754.
31. A. M. Lincoff, *et al.*, Evacetrapib and cardiovascular outcomes in high-risk vascular disease, *N. Engl. J. Med.*, 2017, **376**, 1933–1942.
32. P. M. Ridker, *et al.*, Cardiovascular efficacy and safety of bococizumab in high-risk patients, *N. Engl. J. Med.*, 2017, **376**, 1527–1539.
33. M. L. O'Donoghue, *et al.*, Effect of losmapimod on cardiovascular outcomes in patients hospitalized with acute myocardial infarction: a randomized clinical trial, *JAMA*, 2016, **315**, 1591–1599.
34. C. M. Schooling, *et al.*, Disconnect Between Genes Associated With Ischemic Heart Disease and Targets of Ischemic Heart Disease Treatments, *EBioMedicine*, 2018, **28**, 311–315.
35. J. M. Gregson, *et al.*, Genetic invalidation of Lp-PLA2 as a therapeutic target: Large-scale study of five functional Lp-PLA2-lowering alleles, *Eur. J. Prev. Cardiol.*, 2017, **24**, 492–504.
36. H. Moses, *et al.*, The anatomy of medical research: US and international comparisons, *JAMA*, 2015, **313**, 174–189.
37. C. P. Nelson, *et al.*, Association analyses based on false discovery rate implicate new loci for coronary artery disease, *Nat. Genet.*, 2017, **49**, 1385–1391.
38. Antiplatelet Trialists' Collaboration, Collaborative overview of randomised trials of antiplatelet therapy Prevention of death, myocardial

infarction, and stroke by prolonged antiplatelet therapy in various categories of patients, *BMJ*, 1994, **308**, 81.

39. C. Baigent, C. Sudlow, R. Collins and R. Peto, Collaborative meta-analysis of randomised trials of antiplatelet therapy for prevention of death, myocardial infarction, and stroke in high risk patients, *BMJ*, 2002, **324**, 71–86.
40. S. D. Fihn, *et al.*, ACC/AHA/AATS/PCNA/SCAI/STS Focused Update of the Guideline for the Diagnosis and Management of Patients With Stable Ischemic Heart Disease A Report of the American College of Cardiology/American Heart Association Task Force on Practice Guidelines, and the American Association for Thoracic Surgery, Preventive Cardiovascular Nurses Association, Society for Cardiovascular Angiography and Interventions, and Society of Thoracic Surgeons, *Circulation*, 2014, **130**, 1749–1767.
41. F. J. Neumann, *et al.*, 2019 ESC Guidelines for the diagnosis and management of chronic coronary syndromes, *Eur. Heart J.*, 2020, **41**, 407–477.
42. R. Collins, *et al.*, Aspirin in the primary and secondary prevention of vascular disease: collaborative meta-analysis of individual participant data from randomised trials, *The Lancet*, 2009, **373**, 1849–1860.
43. M. Gent, A randomised, blinded, trial of clopidogrel versus aspirin in patients at risk of ischaemic events (CAPRIE), *Lancet*, 1996, **348**, 1329–1339.
44. A. Colombo, *et al.*, Second-generation drug-eluting stent implantation followed by 6- versus 12-month dual antiplatelet therapy, *J. Am. Coll. Cardiol.*, 2014, **64**, 2086–2097.
45. F. Costa, J. G. Tijssen, S. Ariotti, S. Giatti, E. Moscarella, P. Guastaroba, R. De Palma, G. Andò, G. Oreto, F. Zijlstra and M. Valgimigli, Incremental value of the CRUSADE, ACUITY, and HAS-BLED risk scores for the prediction of hemorrhagic events after coronary stent implantation in patients undergoing long or short duration of dual antiplatelet therapy, *J. Am. Heart Assoc.*, 2015, **4**(12), e002524.
46. G. Giustino, *et al.*, Duration of dual antiplatelet therapy after drug-eluting stent implantation: A systematic review and meta-analysis of randomized controlled trials, *J. Am. Coll. Cardiol.*, 2015, **65**, 1298–1310.
47. E. P. Navarese, F. Andreotti, V. Schulze, M. Kołodziejczak, A. Buffon, M. Brouwer, F. Costa, M. Kowalewski, G. Parati, G. Y. Lip, M. Kelm and M. Valgimigli, Optimal duration of dual antiplatelet therapy after percutaneous coronary intervention with drug eluting stents: meta-analysis of randomised controlled trials, *BMJ*, 2015, **350**, h1618. Erratum in: *BMJ*, 2016, **355**, i5600.
48. T. Palmerini, *et al.*, Mortality in patients treated with extended duration dual antiplatelet therapy after drug-eluting stent implantation: A

pairwise and Bayesian network meta-analysis of randomised trials, *Lancet*, 2015, **385**, 2371–2382.

49. M. Gilard, *et al.*, 6- versus 24-month dual antiplatelet therapy after implantation of drug-eluting stents in patients nonresistant to aspirin: the randomized, multicenter ITALIC trial, *J. Am. Coll. Cardiol.*, 2015, **65**, 777–786.
50. H. C. Gwon, *et al.*, Six-month versus 12-month dual antiplatelet therapy after implantation of drug-eluting stents: the efficacy of xience/promus versus cypher to reduce late loss after stenting (EXCELLENT) randomized, multicenter study, *Circulation*, 2012, **125**, 505–513.
51. S. Schulz-Schüpke, *et al.*, ISAR-SAFE: A randomized, double-blind, placebo-controlled trial of 6 vs. 12 months of clopidogrel therapy after drug-eluting stenting, *Eur. Heart J.*, 2015, **36**, 1252–1263.
52. M. Valgimigli, *et al.*, Short- versus long-term duration of dual-antiplatelet therapy after coronary stenting: A randomized multicenter trial, *Circulation*, 2012, **125**, 2015–2026.
53. F. Feres, *et al.*, Three vs twelve months of dual antiplatelet therapy after zotarolimus-eluting stents: The OPTIMIZE randomized trial, *JAMA*, 2013, **310**, 2510–2522.
54. B. K. Kim, *et al.*, A new strategy for discontinuation of dual antiplatelet therapy: The reset trial (real safety and efficacy of 3-month dual antiplatelet therapy following endeavor zotarolimus-eluting stent implantation), *J. Am. Coll. Cardiol.*, 2012, **60**, 1340–1348.
55. M. P. Bonaca, D. L. Bhatt, M. Cohen, P. G. Steg, R. F. Storey, E. C. Jensen, G. Magnani, S. Bansilal, M. P. Fish, K. Im, O. Bengtsson, T. Oude Ophuis, A. Budaj, P. Theroux, M. Ruda, C. Hamm, S. Goto, J. Spinar, J. C. Nicolau, R. G. Kiss, S. A. Murphy, S. D. Wiviott, P. Held, E. Braunwald and M. S. Sabatine, Long-term use of ticagrelor in patients with prior myocardial infarction, *N. Engl. J. Med.*, 2015, **372**, 1791–1800.
56. J. W. Eikelboom, *et al.*, Rivaroxaban with or without aspirin in stable cardiovascular disease, *N. Engl. J. Med.*, 2017, **377**, 1319–1330.
57. J. L. Mega, *et al.*, Rivaroxaban in patients with a recent acute coronary syndrome, *N. Engl. J. Med.*, 2012, **366**, 9–19.
58. K. A. Fiedler, *et al.*, Duration of triple therapy in patients requiring oral anticoagulation after drug-eluting stent implantation: The ISAR-TRIPLE trial, *J. Am. Coll. Cardiol.*, 2015, **65**, 1619–1629.
59. M. Packer, *et al.*, The effect of carvedilol on morbidity and mortality in patients with chronic heart failure, *N. Engl. J. Med.*, 1996, **334**, 1349–1355.
60. A. Hjalmarson, *et al.*, Effects of controlled-release metoprolol on total mortality, hospitalizations, and well-being in patients with heart failure: The metoprolol CR/XL randomized intervention trial in congestive heart failure (MERIT-HF), *JAMA*, 2000, **283**, 1295–1302.

61. S. Bangalore, *et al.*, β-Blocker use and clinical outcomes in stable outpatients with and without coronary artery disease, *JAMA*, 2012, **308**, 1340–1349.
62. A. A. Motivala, *et al.*, Predictors, Trends, and Outcomes (Among Older Patients ≥65 Years of Age) Associated With Beta-Blocker Use in Patients With Stable Angina Undergoing Elective Percutaneous Coronary Intervention: Insights From the NCDR Registry, *JACC*, 2016, **9**, 1639–1648.
63. T. A. Rousan, S. T. Mathew and U. Thadani, Drug Therapy for Stable Angina Pectoris, *Drugs*, 2017, **77**, 265–284.
64. W. W. Klein, G. Jackson and L. Tavazzi, Efficacy of monotherapy compared with combined antianginal drugs in the treatment of chronic stable angina pectoris: A meta-analysis, *Coron. Artery Dis.*, 2002, **13**, 427–436.
65. S. D. Fihn, *et al.*, 2012 ACCF/AHA/ACP/AATS/PCNA/SCAI/STS guideline for the diagnosis and management of patients with stable ischemic heart disease, *J. Am. Coll. Cardiol.*, 2012, **60**, e44–e164.
66. J. Belsey, I. Savelieva, A. Mugelli and A. J. Camm, Relative efficacy of antianginal drugs used as add-on therapy in patients with stable angina: A systematic review and meta-analysis, *Eur. J. Prev. Cardiol.*, 2015, **22**, 837–848.
67. J. C. Tardif, P. Ponikowski and T. Kahan, Efficacy of the If current inhibitor ivabradine in patients with chronic stable angina receiving beta-blocker therapy: A 4-month, randomized, placebo-controlled trial, *Eur. Heart J.*, 2009, **30**, 540–548.
68. P. Ong, A. Athanasiadis and U. Sechtem, Pharmacotherapy for coronary microvascular dysfunction, *Eur. Heart J.: Cardiovasc. Pharmacother.*, 2015, **1**, 65–71.
69. JCS Joint Working Group, Guidelines for diagnosis and treatment of patients with vasospastic angina (Coronary Spastic Angina) (JCS 2013), *Circ. J.*, 2014, **78**(11), 2779–2801.
70. C. Baigent, *et al.*, Efficacy and safety of more intensive lowering of LDL cholesterol: A meta-analysis of data from 170000 participants in 26 randomised trials, *Lancet*, 2010, **376**, 1670–1681.
71. F. Mach, *et al.*, 2019 ESC/EAS guidelines for the management of dyslipidaemias: Lipid modification to reduce cardiovascular risk, *Eur. Heart J.*, 2020, **41**, 111–188.
72. M. S. Sabatine, *et al.*, Evolocumab and clinical outcomes in patients with cardiovascular disease, *N. Engl. J. Med.*, 2017, **376**, 1713–1722.
73. G. G. Schwartz, *et al.*, Alirocumab and cardiovascular outcomes after acute coronary syndrome, *N. Engl. J. Med.*, 2018, **379**, 2097–2107.
74. C. K. Chow, *et al.*, Association of diet, exercise, and smoking modification with risk of early cardiovascular events after acute coronary syndromes, *Circulation*, 2010, **121**, 750–758.

75. J. A. Critchley and S. Capewell, Mortality Risk Reduction Associated with Smoking Cessation in Patients with Coronary Heart Disease: A Systematic Review, *JAMA*, 2003, **290**, 86–97.
76. N. P. Johnson, *et al.*, Prognostic value of fractional flow reserve: Linking physiologic severity to clinical outcomes, *J. Am. Coll. Cardiol.*, 2014, **64**, 1641–1654.
77. E. J. Velazquez, *et al.*, Coronary-artery bypass surgery in patients with ischemic cardiomyopathy, *N. Engl. J. Med.*, 2016, **374**, 1511–1520.
78. H. Gada, *et al.*, Meta-analysis of trials on mortality after percutaneous coronary intervention compared with medical therapy in patients with stable coronary heart disease and objective evidence of myocardial ischemia, *Am. J. Cardiol.*, 2015, **115**, 1194–1199.
79. L. J. Shaw, *et al.*, Comparative definitions for moderate-severe ischemia in stress nuclear, echocardiography, and magnetic resonance imaging, *JACC*, 2014, **7**, 593–604.
80. S. Sen, *et al.*, Development and validation of a new adenosine-independent index of stenosis severity from coronary wave-intensity analysis: results of the ADVISE (ADenosine Vasodilator Independent Stenosis Evaluation) study, *J. Am. Coll. Cardiol.*, 2012, **59**, 1392–1402.
81. J. E. Davies, *et al.*, Use of the instantaneous wave-free ratio or fractional flow reserve in PCI, *N. Engl. J. Med.*, 2017, **376**, 1824–1834.
82. M. Götberg, *et al.*, Instantaneous wave-free ratio versus fractional flow reserve to guide PCI, *N. Engl. J. Med.*, 2017, **376**, 1813–1823.
83. T. Engstrøm, *et al.*, Complete revascularisation versus treatment of the culprit lesion only in patients with ST-segment elevation myocardial infarction and multivessel disease (DANAMI-3—PRIMULTI): An open-label, randomised controlled trial, *Lancet*, 2015, **386**, 665–671.
84. F. M. Zimmermann, *et al.*, Fractional flow reserve-guided percutaneous coronary intervention vs. medical therapy for patients with stable coronary lesions: Meta-analysis of individual patient data, *Eur. Heart J.*, 2019, **40**, 180–186.
85. P. C. Smits, *et al.*, Fractional flow reserve-guided multivessel angioplasty in myocardial infarction, *N. Engl. J. Med.*, 2017, **376**, 1234–1244.
86. P. Xaplanteris, *et al.*, Five-year outcomes with PCI guided by fractional flow reserve, *N. Engl. J. Med.*, 2018, **379**, 250–259.
87. B. De Bruyne, *et al.*, Fractional flow reserve-guided PCI versus medical therapy in stable coronary disease, *N. Engl. J. Med.*, 2012, **367**, 991–1001.
88. B. De Bruyne, *et al.*, Fractional flow reserve-guided PCI for stable coronary artery disease, *N. Engl. J. Med.*, 2014, **371**, 1208–1217.
89. J. Adjedj, *et al.*, Significance of intermediate values of fractional flow reserve in patients with coronary artery disease, *Circulation*, 2016, **133**, 502–508.
90. J. M. Ahn, *et al.*, Fractional Flow Reserve and Cardiac Events in Coronary Artery Disease: Data from a Prospective IRIS-FFR Registry

(Interventional Cardiology Research Incooperation Society Fractional Flow Reserve), *Circulation*, 2017, **135**, 2241–2251.
91. N. H. J. Pijls, *et al.*, Percutaneous Coronary Intervention of Functionally Nonsignificant Stenosis. 5-Year Follow-Up of the DEFER Study, *J. Am. Coll. Cardiol.*, 2007, **49**, 2105–2111.
92. R. Petraco, *et al.*, Fractional flow reserve-guided revascularization: Practical implications of a diagnostic gray zone and measurement variability on clinical decisions, *JACC*, 2013, **6**, 222–225.
93. J. Escaned, *et al.*, Prospective assessment of the diagnostic accuracy of instantaneous wave-free ratio to assess coronary stenosis relevance: Results of ADVISE II international, multicenter study (ADenosine vasodilator independent stenosis evaluation II), *JACC*, 2015, **8**, 824–833.
94. S. J. Park, *et al.*, Trial of everolimus-eluting stents or bypass surgery for coronary disease, *N. Engl. J. Med.*, 2015, **372**, 1204–1212.
95. T. Mäkikallio, *et al.*, Percutaneous coronary angioplasty versus coronary artery bypass grafting in treatment of unprotected left main stenosis (NOBLE): a prospective, randomised, open-label, non-inferiority trial, *Lancet*, 2016, **388**, 2743–2752.
96. S. J. Park, *et al.*, Randomized trial of stents versus bypass surgery for left main coronary artery disease, *N. Engl. J. Med.*, 2011, **364**, 1718–1727.
97. G. W. Stone, *et al.*, Everolimus-eluting stents or bypass surgery for left main coronary artery disease, *N. Engl. J. Med.*, 2016, **375**, 2223–2235.
98. J. M. Ahn, *et al.*, Comparison of Stenting Versus Bypass Surgery According to the Completeness of Revascularization in Severe Coronary Artery Disease: Patient-Level Pooled Analysis of the SYNTAX, PRECOMBAT, and BEST Trials, *JACC*, 2017, **10**, 1415–1424.
99. S. J. Head, *et al.*, Mortality after coronary artery bypass grafting versus percutaneous coronary intervention with stenting for coronary artery disease: a pooled analysis of individual patient data, *Lancet*, 2018, **391**, 939–948.
100. F. W. Mohr, *et al.*, Coronary artery bypass graft surgery versus percutaneous coronary intervention in patients with three-vessel disease and left main coronary disease: 5-year follow-up of the randomised, clinical SYNTAX trial, *Lancet*, 2013, **381**, 629–638.
101. F. J. Neumann, *et al.*, ESC/EACTS Guidelines on myocardial revascularization, *Eur. Heart J.*, 2019, **40**, 87–165.
102. R. S. B. Beanlands, *et al.*, F-18-Fluorodeoxyglucose Positron Emission Tomography Imaging-Assisted Management of Patients With Severe Left Ventricular Dysfunction and Suspected Coronary Disease. A Randomized, Controlled Trial (PARR-2), *J. Am. Coll. Cardiol.*, 2007, **50**, 2002–2012.
103. R. O. Bonow, *et al.*, Myocardial viability and survival in ischemic left ventricular dysfunction, *N. Engl. J. Med.*, 2011, **364**, 1617–1625.
104. P. G. Thrane, *et al.*, 16-year follow-up of the Danish Acute Myocardial Infarction 2 (DANAMI-2) trial: Primary percutaneous coronary

intervention vs. fibrinolysis in ST-segment elevation myocardial infarction, *Eur. Heart J.*, 2020, **41**, 847–854.

105. I. Y. Elgendy, A. N. Mahmoud, D. J. Kumbhani, D. L. Bhatt and A. A. Bavry, Complete or Culprit-Only Revascularization for Patients With Multivessel Coronary Artery Disease Undergoing Percutaneous Coronary Intervention: A Pairwise and Network Meta-Analysis of Randomized Trials, *JACC*, 2017, **10**, 315–324.
106. A. H. Gershlick, *et al.*, Randomized trial of complete versus lesion-only revascularization in patients undergoing primary percutaneous coronary intervention for STEMI and multivessel disease: The CvLPRIT trial, *J. Am. Coll. Cardiol.*, 2015, **65**, 963–972.
107. S. R. Mehta, *et al.*, Complete revascularization with multivessel PCI for myocardial infarction, *N. Engl. J. Med.*, 2019, **381**, 1411–1421.
108. D. S. Wald, *et al.*, Randomized trial of preventive angioplasty in myocardial infarction, *N. Engl. J. Med.*, 2013, **369**, 1115–1123.
109. J. H. Kwon, *et al.*, Protective effect of heat shock protein 27 using protein transduction domain-mediated delivery on ischemia/reperfusion heart injury, *Biochem. Biophys. Res. Commun.*, 2007, **363**, 399–404.
110. E. Toyota, T. Matsunaga and W. M. Chilian, Myocardial angiogenesis, *Mol. Cell. Biochem.*, 2004, **264**, 35–44.
111. S. Nagai, *et al.*, Protection of hepatic cells from apoptosis induced by ischemia/reperfusion injury by protein therapeutics, *Hepatol. Res.*, 2007, **37**, 133–142.
112. M. Arakawa, *et al.*, Transduction of anti-cell death protein FNK protects isolated rat hearts from myocardial infarction induced by ischemia/reperfusion, *Life Sci.*, 2007, **80**, 2076–2084.
113. S. L. M. A. Beeres, D. E. Atsma, J. van Ramshorst, M. J. Schalij and J. J. Bax, Cell therapy for ischaemic heart disease, *Heart*, 2008, **94**, 1214–1226.
114. C. A. Mack, *et al.*, Biologic bypass with the use of adenovirus-mediated gene transfer of the complementary deoxyribonucleic acid for vascular endothelial growth factor 121 improves myocardial perfusion and function in the ischemic porcine heart, *J. Thorac. Cardiovasc. Surg.*, 1998, **115**, 168–177.
115. F. J. Giordano, *et al.*, Intracoronary gene transfer of fibroblast growth factor-5 increases blood flow and contractile function in an ischemic region of the heart, *Nat. Med.*, 1996, **2**, 534–539.
116. D. W. Losordo, *et al.*, Phase 1/2 placebo-controlled, double-blind, dose-escalating trial of myocardial vascular endothelial growth factor 2 gene transfer by catheter delivery in patients with chronic myocardial ischemia, *Circulation*, 2002, **105**, 2012–2018.
117. P. R. Vale, *et al.*, Left ventricular electromechanical mapping to assess efficacy of phVEGF(165) gene transfer for therapeutic angiogenesis in chronic myocardial ischemia, *Circulation*, 2000, **102**, 965–974.

118. J. F. Symes, *et al.*, Gene therapy with vascular endothelial growth factor for inoperable coronary artery disease, *Ann. Thorac. Surg.*, 1999, **68**, 830–836.
119. T. K. Rosengart, *et al.*, Angiogenesis gene therapy: Phase I assessment of direct intramyocardial administration of an adenovirus vector expressing VEGF121 cDNA to individuals with clinically significant severe coronary artery disease, *Circulation*, 1999, **100**, 468–474.
120. M. C. Manning, K. Patel and R. T. Borchardt, Stability of protein pharmaceuticals, *Pharm. Res.*, 1989, **6**, 903–918.
121. A. Prokop, E. Kozlov, G. Carlesso and J. M. Davidson, in *Advances in Polymer Science*, 2002, vol. **160**, pp. 119–173.
122. K. Y. Lee and S. H. Yuk, Polymeric protein delivery systems, *Prog. Polym. Sci.*, 2007, **32**, 669–697.
123. K. Y. Lee and D. J. Mooney, Hydrogels for tissue engineering, *Chem. Rev.*, 2001, **101**, 1869–1879.
124. S. E. Sakiyama-Elbert and J. A. Hubbell, Functional biomaterials: Design of novel biomaterials, *Annu. Rev. Mater. Sci.*, 2001, **31**, 183–201.
125. A. K. Jain, N. K. Mehra and N. K. Swarnakar, Role of Antioxidants for the Treatment of Cardiovascular Diseases: Challenges and Opportunities, *Curr. Pharm. Des.*, 2015, **21**, 4441–4455.
126. X. Ye and D. Yang, Recent advances in biological strategies for targeted drug delivery, *Cardiovasc. Hematol. Disord. Drug. Targets*, 2009, **9**, 206–221.
127. P. Ehrlich, in *Collected studies on immunity*, John Wiley, 1906, pp. 442–447.
128. J. E. Vance and R. Steenbergen, Metabolism and functions of phosphatidylserine, *Prog. Lipid Res.*, 2005, **44**, 207–234.
129. H. Takahama, *et al.*, Prolonged targeting of ischemic/reperfused myocardium by liposomal adenosine augments cardioprotection in rats, *J. Am. Coll. Cardiol.*, 2009, **53**, 709–717.
130. W. Sang, Z. Zhang, Y. Dai and X. Chen, Recent advances in nanomaterial-based synergistic combination cancer immunotherapy, *Chem. Soc. Rev.*, 2019, **48**, 3771–3810.
131. Z. Zhang, L. Xie, Y. Ju and Y. Dai, Recent Advances in Metal-Phenolic Networks for Cancer Theranostics, *Small*, 2021, **17**, e2100314.
132. Y. M. H. Matsumura, A new concept for macromolecular therapeutics in cancer chemotherapy: mechanism of tumoritropic accumulation of proteins and the antitumor agent smancs, *Cancer Res.*, 1986, **46**(12 Pt 1), 6387–6392.
133. World Health Organization, *Atlas of heart disease and stroke*, World Health Organization, Switzerland, 2004, Available from: https://iris.who.int/handle/10665/43007. Accessed 1 Aug 2024.
134. K. Drexler, in *Engines of creation: the coming era of nanotechnology*, Bantam Doubleday Dell, New York, NY, 1986.

135. D. D. Verma, W. C. Hartner, V. Thakkar, T. S. Levchenko and V. P. Torchilin, Protective effect of coenzyme Q10-loaded liposomes on the myocardium in rabbits with an acute experimental myocardial infarction, *Pharm. Res.*, 2007, **24**, 2131–2137.
136. Y. T. Ko, W. C. Hartner, A. Kale and V. P. Torchilin, Gene delivery into ischemic myocardium by double-targeted lipoplexes with anti-myosin antibody and TAT peptide, *Gene Ther.*, 2009, **16**, 52–59.
137. D. M. Yellon and D. J. Hausenloy, Myocardial reperfusion injury, *N. Engl. J. Med.*, 2007, **357**, 1121–1135.
138. B. S. Juránek, Controversy of free radical hypothesis: reactive oxygen species—cause or consequence of tissue injury?, *Gen. Physiol. Biophys.*, 2005, **24**, 263–278.
139. M. H. Huang, K. K. Poh, H. C. Tan, F. G. Welt and C. Y. Lui, Therapeutic synergy and complementarity for ischemia/reperfusion injury: β1-adrenergic blockade and phosphodiesterase-3 inhibition, *Int. J. Cardiol.*, 2016, **214**, 374–380.
140. D. N. Granger and P. R. Kvietys, Reperfusion injury and reactive oxygen species: The evolution of a concept, *Redox Biol.*, 2015, **6**, 524–551.
141. C. Huang, *et al.*, Biodegradable Redox-Responsive AIEgen-Based-Covalent Organic Framework Nanocarriers for Long-Term Treatment of Myocardial Ischemia/Reperfusion Injury, *Small*, 2022, e2205062.
142. T. Ishihara, *et al.*, Therapeutic effect of lecithinized superoxide dismutase against colitis, *J. Pharmacol. Exp. Ther.*, 2009, **328**, 152–164.
143. V. V. Shuvaev, *et al.*, Targeted detoxification of selected reactive oxygen species in the vascular endothelium, *J. Pharmacol. Exp. Ther.*, 2009, **331**, 404–411.
144. J. Nong, P. M. Glassman and V. R. Muzykantov, Targeting vascular inflammation through emerging methods and drug carriers, *Adv. Drug. Deliv. Rev.*, 2022, **184**, 114180.
145. M. D. Howard, C. F. Greineder, E. D. Hood and V. R. Muzykantov, Endothelial targeting of liposomes encapsulating SOD/catalase mimetic EUK-134 alleviates acute pulmonary inflammation, *J. Controlled Release*, 2014, **177**, 34–41.
146. Y. Zhang, *et al.*, Reactive oxygen species scavenging and inflammation mitigation enabled by biomimetic prussian blue analogues boycott atherosclerosis, *J. Nanobiotechnol.*, 2021, **19**, 161.
147. Y. Cheng, *et al.*, Mitochondria-targeted antioxidant delivery for precise treatment of myocardial ischemia-reperfusion injury through a multistage continuous targeted strategy, *Nanomedicine*, 2019, **16**, 236–249.
148. T. Wu, *et al.*, Aortic plaque-targeted andrographolide delivery with oxidation-sensitive micelle effectively treats atherosclerosis via simultaneous ROS capture and anti-inflammation, *Nanomedicine*, 2018, **14**, 2215–2226.
149. M. C. C. Ferrer, *et al.*, Icam-1 targeted nanogels loaded with dexamethasone alleviate pulmonary inflammation, *PLoS One*, 2014, **9**, e102329.

150. B. S. Ding, *et al.*, Prophylactic thrombolysis by thrombin-activated latent prourokinase targeted to PECAM-1 in the pulmonary vasculature, *Blood*, 2008, **111**, 1999–2006.
151. C. F. Greineder, *et al.*, Vascular immunotargeting to endothelial determinant ICAM-1 enables optimal partnering of recombinant scFv-thrombomodulin fusion with endogenous cofactor, *PLoS One*, 2013, e80110.
152. C. Garnacho, S. M. Albelda, V. R. Muzykantov and S. Muro, Differential intra-endothelial delivery of polymer nanocarriers targeted to distinct PECAM-1 epitopes, *J. Controlled Release*, 2008, **130**, 226–233.
153. S. Muro, *et al.*, Slow intracellular trafficking of catalase nanoparticles targeted to ICAM-1 protects endothelial cells from oxidative stress, *Am. J. Physiol. Cell. Physiol.*, 2003, **285**, C1339–C1347.
154. S. Perkowski, *et al.*, Gene expression profiling of the early pulmonary response to hyperoxia in mice, *Am. J. Respir. Cell. Mol. Biol.*, 2003, **28**, 682–696.
155. S. Kessner, A. Krause, U. Rothe and G. Bendas, Investigation of the cellular uptake of E-Selectin-targeted immunoliposomes by activated human endothelial cells, *Biochim. Biophys. Acta*, 2001, **1514**, 177–190.
156. B. Bénéteau-Burnat and B. Baudin, Angiotensin-converting enzyme: clinical applications and laboratory investigations on serum and other biological fluids, *Crit. Rev. Clin. Lab. Sci.*, 1991, **28**, 337–356.
157. D. P. McIntosh, X. Y. Tan, P. Oh and J. E. Schnitzer, Targeting endothelium and its dynamic caveolae for tissue-specific transcytosis in vivo: a pathway to overcome cell barriers to drug and gene delivery, *Proc. Natl. Acad. Sci. US A*, 2002, **99**, 1996–2001.
158. R. O. Hynes, Integrins: bidirectional, allosteric signaling machines, *Cell*, 2002, **110**(6), 673–687.
159. M. Chorny, E. Hood, R. J. Levy and V. R. Muzykantov, Endothelial delivery of antioxidant enzymes loaded into non-polymeric magnetic nanoparticles, *J. Controlled Release*, 2010, **146**, 144–151.
160. S. Kartha, L. Yan, C. L. Weisshaar, M. E. Ita, V. V. Shuvaev, V. R. Muzykantov, A. Tsourkas, B. A. Winkelstein and Z. Cheng, Superoxide Dismutase-Loaded Porous Polymersomes as Highly Efficient Antioxidants for Treating Neuropathic Pain, *Adv. Healthc. Mater.*, 2017, **6**(17), DOI: 10.1002/adhm.201700500.
161. E. A. Simone, *et al.*, Loading PEG-catalase into filamentous and spherical polymer nanocarriers, *Pharm. Res.*, 2009, **26**, 250–260.
162. R. Atoui, D. Shum-Tim and R. C. Chiu, Myocardial regenerative therapy: immunologic basis for the potential "universal donor cells", *Ann. Thorac. Surg.*, 2008, **86**, 327–334.
163. J. Rehman, *et al.*, Secretion of angiogenic and antiapoptotic factors by human adipose stromal cells, *Circulation*, 2004, **109**, 1292–1298.
164. Z. Tao, *et al.*, Coexpression of VEGF and angiopoietin-1 promotes angiogenesis and cardiomyocyte proliferation reduces apoptosis in

porcine myocardial infarction (MI) heart, *Proc. Natl. Acad. Sci. U. S. A.*, 2011, **108**, 2064–2069.

165. L. Zhou, *et al.*, VEGF165 and angiopoietin-1 decreased myocardium infarct size through phosphatidylinositol-3 kinase and Bcl-2 pathways, *Gene Ther.*, 2005, **12**, 196–202.
166. S. M. Samuel, *et al.*, Coadministration of adenoviral vascular endothelial growth factor and angiopoietin-1 enhances vascularization and reduces ventricular remodeling in the infarcted myocardium of type 1 diabetic rats, *Diabetes*, 2010, **59**, 51–60.
167. A. Paul, M. Nayan, A. A. Khan, D. Shum-Tim and S. Prakash, Angiopoietin-1-expressing adipose stem cells genetically modified with baculovirus nanocomplex: investigation in rat heart with acute infarction, *Int. J. Nanomed.*, 2012, **7**, 663–682.
168. S. Qi, *et al.*, Cellular Internalization-Induced Aggregation of Porous Silicon Nanoparticles for Ultrasound Imaging and Protein-Mediated Protection of Stem Cells, *Small*, 2019, **15**, 1804332.
169. Y. Liu, Z. Zhao and M. Li, Overcoming the cellular barriers and beyond: Recent progress on cell penetrating peptide modified nanomedicine in combating physiological and pathological barriers, *Asian J. Pharm. Sci.*, 2022, **17**, 523–543.
170. J. W. Myerson, *et al.*, Non-affinity factors modulating vascular targeting of nano- and microcarriers, *Adv. Drug Deliv. Rev.*, 2016, **99**, 97–112.
171. H. S. Choi, *et al.*, Renal clearance of quantum dots, *Nat. Biotechnol.*, 2007, **25**, 1165–1170.
172. R. A. Petros and J. M. DeSimone, Strategies in the design of nanoparticles for therapeutic applications, *Nat. Rev. Drug. Discovery*, 2010, **9**, 615–627.
173. A. C. Anselmo and S. Mitragotri, Cell-mediated delivery of nanoparticles: taking advantage of circulatory cells to target nanoparticles, *J. Controlled Release*, 2014, **190**, 531–541.
174. S. Muro, *et al.*, Control of endothelial targeting and intracellular delivery of therapeutic enzymes by modulating the size and shape of ICAM-1-targeted carriers, *Mol. Ther.*, 2008, **16**, 1450–1458.
175. P. Kolhar, *et al.*, Using shape effects to target antibody-coated nanoparticles to lung and brain endothelium, *Proc. Natl. Acad. Sci. U. S. A.*, 2013, **110**, 10753–10758.
176. V. V. Shuvaev, *et al.*, Endothelial targeting of antibody-decorated polymeric filomicelles, *ACS Nano*, 2011, **5**, 6991–6999.
177. M. Miragoli, *et al.*, Inhalation of peptide-loaded nanoparticles improves heart failure, *Sci. Transl. Med.*, 2018, **10**, eaan6205.
178. P. L. Ma, M. Lavertu, F. M. Winnik and M. D. Buschmann, Stability and binding affinity of DNA/chitosan complexes by polyanion competition, *Carbohydr. Polym.*, 2017, **176**, 167–176.
179. Z. M. Binsalamah, A. Paul, A. A. Khan, S. Prakash and D. Shum-Tim, Intramyocardial sustained delivery of placental growth factor using

nanoparticles as a vehicle for delivery in the rat infarct model, *Int. J. Nanomed.*, 2011, **6**, 2667–2678.
180. Y. Nakano, *et al.*, Nanoparticle-Mediated Delivery of Irbesartan Induces Cardioprotection from Myocardial Ischemia-Reperfusion Injury by Antagonizing Monocyte-Mediated Inflammation, *Sci. Rep.*, 2016, **6**, 29601.
181. M. M. Nguyen, N. C. Gianneschi and K. L. Christman, Developing injectable nanomaterials to repair the heart, *Curr. Opin. Biotechnol.*, 2015, **34**, 225–231.
182. A. Ishikita, T. Matoba, G. Ikeda, J. Koga, Y. Mao, K. Nakano, D. Takeuchi, J. Sadoshima and K. Egashira, Nanoparticle-Mediated Delivery of Mitochondrial Division Inhibitor 1 to the Myocardium Protects the Heart From Ischemia-Reperfusion Injury Through Inhibition of Mitochondria Outer Membrane Permeabilization: A New Therapeutic Modality for Acute Myocardial Infarction, *J. Am. Heart Assoc.*, 2016, **5**(7), e003872.
183. G. Ikeda, *et al.*, Nanoparticle-Mediated Targeting of Cyclosporine A Enhances Cardioprotection Against Ischemia-Reperfusion Injury Through Inhibition of Mitochondrial Permeability Transition Pore Opening, *Sci. Rep.*, 2016, **6**, 20467.
184. J. Kim, L. Cao, D. Shvartsman, E. A. Silva and D. J. Mooney, Targeted delivery of nanoparticles to ischemic muscle for imaging and therapeutic angiogenesis, *Nano. Lett.*, 2011, **11**, 694–700.
185. P. M. Ridker, *et al.*, Antiinflammatory Therapy with Canakinumab for Atherosclerotic Disease, *N. Engl. J. Med.*, 2017, **377**, 1119–1131.
186. S. Singha, *et al.*, Peptide-MHC-based nanomedicines for autoimmunity function as T-cell receptor microclustering devices, *Nat. Nanotechnol.*, 2017, **12**, 701–710.
187. X. Clemente-Casares, *et al.*, Expanding antigen-specific regulatory networks to treat autoimmunity, *Nature*, 2016, **530**, 434–440.
188. R. Feiner, *et al.*, Engineered hybrid cardiac patches with multifunctional electronics for online monitoring and regulation of tissue function, *Nat. Mater.*, 2016, **15**, 679–685.
189. K. Cheng, *et al.*, Magnetic antibody-linked nanomatchmakers for therapeutic cell targeting, *Nat. Commun.*, 2014, **5**, 4880.
190. D. B. Cowan, *et al.*, Intracoronary Delivery of Mitochondria to the Ischemic Heart for Cardioprotection, *PLoS ONE*, 2016, **11**, e0160889.
191. D. Lee, *et al.*, Hydrogen peroxide-responsive copolyoxalate nanoparticles for detection and therapy of ischemia-reperfusion injury, *J. Controlled Release*, 2013, **172**, 1102–1110.
192. M. Howard, B. J. Zern, A. C. Anselmo, V. V. Shuvaev, S. Mitragotri and V. Muzykantov, Vascular targeting of nanocarriers: perplexing aspects of the seemingly straightforward paradigm, *ACS Nano*, 2014, **8**, 4100–4132.
193. H. Parhiz, *et al.*, Unintended effects of drug carriers: Big issues of small particles, *Adv. Drug Delivery Rev.*, 2018, **130**, 90–112.

CHAPTER 4

Cardiovascular Nanomedicine: Mode of Action and Applications

FOUZIA PERVEEN MALIK*[a], SUMAIYA AZAM[a], AQSA ALEEM[a] AND JAHANZAIB NOWSHERWAN[b]

[a]School of Interdisciplinary Engineering and Sciences (SINES), NUST, H-12, Islamabad 44000, Pakistan; [b]Inorganic Chemistry Department, Department of Chemistry, University of Oxford, Oxford, UK
*E-mail: fouzia@sines.nust.edu.pk

4.1 Introduction

Cardiovascular diseases (CVDs) are a broad class of chronic heart illnesses and vascular system disorders, primarily including congenital heart disease, cerebrovascular disease, peripheral arterial disorders, coronary heart disease, deep vein thrombosis, and pulmonary embolism. CVDs are a global health concern and a primary cause of mortality and morbidity worldwide. In 2019, CVDs caused 17.9 million deaths, representing 32% of all global deaths, according to the World Health Organization. People suffering from CVDs are at a greater risk of comorbid conditions, such as diabetes, depression, or obesity.[1]

Over the past twenty years, tremendous research focus has been on nanotechnology, which involves the study of synthesizing systems, devices, and structures on length scales of about 1–100 nm. The medicinal

Nanoscience & Nanotechnology Series No. 64
Cardiovascular Nanomedicine
Edited by Roger Narayan and Tanveer Tabish
© The Royal Society of Chemistry 2024
Published by the Royal Society of Chemistry, www.rsc.org

application of nanotechnology is termed as nanomedicine. Nanodrugs consist of both biological and non-biological medical species. Because of their multi-functionality, zero-dimensional particles with dimensions under 100 nm have potential applications in the treatment of diseases in the modern era, and many of these nanotechnology-based drugs have been approved by the Food and Drug Administration (FDA). The first nanomedicine, Doxil®, was used in 1995 for treating Kaposi's sarcoma and ovarian cancer.[2]

Nanomedicine, being a research hotspot, has taken the world by storm, as it is well suited to address challenges in pharmaceutics. Due to the unique chemical, structural, magnetic, mechanical, and biological properties of nanoparticles (NPs), it is possible to synthesize precise nanoparticles as drug carriers for controlled and targeted delivery of therapeutic substances to the desired sites, diagnostic applications, and biomarker detection. Moreover, these nanomedicines are currently being used in imaging techniques, designing of medical devices, stem cell applications, and wound healing. The therapeutic effects of nanoparticles depend on their physicochemical properties, such as their size, shape, distribution, and composition. Furthermore, these properties influence the stability of synthesized NPs, drug loading, drug release, and interaction of these nanoparticles with biological systems. In addition to particle size, shape is also an important parameter to consider, as it affects the intracellular release of the therapeutic compound.[3]

Applications of pharmaceutical nanotechnology in cardiovascular diseases are gaining momentum and aim to establish innovative solutions for current challenges in cardiovascular diseases. Nanopharmaceutics emphasizes site-specific drug delivery along with a diagnosis of cardiovascular disorders at early stages, which provides a comprehensive understanding of disease mechanisms, treatment processes, and their control at initial stages. This advanced treatment modality has advantages, such as higher safety, improved bioavailability, enhanced metabolic stability, high membrane permeability, increased half-life of therapeutics and thus a longer duration of action, precise distribution, and sustainable delivery of drug payloads specifically to diseased cells, avoiding non-specific uptake. Other benefits of nanodrugs include increased efficacy, improved safety of dosage, biodegradability, high tolerability profiles, and minimum side effects.[4]

4.1.1 Nanomaterials as Nanocarriers

Nanomaterials are a broad class of materials consisting of very small particles having diameters not larger than 100 nm. Because of their unique properties, these materials have broad applications in medicine and pharmaceutics. A high surface-area-to-volume ratio is responsible for the high drug-loading ability of nanomaterials; thus, in the modern era, these materials are an excellent choice as drug vehicles. Nanomaterials are broadly classified into different categories based on their morphology, size, and chemical characteristics. They are used in the treatment of cardiovascular diseases and are recognized as efficient and safer treatment modalities.

4.1.1.1 Micelles

Ordered arrangements of amphiphiles (molecules with hydrophobic tails and hydrophilic heads) form self-assembled structures termed micelles. In a micelle, the hydrophobic tails of amphiphiles arrange together to form the hydrophobic core and the hydrophilic head regions face outward in the aqueous environment, as represented in Figure 4.1A. Depending on the concentration and composition, micelles have a size ranging from 20 Å to 200 Å and form aggregates in the solvent. Usually, micelles are used as nanocarriers for targeted drug delivery by making drugs biocompatible and increasing their half-life. The efficiency of drug delivery depends on the compatibility between the drug and the inner core of the micelle.

The hydrophilic head region renders micelles water soluble and helps in intravenous delivery, whereas the unique core structure (hydrophobic) of micelles acts as a container for therapeutic agents, allowing targeted delivery of the loaded drug to the site of interest. This dynamic design of polymeric micelles makes them biocompatible and highly stable *in vivo* and *in vitro* and allows them to dissolve a diverse variety of poorly soluble pharmaceutical agents.[6]

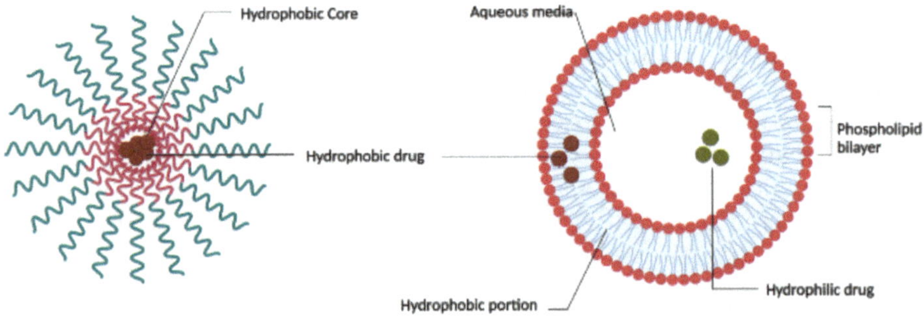

Figure 4.1 Schematic representation of a micelle and a liposome. (A) Pharmaceutical agent is encapsulated in the hydrophobic core of the micelle. (B) Depiction of a liposome containing hydrophilic drugs in its core and hydrophobic drugs in the lipid bilayer.[5] Reproduced from ref. 5, https://doi.org/10.3390/pharmaceutics14102195, under the terms of the CC BY 4.0 license, https://creativecommons.org/licenses/by/4.0/.

4.1.1.2 Liposomes

Liposomes are aggregates of amphiphiles, mainly consisting of a phospholipid bilayer, especially cholesterol and phosphatidylcholine. Liposomes have a size range of 50–500 nm, similar to micelles, but they have quite dissimilar structures. Liposomes have a hydrophilic core, which is surrounded by the hydrophobic membrane (lipid bilayer), as shown in Figure 4.1B. Owing to their structural property and biocompatibility, liposomes can be used as drug vehicles and diagnostic agents. The aqueous interior of liposome can store hydrophilic particles and the hydrophobic part can accommodate therapeutic molecules. The lipid bilayer can fuse with other bilayers, such as the cell membrane, thus releasing the entrapped pharmaceutical agent to the desired action site and limiting the attachment of drugs to non-target cells. Liposomes can also be used for the encapsulation of contrast-generating materials, which facilitates an early diagnosis and a better understanding of disease progression.

As a drug carrier, liposomes possess unique properties, *e.g.*, sustained release, biodegradability, particle size, possibility of surface functionalization, improved drug efficacy, stability, prolonged half-life of drugs, surface modifications, improved circulation time, and better distribution in biological systems. The aforementioned properties of liposomes make liposomal nanoformulated drugs one of the most successful drug delivery systems that overcome the elevated side effects of many conventional drugs.[7]

4.1.1.3 Dendrimers

Dendrimers are globular, three-dimensional nanomacromolecules that exhibit a high degree of molecular geometry. Dendrimers are radially symmetric molecules having three main structural components: a

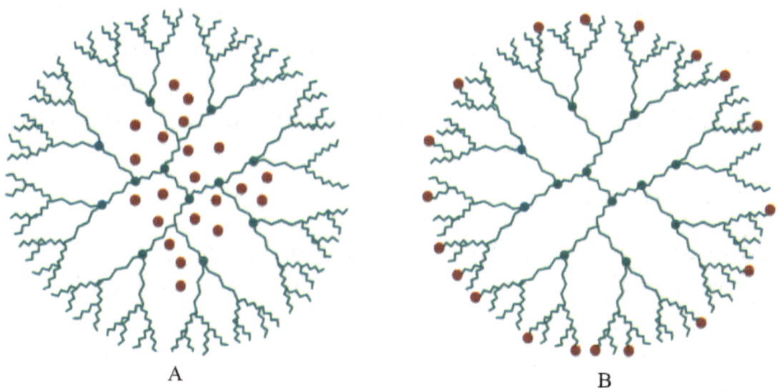

Figure 4.2 Strategies for the interaction between a pharmaceutical agent and dendrimer: (A) encapsulation inside the core and (B) electrostatic interaction or covalent conjugation.[8] Reproduced from ref. 8 with permission from Elsevier, Copyright 2008.

multivalent shell, branching layers (generation), and a well-defined interior core, as indicated in Figure 4.2. This well-specified and uniform structure has a profound influence on the chemical and physical properties of these highly ordered molecules. Dendrimers have diverse industrial and biomedical applications because of their unique characteristics, such as reduced toxicity, high permeability, biodegradation, solubility, nanoscale size, chemical stability, and the presence of internal cavities.

Biomedical applications of dendrimers include solubilization, encapsulation, complexation, site-specific drug delivery, biopharmaceuticals, and contrast agents. Dendrimers with hydrophilic periphery and hydrophobic core have been shown to possess a micelle-like behavior. Hydrophobic drugs with poor solubility are entrapped in the hydrophobic core of dendrimers, where they can be solubilized, thus offering improved drug bioavailability. By modifying the physicochemical properties of dendrons or dendrimer generation, it is possible to control the drug-carrying and drug-releasing capacities of dendrimers. Designing an appropriate dendrimer is the major step for drug accommodation, as the entrapment of the drug is dependent on the structures of both the dendrimer and the drug. The association between a dendrimer and a pharmaceutical agent may occur in different ways, either through covalent conjugation or through non-covalent interactions.

The inherent architecture and empty internal cavities of dendrimers allow the direct encapsulation of drug molecules inside the core through non-covalent interactions. The internal cavities (hydrophobic in nature) of dendrimers interact with poorly soluble drugs *via* hydrophobic interactions. Drug encapsulation provides advantages, such as rapid preparation without disturbing the pharmacological activity of drugs, reduced possible drug toxicity, stability, high solubility, and preservation of the pharmacological properties of pharmaceutical agents. Drug loading can also be achieved through covalent conjugation, as dendrimers possess a large number of surface functional groups. Dendrimer–drug conjugation offers advantages such as selective and controlled drug release that cannot be obtained through simple encapsulation of a pharmaceutical or electrostatic interaction between the drug and dendrimer. Other advantages of drug conjugation include optimum biodistribution and tissue targeting.[8]

4.1.1.4 Quantum Dots

Having diameters less than 10 nm, quantum dots (QDs) are tiny particles possessing intense fluorescent properties. First, as reported by the physicist Alexei Ekimov, in the 1980s, QDs have narrow tunable emission spectra and broadband absorbance. These tiny particles have characteristics, including high quantum yield, good chemical and photostability, and size-tuneable light emission, which make them versatile in applications such as LED technology, solar cells, and biomedical applications, including imaging, drug delivery, and cancer photodynamic therapy. QDs are mainly used for capturing fluorescence tomography images, biological optical detection, and cellular and intracellular targeting. QDs can be used as traceable drug carriers because of their ultra-fine particle size, ease of fabrication, and ability to conjugate with a variety of therapeutic agents and optical properties.

QDs having water-soluble capping stabilizers, such as thioglycolic acid (TGA), mercaptoethylamine, and polyethylene glycol (PEG) polymer, conjugate readily with a pharmaceutical agent *via* electrostatic interactions or covalent bonds. A better understanding of the mechanistic pathways of delivery of drugs and investigation of properties, such as site-specific targeting and effective drug release within molecules, is provided by tracking the fluorescence signal of QDs. Moreover, sophisticated drug delivery systems that assemble treatment modalities and QDs into a single construct can significantly enhance therapeutic effects and minimize side effects. Certain small or large compounds, such as peptides and DNA that have significant medical applications, have been studied successfully using fluorescent imaging techniques based on QDs. These tiny particles have high drug-loading efficiency, targeted and controlled drug release, prolonged blood circulation time, and integration of multiple targeting ligands on the surface.[9]

4.1.1.5 Nanocrystals

Nanocrystals are crystalline materials with at least one dimension less than 100 nm (size range is 1–1000 nm). Nanocrystals have received great attention as a viable drug delivery platform for the effective treatment of various diseases because of their physical stability, relative ease of production, high drug-loading properties, low particle size, and site-specific targeting. Formulation of drugs having poor solubility into nanocrystals results in enhanced saturation solubility, satisfactory therapeutic concentrations at low doses, and improved biodistribution. All of these properties, regardless of the drug's administration route, promise an improved bioavailability of poorly soluble drugs. The size, crystallinity, surface charge, and shape of nanocrystals and the type of stabilizers used in nanocrystal formulation play a crucial role in medicinal applications. Some commonly used stabilizers in nanocrystal drug formulation are sodium lauryl sulfate, polyvinyl alcohol, polyvinylpyrrolidone (PVP), trioctylphosphine oxide (TOPO), and hydroxypropyl methylcellulose. Iron oxide (IO; 11–16 nm), copper nanocrystals (2–10 nm), gold nanoparticles (AuNPs; 1–4 nm), and chitosan/$LaF_3 : Eu^{3+}$ (13–18 nm) are some common examples of nanocrystals. TriCor is the nanocrystal drug available in the market for treating hypercholesterolemia.[10]

4.1.1.5.1 Iron Oxide. IO nanoparticles have unique chemical and magnetic properties, which are responsible for their biomedical applications. They are also known as superparamagnetic iron oxide (SPIO) nanoparticles (SPIONs) because of their exceptional magnetic properties. At present, SPIO nanoparticle-based materials are widely used in biomedical applications. SPIO nanoparticles have a size in the range of 10–100 nm, and ultra-small superparamagnetic iron oxide (USPIO) nanoparticles are less than 50 nm in size. IO nanoparticles are clinically approved as a biocompatible system for drug delivery and as an MRI contrast medium for the non-invasive monitoring of cells in living organisms. They have low toxicity, prolonged blood retention time, and timely biodegradability, and they can be modified by suitable moieties, such as inorganic molecules, polymeric stabilizers, poloxamers, and PEG derivatives. Dextran-coated iron oxide nanoparticles have been clinically approved and are commonly used as MRI contrast agents; primarily, these designed nanoparticles play a significant role in the identification of atherosclerotic plaques and tracking of mesenchymal stem cells.[11]

4.1.1.6 Nanoparticles

Nanoparticles are defined as particles of matter having diameters ranging between 1 and 100 nm. These ultrafine particles can be 0D, 1D, 2D, or 3D, depending on their overall shape. Through nanotechnology, the size of particles and devices can be controlled. Nanoengineered devices or nanodevices are much less invasive than existing medical devices and have sizes similar to those of biological macromolecules, such as enzymes and receptors, *e.g.*, hemoglobin, which is approximately 5 nm in diameter. Due to their small size, these nanoconstructs and nanodrugs can interact readily with biomolecules on the cell surface and within the cell, thus providing a better understanding of processes that derive cellular systems.[12]

4.1.1.6.1 Gold Nanoparticles. Gold nanoparticles (size ranging from 5 to 400 nm) have significant applications in bionanotechnology, pharmaceuticals, and catalysis owing to their exceptionally unique and tunable optical properties. As these tiny materials possess remarkable biocompatibility and high chemical and physical stability, they are especially used in the pharmaceutical sector. Besides being feasible to synthesize, the versatile surface chemistry of these NPs allows for surface modification, allowing one to coat AuNPs with a variety of substances, including biomolecules, small molecules, and polymers, or to functionalize them with biologically active organic molecules. Moreover, fluorophores and gadolinium chelates can be attached to their surface to yield greater contrast for MRI and fluorescence methods.[13]

Presently, these superior NPs are used as contrast agents for the majority of radiological procedures, including computed tomography (CT) scans, intravascular ultrasound (IVUS), and X-ray imaging. In cardiovascular applications, however, AuNPs are potential candidates due to their use as contrast agents in CT scans. To deliver large quantities of radio-opaque materials, required for CT scans, the use of AuNPs is considered to be a relatively straightforward approach.[14]

4.1.1.7 Microbubbles (MBs) and Nanobubbles (NBs)

Microbubbles (diameter: 1–999 μm) are small core–shell bubbles, *i.e.*, they are composed of a gas core protected by a shell, and nanobubbles are even smaller in size with diameters in the range of 1–999 nm. There are many types of microbubbles and nanobubbles depending on the composition of the core and shell.[15] MBs and NBs have established significant roles in theranostics, which aims to diagnose, treat, and confirm the successful treatment of a disease by using a single approach. In the diagnosis, MBs and NBs are used as ultrasound contrasting agents (UCAs), while in the treatment, they are used for targeted drug delivery.

Bubbles can be categorized as first- and second-generation agents. The first intravenous UCAs, capable of left heart contrast first-generation agents, *e.g.*, Albunex and Levovist (Table 4.1), use air as their core, which is stabilized by a shell material composed of proteins, lipids, or fatty acids. The lifetime of the 1st-generation MBs and NBs after induction *in vivo* is very short due to the solubility of the air core in the blood.

In second-generation MBs and NBs, *e.g.*, SonoVue (Table 4.1), the air core has been replaced with gases that show less solubility in blood. These gases enhance the stability and circulation time of the MBs and NBs within the body *in vivo*.[17]

A second-generation UCA, *e.g.*, Definity, has a gas core of octafluoropropane with a shell material of phospholipid, with perflutren gas (C_3F_8) as the active substance, as shown in Figure 4.3.

Table 4.1 Overview of UCAs and their commercial availability.[16] Reproduced from ref. 16 with permission from Taylor & Francis, Copyright 2012.

Name of the UCAs	Material of core	Material of shell	Availability
1st generation			
Echovist	Air	Galactose matrix	Europe
Levovist	Air	Palmitic acid	Europe
Albunex	Air	Albumin	Available and approved by the FDA
2nd generation			
Definity	Octafluoropropane	Phospholipid	Available
Imagnet	Perfluorohexane	Phospholipid	Available
SonoVue	Sulfur hexafluoride	Phospholipid	Yet to be approved by the FDA and is used in Europe

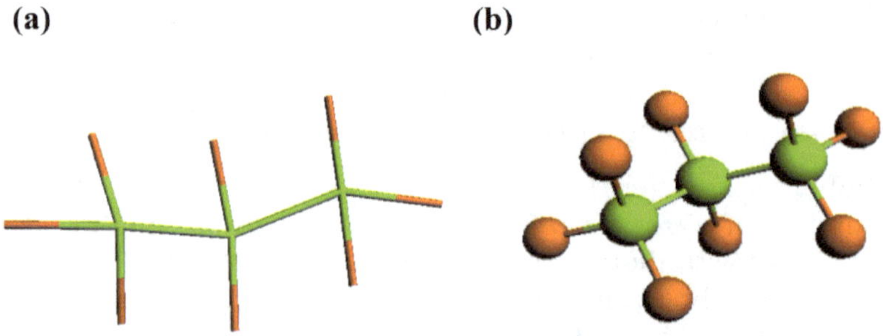

Figure 4.3 Structural formula of the active substance (perflutren gas) in Definity® [C_3F_8]. (a) Shows the chemical structure. (b) 3D orientation of the octafluoropropane (green carbon atoms and orange fluorine atoms).

4.1.2 Nanomedicine in Targeted Drug Delivery

4.1.2.1 Therapeutic Role of Nanomedicine

4.1.2.1.1 Fenofibrate (TriCor). Hyperlipidemia and hypercholesterolemia (high cholesterol levels) are associated with many diseases and are also major cardiovascular risk factors. Our liver maintains a balanced cholesterol level by producing enough cholesterol to meet the body's needs and by getting rid of the excess. High-density lipoprotein (HDL) or good cholesterol transports cholesterol to the liver and functions to remove excess cholesterol from the blood. Another type of lipoprotein, called low-density lipoproteins (LDLs) or bad cholesterol, is responsible for heart diseases as it can cling to arterial walls, consequently narrowing the arteries and blocking them. Unbalanced cholesterol level plays a detrimental role in the pathogenesis of atherosclerosis. High cholesterol level forms fatty deposits in blood vessels, known as plaques. The growth of these plaques results in the narrowing of arteries and thus reduces the blood flow through them. When this plaque grows large enough, it becomes brittle or inflamed and may rupture, resulting in a blood clot. Now, this clot can worsen the condition by narrowing the artery or completely blocking it.[18]

TriCor (nanocrystal) was approved in 2004 by the FDA. It is used for treating hypercholesterolemia and hyperglyceridemia. TriCor reduces the plaque on arterial walls, thus preventing the development of atherosclerosis, which is the underlying cause of heart attacks and strokes. Lipoprotein is responsible for transporting cholesterol throughout the body. We can distinguish these lipoproteins based on their density, which is directly correlated with the amount of either apolipoprotein A-1 (ApoAI; a major structural protein component of HDL) or apolipoprotein B [ApoB; a major constituent of atherogenic particles, very-low-density lipoprotein (VLDL) and LDL]. VLDL and LDL are associated with a greater risk of coronary heart disease. A higher level of HDL may lessen the risk of cardiovascular diseases because of its potential role in reverse cholesterol transport (a mechanism that works by removing the excess cholesterol from peripheral tissues and transporting it back to the liver).

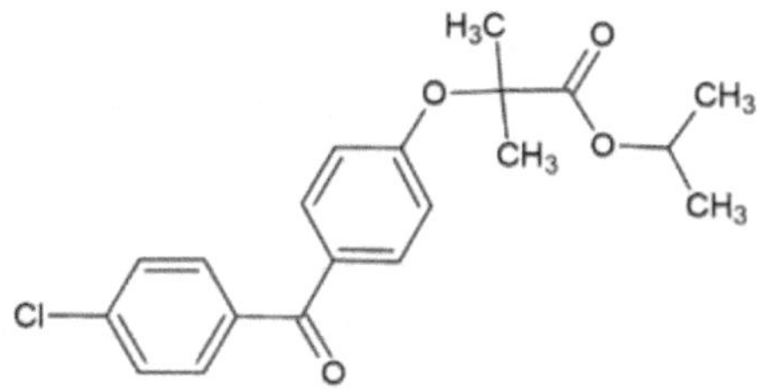

Propan-2-yl 2-[4-(4 chlorobenzoyl)phenoxy]-2-methylpropanoate.

Fenofibrate works by reducing elevated levels of LDL, triglycerides, and apolipoprotein B and by increasing HDL. When administered in the body, fenofibrate is hydrolyzed into fenofibric acid. Fenofibric acid binds with peroxisome proliferator-activated receptor-alpha (PPAR-α; a ligand-activated transcriptional factor that is responsible for regulating gene expression and critical for lipoprotein and lipid metabolism). This activated receptor regulates the gene expression after dimerization with retinoid X receptor (RXR). The PPAR-α/RXR heterodimer binds to PPAR response elements (PPREs), thereby acting as a transcriptional regulator. This complex results in:

- upregulation of the lipoprotein lipase (LPL) enzyme, promoting the degradation of circulating triglycerides into free fatty acids in the bloodstream. This increased lipolysis results in the lowering of triglyceride-rich VLDL within the body (more degradation, less production);
- stimulating the production of Apo-A1 and Apo-A2 (major components of HDL), consequently producing more HDL;
- suppressing the synthesis of apolipoprotein C-III (an inhibitor of the lipoprotein lipase activity), which reduces the lipolysis of triglyceride-rich molecules; and
- suppressing the synthesis of apolipoprotein C-III (an inhibitor of the lipoprotein lipase activity), which reduces the lipolysis of triglyceride-rich molecules.[19]

All these processes decrease circulating triglycerides, total cholesterol VLDL, triglyceride-rich LDL, and apolipoprotein B, while increasing HDL.

4.1.2.1.2 Probucol (PBC). Probucol (Lorelco) is a powerful antioxidant that is commonly used to treat high cholesterol levels in the blood. PBC works by promoting the catabolic rate of LDL, thus lowering the cholesterol level in the bloodstream. PBC, being a powerful antioxidant, inhibits the oxidation of cholesterol in LDLs, which slows down the process of foam cell formation. These cells play a crucial role in the pathogenesis of atherosclerosis. The efficiency of this drug, however, is reliant on its solubility, and it has approximately negligible solubility (about 5 ng ml^{-1}) in water. This hydrophobic medication has a low rate of absorption in the gastrointestinal tract, which considerably reduces its bioavailability. Thus, to achieve the therapeutic plasma concentration, a higher dose is needed, which may have some associated undesirable side effects. Therefore, it is necessary to develop novel strategies to enhance the efficacy of this antioxidant. Nanoformulation by cogrinding or wet-grinding can enhance its water solubility and drug absorption *in vivo*. The wet-grinding strategy, however, results in drug particles with submicron size, while the cogrinding approach provides nanoparticles (below 100 nm) with heterogeneous sizes. By integrating spray drying and cogrinding, a new approach has been demonstrated. The bioavailability, solubility, effectiveness, and dissolution rate can be improved by decreasing the particle size of drugs.

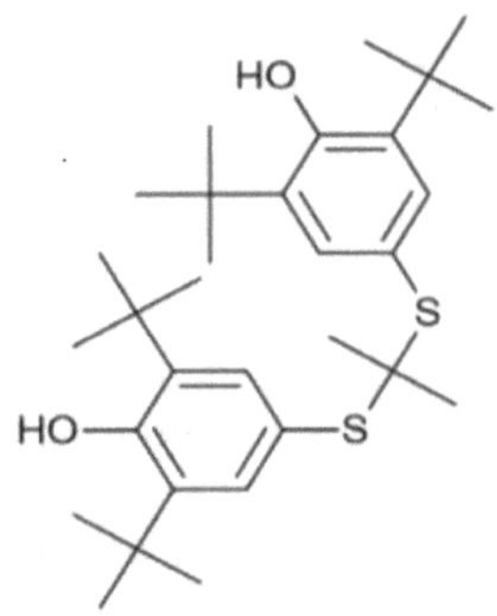

2,6-Di-*tert*-butyl-4-[2-(3,5-di-*tert*-butyl-4-hydroxyphenyl)sulfanylpropan-2-ylsulfanyl]phenol.

The first step of this novel technique is to prepare a coground mixture of PBC, sodium dodecyl sulfate (SDS), and methacrylic acid-methyl methacrylate copolymer (MMC) (1 : 1 : 5). The particle size measurement of this ternary component confirmed that PBC is present as a colloidal nanoparticle. The presence of primary nanoparticles in a ternary ground mixture is confirmed by microscopic techniques, such as scanning electron microscopy (SEM) and atomic force microscopy (AFM). The second step, as represented in Figure 4.4, involves the suspension of the ternary component mixture in distilled water and the gradual addition of 1.7% NH_4OH with slow agitation for an hour using a magnetic stirrer. Finally, pharmaceutical granules are formed by spray coating the ternary component mixture onto carrier particles.

It is necessary to examine the ability of the newly synthesized nanoparticle mixture to permeate the cell membrane, as the effectiveness of PBC

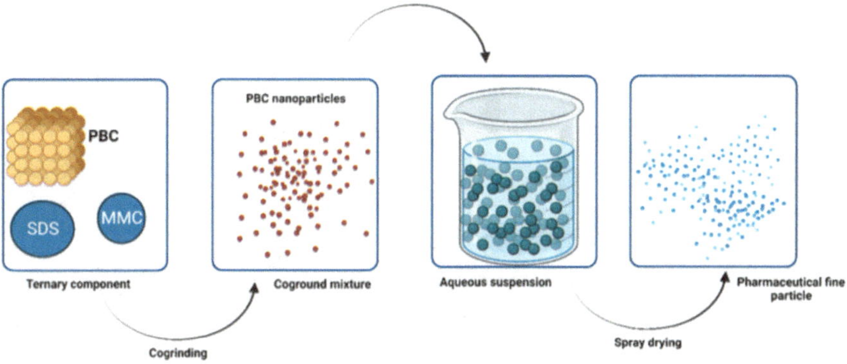

Figure 4.4 An illustration showing the process for synthesizing probucol containing fine pharmaceutical particles.[20] Reproduced from ref. 20 with permission from American Chemical Society, Copyright 2009.

is reliant on this ability. Caco-2 cells are commonly used to examine drug absorption in humans. The permeated amounts of probucol were measured from the mixture of PBC/MMC/SDS with and without grinding. The experimental results revealed that the fine particles of PBC pass through the cell membrane, while the ungrounded mixture of PBC failed to do so, even after 1 hour (Fukami *et al.* 2009). Furthermore, the prepared PBC nanoparticles exhibited improved oral bioavailability and efficient absorption through the mucous membranes of the human small intestine. These effects are significant in the treatment of myocardial infarction by the inhibition of LDL oxidation and its deposition in plaque. Thus, this novel approach of cogrinding followed by spraying has enhanced the solubility and bioavailability of the hydrophobic PBC drug.[20]

Some reported nanocarriers and nanodrugs for treating various cardiovascular diseases are listed in Table 4.2.

4.1.2.2 Theranostic Roles of Nanomedicine

4.1.2.2.1 Applications of Microbubbles and Nanobubbles. In the clinical management of diseases, there are three major steps, *i.e.*, diagnosis, treatment, and confirmation of successful treatment. At each stage, there are approaches that are used for an effective result, but this also prolongs and complicates the management process. A novel approach that is highly in demand is "theranostics", which involves adopting a single approach/technique for diagnosis, treatment, and confirmation of a successful cure of a disease. In this regard, MBs and NBs can play a vital role as they can be used in diagnosis [as UCAs in treatment (in targeted drug delivery)] and in confirmation that drugs have been delivered [*e.g.*, continuous ultrasound imaging (UI) of target site could confirm the drug delivery].[33]

Table 4.2 Different nanomedicine approaches toward CVDs.[21] Reproduced from ref. 21, https://doi.org/10.3390/ph15040441, under the terms of the CC BY 4.0 license, https://creativecommons.org/licenses/by/4.0/.

CVDs	Nanocarrier/drug	Targeted areas	Inferences	Ref.
Myocardial infarction	Magnetic nanoparticles	Extracellular matrix—metalloproteinase inducer	*In vivo* visualization and regression of acute myocardial infarction	22
Angina pectoris	Ivabradine/polymeric nanoparticles	Funny channels of SA node	Improved permeability and anti-fungal effects	23
Thrombosis	tPA/chitosan magnetic nanoparticles	Fibrin clot	Increased efficacy of drugs	24
Thrombosis	Fibrin-targeted H_2O_2-responsive nanoparticles	Fibrin	Inhibits the formation of thrombus	25
Atherosclerosis	Prednisolone/liposomes	Atherosclerotic macrophages	No anti-inflammatory effects were observed	26
Atherosclerosis	Fumagillin/paramagnetic nanoparticles	Endothelial $\alpha v\beta 3$ integrin	Quantification and inhibition of angiogenesis	27
Hyperlipidemia	Simvastatin/chitosan NPs	HMG-CoA reductase enzyme	Improved hypolipidemic effect	28
Hyperlipidemia	Lovastatin/hyaluronic acid-reconstituted high-density lipoprotein	Atherosclerotic lesions	Greater atheroprotective efficacy	29
Hyperlipidemia	Isradipine/invasomes	L-type calcium channels of vascular smooth muscle and myocardium	Enhanced antihypertensive activity	30
Pulmonary arterial hypertension	Fasudil/liposomes	Rho-kinase receptors	Prolonged vasodilatory effects	31
Atherosclerosis	PLGA nanoparticles	Plaque-targeted peptides PP1 and cRGD	Diagnosis and therapy of atherosclerotic plaques	32

HMG-CoA, hydroxymethylglutaryl-coenzyme A; PLGA, poly (lactic-*co*-glycolic acid); PP1, protein phosphatase-1; SA, sinoatrial node; cRGD, cyclic arginine-glycine-aspartic acid; tPA, tissue plasminogen activator.

4.1.2.2.1.1 As Ultrasound Contrast Agents (Diagnostic Application)

Microbubbles and nanobubbles have established a significant role in medical diagnostics as UCAs.[34] Ultrasound is a mechanical vibration with a high frequency >20 kHz that uses sound waves to generate images and is extensively used in clinical practices for diagnosis and therapeutic applications.

As ultrasound waves travel by collisions and vibrations of molecules, a huge amount of energy is lost in their propagation through body tissues.[35] Imaging the blood flowing in the arteries with conventional ultrasound is challenging, as the standard ultrasound scatters the blood at lower frequencies that fall below the display range, resulting in blood appearing black in images.[36]

Gramiak and Shah first reported the application of MBs and NBs as UCAs to increase the UI in 1968.[37] However, in circulating blood, the *in vivo* decomposition of these air bubbles took place rapidly. MBs and NBs are found to enhance the strength of the signals of ultrasound, and therefore, both have established a significant role in medical diagnosis as UCAs. We can define UCAs as comprising microbubbles and nanobubbles as a gas core encapsulated by a shell, which are greatly echogenic (reflect ultrasound very well) and are used to increase the UI in blood-filled tissues.

The composition of the MB and NB shell is represented in Figure 4.5. MB and NB shell composition is of significant importance, as the material of the shell affects the stability of bubbles *in vivo*. The UCAs can be soft- and hard-shell bubbles. Soft-shell MBs and NBs are those in which a soft-shell material is used, such as lipids, phospholipids, and surfactants. However, hard-shell bubbles are those in which a hard-shell material is used, such as

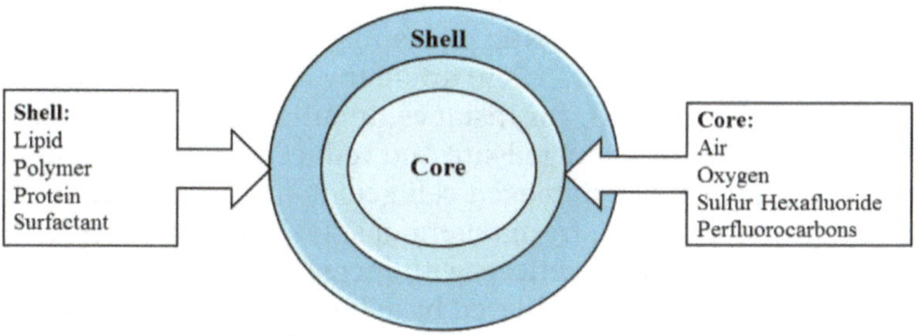

Figure 4.5 Schematic representation of microbubbles and nanobubbles and their core–shell composition. Core material (composed of air or gases) is encapsulated by a shell (composed of lipids, polymers, proteins, or surfactants).[38] Reproduced from ref. 38 with permission from Future Science, Copyright 2016.

polymers, proteins, or inorganic compounds. Studies suggest that soft-shell bubbles are more echogenic but less stable compared to hard-shell bubbles due to their shell compositions.[39]

4.1.2.2.1.2 Applications of Microbubbles and Nanobubbles in Vascular Perfusion Imaging

MBs and NBs have vast applications in echocardiography. They are applied for a better diagnosis of the endocardial border and myocardial contractility. The examination of myocardial perfusion is necessary for diagnosing coronary artery disease and acute myocardial infarction, as it helps to find out the amount of myocardium at risk of necrosis and assess the blood flow in tissues.[40] Myocardial perfusion imaging uses both high and low molecular imaging (MI) ultrasound pulses. The high pulses are used to destroy the MBs and NBs (for better dispersion), and low pulses are used to detect the MBs and NBs. The region affected by myocardial infarction is perfused with smaller amounts of MBs and NBs and appears darker than healthy tissues.[41]

4.1.2.2.1.3 As Tools for Molecular Ultrasound Imaging

The development of MBs and NBs for targeted MI has increased in recent years. They are being prepared with conjugated antibodies or with specific ligands for targeted MI, which increases the retention time within damaged tissues.[42] The conjugation of MBs and NBs with ligands is represented in Figure 4.6; the advantage of this conjugation is that they can be visualized and destroyed by even standard ultrasound machines, and contents of MBs and NBs can be accumulated at the target site.[43] Thus, we can say that ligand-conjugated MBs and NBs are highly in demand for clinical applications, as they are highly organ-specific and can increase drug delivery.[44] Examples of their use in some cardiac applications include the management of coronary artery disease, heart failure, and cardiac remodeling.[45]

4.1.2.2.1.4 Targeted Drug Delivery (Therapeutic Applications)

As described earlier, MBs and NBs play a major role as drug carrier candidates in clinical applications. There are many advantages of targeted drug delivery using MBs and NBs; for instance, it enhances the concentration of drugs at the targeted or damaged site and reduces the interaction of drugs with normal tissues.[46] The damaged cell gets its required drug for healing, and the healthy cell is saved from potential side effects of drugs.[47] Thus, the reduction in toxicity and side effects with increased concentrations of drugs in the affected region can be achieved by using bubbles as carriers.

Studies suggest that the dispersion of bubbles to deliver drugs to the targeted region can be achieved by using radiation force on spheres. Once the dispersion is done, the bubbles are destroyed by applying high MI ultrasound pulses and can be visualized by applying low MI ultrasound pulses. Osei and Al-Asady used ultrasound radiation force to enhance the control of motion of

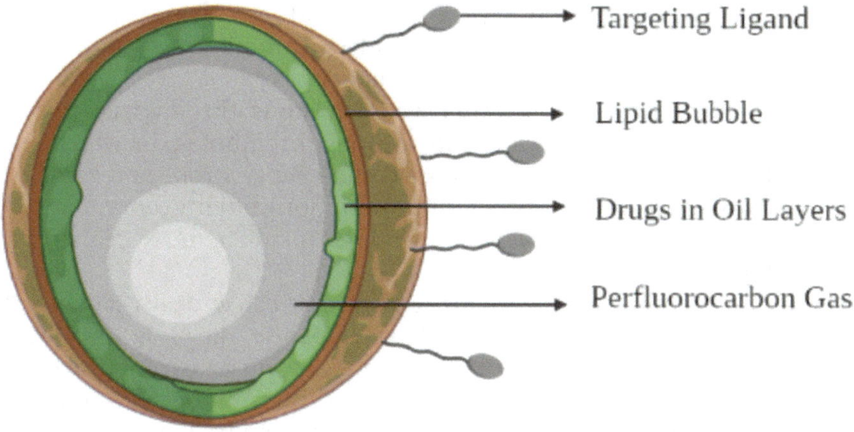

Figure 4.6 Representation of ligand-conjugated MBs and NBs used for drug delivery. Targeting ligand is attached at the outer surface of the bubble, with drugs embedded in the oil layer between the shell and core materials.[16] Reproduced from ref. 16 with permission from Taylor & Francis, Copyright 2012.

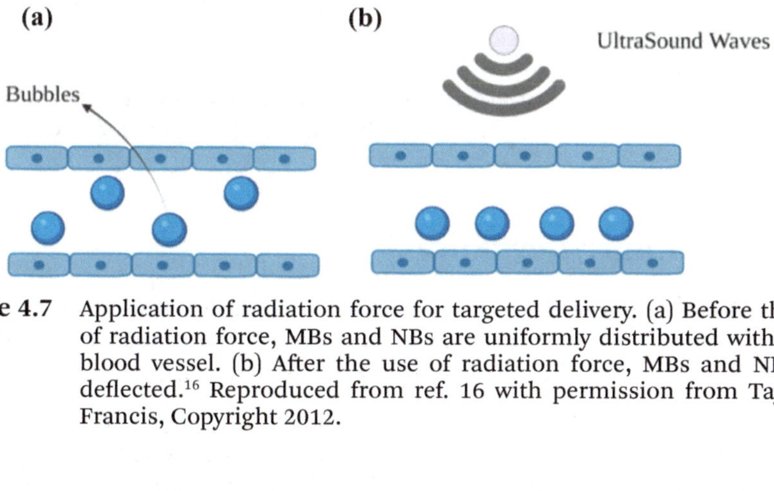

Figure 4.7 Application of radiation force for targeted delivery. (a) Before the use of radiation force, MBs and NBs are uniformly distributed within the blood vessel. (b) After the use of radiation force, MBs and NBs are deflected.[16] Reproduced from ref. 16 with permission from Taylor & Francis, Copyright 2012.

MBs and NBs, as shown in Figure 4.7. They concluded that this technique can be helpful for increasing the targeting of drugs at a specific site.[48]

4.1.2.2.1.5 Role of Microbubbles and Nanobubbles in the Diagnostics of Thrombosis

Thrombosis refers to the formation of blood clots inside blood vessels (a vein or an artery). The formation of malignant blood clots could occur due to the abnormal regulation of the normal hemostatic system (in response to minimizing blood loss due to injury), myocardial infarction, embolism (a clot that breaks apart and begins to propagate in the body), increased blood coagulability, and strokes. An injury to the blood vessel causes disturbances in the blood flow, and the body uses platelets (thrombocytes) and fibrin for the generation of blood clots to prevent blood loss.[49]

Two common types of thrombosis are venous thrombosis, in which blood clotting blocks the vein, and arterial thrombosis, in which blood clotting blocks an artery.[50] Thrombosis could lead to stroke and myocardial infarction. A heart attack (myocardial infarction) often occurs due to the blockage of a coronary artery, which leads to insufficient oxygen supply to the heart, by thrombosis.

Thrombosis is one of the major factors that cause death globally. Yet, its diagnosis is challenging, and its therapeutic range is also limited to produce fruitful treatment. Thus, an early diagnosis of thrombosis is the foremost requirement.[51] UI is the most common technique used for the diagnosis of thrombosis, and its equipment is also available in most hospitals.[52] However, its quality is limited due to its low contrasting properties. The use of MBs and nanobubbles has helped to overcome this challenge.

Li *et al.* worked on various polysaccharide-coated microbubbles using a new, easy one-step polymerization reaction that allows the modification of MBs with different surface groups. They developed fucoidan-functionalized nano-/microparticles as a molecular contrast agent to assess thrombi using MRI. Fucoidan acts as a radiotracer and has already been proven effective for detecting thrombus and heart ischemia.[53] They suggested that, in addition to this, more antithrombotic drugs can be incorporated into the fucoidan MBs, and these could be used for targeted delivery at the thrombus site.[54]

4.2 Summary/Key Points

- CVDs are ever-increasing health concerns, and their medical management is reliant on conventional treatment methodologies, which can result in some potential side effects and toxicities.
- Nanotechnology is an emerging interdisciplinary field that provides innovative solutions to current challenges in CVDs through early disease detection, tissue engineering, imaging diagnostics, and biomarker detection.
- Different classes of nanomaterials, such as micelles, liposomes, dendrimers, quantum dots, iron oxide NPs, MBs, and NBs, have been used for diagnostic and therapeutic purposes in CVDs.
- Microbubbles and nanobubbles are small core bubbles encapsulated by a shell. Bubbles can be categorized as first- and second-generation agents depending on their composition.

- Both MBs and NBs have a significant role in medical diagnostics as UCAs.
- Drug-loaded nanocarriers, MBs and NBs, help in sustained drug release to diseased cells and offer better therapeutic efficacy, avoiding non-specific cell uptake and reducing potential side effects.
- A targeted drug delivery system offers advantages such as drug stabilization, improved bioavailability, extended plasma circulation time, and precise distribution.
- Fenofibrate was approved by the FDA in 2004 for the effective treatment of hypercholesterolemia and hyperlipidemia, both of which are major cardiovascular risk factors. Nanoformulation of PBC enhances the efficiency of this antioxidant drug.
- UI is the most common technique used for the diagnosis of thrombosis, but its quality is limited due to its low contrasting properties. MBs and NBs are vital to overcome this challenge.

References

1. N. Manners, V. Priya, A. K. Mehata, M. Rawat, S. Mohan, H. A. Makeen, M. Albratty, A. Albarrati, A. M. Meraya and M. S. Muthu, Theranostic Nanomedicines for the Treatment of Cardiovascular and Related Diseases: Current Strategies and Future Perspectives, *Pharmaceuticals*, 2022, **15**, 441.
2. M. Iafisco, A. Alogna, M. Miragoli and D. Catalucci, Cardiovascular nanomedicine: the route ahead, *Nanomedicine*, 2019, **14**, 2391–2394.
3. B. Godin, J. H. Sakamoto, R. E. Serda, A. Grattoni, A. Bouamrani and M. Ferrari, Emerging applications of nanomedicine for the diagnosis and treatment of cardiovascular diseases, *Trends Pharmacol. Sci.*, 2010, **31**, 199–205.
4. Z. M. Binsalamah, A. Paul, S. Prakash and D. Shum-Tim, Nanomedicine in cardiovascular therapy: recent advancements, *Expert Rev. Cardiovasc. Ther.*, 2012, **10**, 805–15.
5. M. P. Nikolova, E. M. Kumar and M. S. Chavali, Updates on Responsive Drug Delivery Based on Liposome Vehicles for Cancer Treatment, *Pharmaceutics*, 2022, 2195.
6. M. Chandarana, A. Curtis and C. Hoskins, The use of nanotechnology in cardiovascular disease, *Appl. Nanosci.*, 2018, **8**, 1607–1619.
7. A. Kumar, R. Singla, A. Guliani and S. K. Yadav, Nanoencapsulation for drug delivery, *EXCLI J.*, 2014, 265–286.

8. Y. Cheng, Z. Xu, M. Ma and T. Xu, Dendrimers as drug carriers: Applications in different routes of drug administration, *J. Pharm. Sci.*, 2008, **97**(1), 123–143.
9. Y. Wang and L. Chen, Quantum dots, lighting up the research and development of nanomedicine, *Nanomedicine*, 2011, **7**, 385–402.
10. M. Jarvis, V. Krishnan and S. Mitragotri, Nanocrystals: A perspective on translational research and clinical studies, *Bioeng. Transl. Med.*, 2019, 5–16.
11. N. K. Younis, J. A. Ghoubaira, E. P. Bassil, H. N. Tantawi and A. H. Eid, Metal-based nanoparticles: Promising tools for the management of cardiovascular diseases, *Nanomedicine*, 2021, **36**, 102433.
12. J. Tang, M. E. Lobatto, J. C. Read, A. J. Mieszawska, Z. A. Fayad and W. J. M. Mulder, Nanomedical Theranostics in Cardiovascular Disease, *Curr. Cardiovasc. Imaging Rep.*, 2012, **5**, 19–25.
13. Y. C. Yeh, B. Creran and V. M. Rotello, Gold nanoparticles: preparation, properties, and applications in bionanotechnology, *Nanoscale*, 2012, **4**, 1871–1880.
14. J. H. Ross and R. P. Victor, in *Nanomedicine and the Cardiovascular System*, 2011.
15. I. Lentacker, I. De Cock, R. Deckers, S. C. De Smedt and C. T. W. Moonen, Understanding ultrasound induced sonoporation: definitions and underlying mechanisms, *Adv. Drug Delivery Rev.*, 2014, 49–64.
16. C. H. Lee, Use of Micobubbles and Nanobubbles for Diagnostic Vascular molecular imaging and Therapeutic applications, in *Nanomedicine and the Cardiovascular System*, ed. R. J. Hunter, 2012, pp. 303–323.
17. M. Ward, J. Wu and J. F. Chiu, Experimental study of the effects of optison® concentration on sonoporation in vitro, *Ultrasound Med. Biol.*, 2000, 1169–1175.
18. S. S. Carr, A. J. Hooper, D. R. Sullivan and J. R. Burnett, Non-HDL-cholesterol and apolipoprotein B compared with LDL-cholesterol in atherosclerotic cardiovascular disease risk assessment, *Pathology*, 2019, **51**(2), 148–154.
19. M. Farnier, Update on the clinical utility of fenofibrate in mixed dyslipidemias: mechanisms of action and rational prescribing, *Vasc. Health Risk Manage.*, 2008, **4**(5), 991–1000.
20. T. Fukami, T. Ishii, T. Io, N. Suzuki, T. Suzuki, K. Yamamoto, J. Xu, A. Ramamoorthy and K. Tomono, Nanoparticle Processing in the Solid State Dramatically Increases the Cell Membrane Permeation of a Ccholesterol-Lowering Drug, Probucol, *Mol. Pharmaceutics*, 2009, 1029–1035.
21. N. Manners, V. Priya, A. K. Mehata, M. Rawat, S. Mohan, H. A. Makeen, M. Albratty, A. Albarrati, A. M. Meraya and M. S. Muthu, Theranostic Nanomedicines for the Treatment of Cardiovascular and Related Diseases: Current Strategies and Future Perspectives, *Pharmaceuticals*, 2022, 441.

22. I. Cuadrado, M. J. G. M. Piedras, I. Herruzo, M. D. C. Turpin, B. Castejón, P. Reventun, A. Martin, M. Saura, J. L. Zamorano and C. Zaragoza, EMMPRIN-Targeted Magnetic Nanoparticles for In Vivo Visualization and Regression of Acute Myocardial Infarction, *Theranostics*, 2016, 545–557.
23. V. Sharma, H. K. Dewangan, L. Maurya, K. Vats, H. Verma and S. Singh, Rational design and in-vivo estimation of Ivabradine Hydrochloride loaded nanoparticles for management of stable angina, *J. Drug Delivery Sci. Technol.*, 2019, **54**, 101337.
24. J. P. Chen, P. C. Yang, Y. H. Ma and T. Wu, Characterization of chitosan magnetic nanoparticles for in situ delivery of tissue plasminogen activator, *Carbohydr. Polym.*, 2011, **84**(1), 364–372.
25. K. Changsun Kang, S. Gwon, C. Song, P. M. Kang, S. C. Park, J. Jeon, D. W. Hwang and D. Lee, Fibrin-Targeted and H2O2-Responsive Nanoparticles as a Theranostics for Thrombosed Vessels, *ACS Nano*, 2017, 6194–6203.
26. M. E. Lobatto, Z. A. Fayad, S. Silvera, E. Vucic, C. Calcagno, V. Mani, S. D. Dickson, K. Nicolay, M. Banciu, R. M. Schiffelers, J. M. Metselaar, L. van Bloois, H.-S. Wu, J. T. Fallon, J. H. Rudd, V. Fuster, E. A. Fisher, G. Storm and W. J. M. Mulder, Multimodal Clinical Imaging to Longitudinally Assess a Nanomedical Anti-Inflammatory Treatment in Experimental Atherosclerosis, *Mol. Pharmaceutics*, 2010, 2020–2029.
27. P. M. Winter, A. M. Neubauer, S. D. Caruthers, T. D. Harris, J. D. Robertson, T. A. Williams, A. H. Schmieder, G. Hu, J. S. Allen, E. K. Lacy, H. Zhang., S. A. Wickline and G. M. Lanza, Endothelial $\alpha v\beta 3$ Integrin–Targeted Fumagillin Nanoparticles Inhibit Angiogenesis in Atherosclerosis, *Arterioscler. Thromb. Vasc. Biol.*, 2006, 2103–2109.
28. N. Selvasudha and K. Koumaravelou, The multifunctional synergistic effect of chitosan on simvastatin loaded nanoparticulate drug delivery system, *Carbohydr. Polym.*, 2017, **163**, 70–80.
29. L. Liu, H. He, M. Zhang, S. Zhang, W. Zhang and J. Liu, Hyaluronic acid-decorated reconstituted high density lipoprotein targeting atherosclerotic lesions, *Biomaterials*, 2014, **35**(27), 8002–8014.
30. G. R. Qadri, A. Ahad, M. Aqil, S. S. Imam and A. Ali, Invasomes of isradipine for enhanced transdermal delivery against hypertension: formulation, characterization, and in vivo pharmacodynamic study, *Artif. Cells, Nanomed., Biotechnol.*, 2016, 139–145.
31. V. Gupta, N. Gupta, I. H. Shaik, R. Mehvar, I. F. McMurtry, M. Oka, E. Nozik-Grayck, M. Komatsu and F. Ahsan, Liposomal fasudil, a rho-kinase inhibitor, for prolonged pulmonary preferential vasodilation in pulmonary arterial hypertension, *J. Controlled Release*, 2013, **167**(2), 189–199.
32. B. Gao, J. Xu, J. Zhou, H. Zhang, R. Yang, H. Wang, J. Huang, F. Yan and Y. Luo, Multifunctional pathology-mapping theranostic nanoplatforms

for US/MR imaging and ultrasound therapy of atherosclerosis, *Nanoscale*, 2021, **13**(18), 8623–8638.
33. D. Cosgrove and C. Harvey, Clinical uses of microbubbles in diagnosis and treatment, *Med. Biol. Eng. Comput.*, 2009, 813–826.
34. V. Paefgen, D. Doleschel and F. Kiessling, Evolution of contrast agents for ultrasound imaging and ultrasound-mediated drug delivery, *Front. Pharmacol.*, 2015, 197.
35. F. Kiessling, J. Huppert and M. Palmowski, Functional and molecular ultrasound imaging: concepts and contrast agents, *Curr. Med. Chem.*, 2009, 627–642.
36. A. L. Klibanov, P. T. Rasche, M. S. Hughes, J. K. Wojdyla, K. P. Galen, J. H. Wible and G. H. Brandenburger, Detection of individual microbubbles of an ultrasound contrast agent: Fundamental and pulse inversion imaging, *Acad. Radiol.*, 2002, S279–S281.
37. R. Gramiak and P. M. Shah, Echocardiography of the aortic root, *Invest. Radiol.*, 1968, **3**, 356–366.
38. R. Cavalli, M. Soster and M. Argenziano, Nanobubbles: a promising efficient tool for therapeutic delivery, *Ther. Delivery*, 2016, **7**, 117–138.
39. D. L. Miller, M. A. Averkiou, A. A. Brayman, E. C. Everbach, C. K. Holland, J. H. Wible and J. Wu, Bioeffects considerations for diagnostic ultrasound contrast agents, *J. Ultrasound Med.*, 2008, 611–632.
40. D. Cosgrove and C. Harvey, Clinical uses of microbubbles in diagnosis and treatment, *Med. Biol. Eng. Comput.*, 2009, **47**, 813–826.
41. J. M. Jones and W. J. Koch, Gene therapy approaches to cardiovascular disease, *Mol. Med.*, 2005, **112**, 15–35.
42. J. J. Rychak, J. Graba, A. M. Y. Cheung, B. S. Mystry, J. R. Lindner, R. S. Kerbel and F. S. Foster, Microultrasound molecular imaging of vascular endothelial growth factor receptor 2 in a mouse model of tumor angiogenesis, *Mol. Imaging.*, 2007, **6**, 289–296.
43. I. M. Verma and N. Somia, Gene therapy -- promises, problems and prospects, *Nature*, 1997, **389**, 239–242.
44. E. G. Schutt, D. H. Klein, R. M. Mattrey and J. G. Riess, Injectable microbubbles as contrast agents for diagnostic ultrasound imaging: The key role of perfluorochemicals, *Angew. Chem., Int. Ed.*, 2003, 3218–3235.
45. E. C. Unger, T. Porter, W. Culp, R. Labell, T. Matsunaga and R. Zutshi, Therapeutic applications of lipid-coated microbubbles, *Adv. Drug Delivery Rev.*, 2004, **56**, 1291–1314.
46. K. W. Ferrara, Driving delivery vehicles with ultrasound, *Adv. Drug Delivery Rev.*, 2008, **60**, 1097–1102.
47. P. Marmottant and S. Hilgenfeldt, Controlled vesicle deformation and lysis by single oscillating bubbles, *Nature*, 2003, 153–156.
48. E. Osei and A. Al-Asady, A review of ultrasound-mediated microbubbles technology for cancer therapy: A vehicle for chemotherapeutic drug delivery, *J. Radiother. Prac.*, 2019, 8.

49. B. Furie and B. C. Furie, Mechanisms of thrombus formation, *N. Engl. J. Med.*, 2008, **359**(9), 938–949.
50. E. S. R. Hughes, Venous obstruction in the upper extremity; Paget-Schroetter's syndrome; a review of 320 cases, *Surg., Gynecol. Obstet.*, 1949, **88**(2), 89–127.
51. N. Mackman, Triggers, targets and treatments for thrombosis, *Nature*, 2008, 914–918.
52. J. Brangsch, C. Reimann, F. Collettini, R. Buchert, R. M. Botnar and M. R. Makowski, Molecular Imaging of Abdominal Aortic Aneurysms, *Trends Mol. Med.*, 2017, 150–164.
53. M. Suzuki, L. Bachelet-Violette, F. Rouzet, A. Beilvert, G. Autret, M. Maire, C. Menager, L. Louedec, C. Choqueux, P. Saboural, O. Haddad, C. Chauvierre, F. Chaubet, J.-B. Michel, J. M. Serfaty and D. Letourneur, Ultrasmall superparamagnetic iron oxide nanoparticles coated with fucoidan for molecular MRI of intraluminal thrombus, *Nanomedicine*, 2015, 73–87.
54. B. Li, R. Aid-Launais, M. N. Labour, A. Zenych, M. Juenet, C. Choqueux, V. Ollivier, O. Couture, D. Letourneur and C. Chauvierre, Functionalized polymer microbubbles as new molecular ultrasound contrast agent to target P-selectin in thrombus, *Biomaterials*, 2019, 139–150.

CHAPTER 5

Novel Nanotherapies for the Treatment of Coronary Atherosclerosis

KURT BRASSINGTON[a,b], YI EE LYE[a], ALEX BOBIK[a,c,d] AND TIN KYAW*[a,b,c]

[a]Inflammation and Cardiovascular Disease Laboratory, Baker Heart and Diabetes Institute, Melbourne, VIC, Australia; [b]Baker Department of Cardiometabolic Health, University of Melbourne, Parkville, VIC, Australia; [c]Centre for Inflammatory Diseases, Department of Medicine, Monash Medical Centre, Clayton, VIC, Australia; [d]Department of Immunology, Monash University, Melbourne, VIC, Australia
*E-mail: Tin.Kyaw@baker.edu.au

5.1 Introduction

Coronary atherosclerosis is the major contributing factor to myocardial infarction (MI), the leading cause of death globally, which is expected to remain the primary cause of death until 2030.[1] Thromboembolic events from rupture and/or erosion of advanced coronary atherosclerotic plaques cause myocardial ischaemia that requires immediate intervention to restore the blood supply. Current atherosclerosis management focuses on reducing plasma lipids, inhibiting platelet activation and preventing the formation of thrombosis. However, despite these preventive measures, along with advanced interventions, there is an increasing trend in mortality

Nanoscience & Nanotechnology Series No. 64
Cardiovascular Nanomedicine
Edited by Roger Narayan and Tanveer Tabish
© The Royal Society of Chemistry 2024
Published by the Royal Society of Chemistry, www.rsc.org

related to MI, suggesting a significant residual risk that may arise from mechanisms independent of plasma lipids.

Arterial inflammation has been shown to promote vulnerable plaque development and subsequent rupture, and systemic anti-inflammatory therapies targeting arterial inflammation have been proposed to reduce MI incidence and mortality.[2] Recent research showed that anti-inflammatory therapies combined with lipid-lowering agents reduce MI mortality; however, the incidence of fatal infection as a result of being immunocompromised is increased in patients chronically treated with systemic anti-inflammatory agents.[3] Thus, there is an urgent need to develop novel therapeutics that specifically target intimal atherosclerotic lesions, without compromising the immune system to ultimately reduce the instance of fatal MI.

Nanotechnology has been integrated with many science and engineering applications, including medicine and biotechnology. Nanomedicine, compared to traditional medicine, has several advantages, such as improved target specificity, reduced accumulation and toxicity in non-target tissues and enhanced stability/solubility, and thus a lower therapeutic dose is required.[4] Nanotechnology-based therapeutic agents are now approved for use in the management of cancers, iron-deficiency anaemia, fungal infection and macular degeneration.[5] The use of nanoparticles (NPs) is currently being investigated in preclinical studies for the treatment of chronic inflammatory diseases,[6] potentially revolutionising the management of atherosclerosis.[7] Nanoparticle-eluting stents containing pitavastatin- and pioglitazone-encapsulated nanoparticles have been shown to reduce local inflammation at the stent site in animal models,[8,9] showcasing the therapeutic potential of anti-inflammatory nanoparticles in the management of atherosclerosis and recurrent MI.

In this chapter, we will review the development of cardiovascular nanomedicine, its potential mechanism of action, point of intervention and various intervention strategies in managing atherosclerosis and effectively reducing MI-associated mortality.

5.2 Immunopathogenesis of Atherosclerosis

Atherosclerosis is a chronic inflammatory disease of medium- and large-sized arteries where arterial injury is initiated by the accumulation of oxidised low-density lipoprotein (oxLDL) cholesterol in the intima. Reversible atherosclerotic lesions are often seen in children and young adults; however, these reversible fatty steaks often manifest silently, eventually transforming into advanced, fibrous lesions that become irreversible and vulnerable to rupture. They are often detected once patients present to clinics for hyperlipidaemia. Therefore, current therapeutic strategies that utilise anti-lipid, anti-platelet and anti-coagulant drugs may not entirely target local arterial inflammation, and thus, novel therapeutics are required to specifically prevent arterial inflammation and reduce MI-related mortality more effectively and accurately.

5.2.1 Initiation of Intimal Injury, Foam Cells and Fatty Steak Formation

Raised plasma LDL cholesterol levels are the key triggers of intimal injury, *i.e.*, proteoglycan-mediated retention and free radical- and enzyme-mediated oxidation in the intimal layers. The accumulation of oxLDL cholesterol disturbs local homeostasis and upregulates endothelial cell expression of leucocyte adhesion and chemoattractant molecules, facilitating the recruitment of circulating immune cells into the intima. Monocytes are the predominant subset among all leucocytes recruited and differentiate locally into macrophages, which become activated in response to immunogenic oxLDL. Over time, cytosolic accumulation of oxLDL transforms macrophages into lipid-laden foam cells, which form macroscopic fatty streaks, the first sign of atherosclerotic lesion development. If oxLDL is cleared and the normal intima restored, fatty streaks can be reversed. Studies have shown that people in their late 20s, 30s and 40s have advanced coronary lesions, which were present in 35%–77% of apparently healthy individuals.[10,11]

5.2.2 Immune Cell Recruitment

Further recruitment of monocytes into growing, lipid-rich lesions is facilitated by chemokines derived from initial macrophages, endothelial cells and smooth muscle cells. In addition to monocyte-derived macrophages, other major immune cell types are present in early and advanced lesions.[12] Both human and mouse studies have identified dendritic cells, neutrophils, mast cells, CD4 T cells, CD8 T cells, natural killer (NK) cells, NK T (NKT) cells and γδ-T cells within lesions,[13] all of which contribute to plaque progression, destabilisation and rupture.[14] It is interesting to note that diphtheria toxin-induced transient depletion of circulating monocytes and tissue-resident macrophages attenuated atherosclerosis,[15] and genetic depletion of lymphocytes also reduced atherosclerosis in atherosclerotic mouse models.[16] Cumulatively, this suggests the reciprocal interaction and importance between myeloid and lymphoid immune cells in immune cell recruitment, immune responses and immune regulation.

Leucocyte recruitment is mediated *via* various factors, including chemokines, adhesion molecules and platelet factors. Chemokine receptors are G-protein-coupled surface receptor proteins predominantly expressed on leucocytes and respond to chemokines and small cytokines with chemotactic properties, which are expressed by not only vascular endothelial cells and smooth muscle cells but also recruited leucocytes, further honing immune cell recruitment to the lesion site. The chemokine and chemokine receptor axis have been shown to be vital to leucocyte migration, driving atherosclerosis development and progression in a cell-specific manner.[17] Chemokines are differentially expressed in various tissues, particularly following injury,[18] highlighting chemokines as a potential therapeutic/drug delivery target. It is also critical to define the importance of cell-specific chemokine receptors in atherosclerosis, as there is conflicting evidence in this area, with C-C chemokine receptor 7 (CCR7) deficiency controversially affecting atherosclerosis in animal studies.[19,20] Additional research is warranted to deepen the understanding of the cell-specific chemokine–chemokine receptor axis and how it can be effectively targeted to halt the progression of atherosclerosis.

Cellular adhesion molecules (CAMs) are cell surface proteins that mediate the binding of cells to other cells or promote anchoring to the extracellular matrix (ECM). CAM expression is upregulated in response to inflammation, infections and cancers, resulting in the recruitment and activation of leucocytes and immune cell proliferation. It is well accepted that atherosclerotic lesions express several adhesion molecules, such as intercellular adhesion molecules (ICAMs), vascular cell adhesion molecules (VCAMs), platelet endothelial cellular adhesion molecules (PECAMs) and selectins (P, E and L). Similar to chemokines, adhesion molecules are produced by leucocytes, vascular endothelial cells and smooth muscle cells. Knocking out the gene(s) responsible for their production or antagonising their function significantly decreases atherosclerotic lesion size and formation, reinforcing their significance in the pathogenesis of atherosclerosis.[21–23] The use of adhesion molecule inhibitors and CAM-targeted drug delivery nanocarriers to reduce leucocyte recruitment and their associated inflammation has been proposed as a novel means to treat atherosclerosis.[24]

5.2.3 Immune Response, Lesion Inflammation and Cell Death

Monocyte-derived macrophages, the major cellular constituent of atherosclerotic lesions, play a critical role in the development and progression of atherosclerosis. They are the first cell type to appear and are continuously present, acting as clear initiators and aggravators of atherosclerosis *via* phagocytosis and endocytosis in their attempt to restore local homeostasis. Atherogenic stimulants are considered to initiate immune responses that promote and/or aggravate atherosclerosis; these stimulants include oxLDL, apolipoprotein B (Apo-B), heat shock proteins and self-antigens from both apoptotic and necrotic cell death within lesions. Scavenger receptors and apoptotic cell recognition receptors are membrane-bound pattern recognition receptors that bind modified self-antigens, and intimal macrophages utilise these receptors to internalise both lipid and non-lipid ligands.[25–27] Deficiency of scavenger receptors CD36 and lectin-like oxidised low-density lipoprotein receptor-1 (LOX-1) diminished foam cell formation *in vitro*[28,29] and reduced aortic atherosclerosis in atherogenic mouse models,[29,30] suggesting foam cell formation[31] and oxLDL clearance[32] as therapeutic targets in atherosclerosis. However, overexpression of scavenger receptors in haemopoietic cells did not reduce atherosclerosis despite increased oxLDL clearance.[33,34] In accordance with genetic knockout models,[29,30] targeting foam cell formation reduces atherosclerosis development,[35] although further research is warranted to determine the potential of foam cell-targeted intervention strategies using non-coding RNAs in atherosclerosis progression.[31] It is important to consider whether foam cells and fatty streaks are in a reversible state as macrophages loaded with atherogenic stimulants trigger the chronic cascade of downstream innate and adaptive immune responses.

The initial foam cell death signalling is unclear, as foam cells can commit oxLDL-induced apoptosis, autophagy, necroptosis and pyroptosis,[36] leading to lesion inflammation, necrosis and foam cell accumulation. Macrophages together with recruited innate and adaptive immune cells are responsible for vicious circles of reciprocal augmentation of lesion inflammation and cell death. Decades of research have shown that chronic subclinical localised inflammation results in the progressive development of mature plaques. This progression is evidenced by various histological characteristics, including lipid-rich acellular necrotic cores, collagen-rich extracellular matrix, a thin fibrosis cap, neovascularisation and intraplaque haemorrhage, and increased proinflammatory responses, such as lesion inflammation and immune cell infiltrates, together with imaging evidence of luminal stenosis and lesion calcification.

5.2.4 Vulnerable Plaques, Plaque Instability and Plaque Rupture

Coronary thromboembolic events lead to the clinical manifestations of angina pectoris, MI and MI-related death. The most common causes of a coronary ischaemic attack are plaque rupture and plaque erosion.[37,38] Plaque rupture is characterised by a disturbed thin fibrous cap, exaggerated lesion inflammation, a large necrotic core and intraplaque haemorrhage. Vulnerable plaques are generally silent (asymptomatic) in nature and spontaneously rupture at any time. The clinically important question is when these rupture-associated thromboembolic events happen. Whilst genetic, demographic and epigenetic factors influence atherosclerosis, chronic diseases, such as hypertension, diabetes, autoimmune diseases and a previous history of ischaemic events, are also well-established risk factors.

The development of lesions into rupture-prone plaques is the key cause of ischaemic injury, as the ruptured lesion or lesion fragments can enter the circulation and lead to arterial blockage. Another key issue is that, upon rupture, there is a drastic increase in immune cell and platelet recruitment, leading to thrombus formation and enhanced local inflammation,[39] placing these thrombosis survivors at a heightened risk of recurrent MI.[40] Plaque vulnerability is largely associated with the activity of both matrix metalloproteinases (MMPs) and collagenases. The activity of MMPs within lesions is conflicting, as they can be beneficial by promoting intimal thickening that aids in plaque stability; however, in the ECM, the presence of MMPs has been shown to drive plaque destabilisation due to the degradation of the surrounding lesion microenvironment.[41–43] The degradation of the ECM *via* MMPs and collagenases is a key contributor to plaque rupture due to reduced strength of the shoulder region as a result of collagen degradation, promoting atheroma destabilisation. It is known that macrophage-derived MMPs are capable of degrading all ECM components,[44] and the modulation of MMP secretion by these cells as a potential therapy for plaque stabilisation has been investigated. A study by Nakashiro *et al.* investigated the effect of monocyte/macrophage differentiation in preventing plaque destabilisation and rupture in atherogenic $ApoE^{-/-}$ mice.[8] The administration of pioglitazone, a peroxisome proliferator-activated receptor gamma (PPARγ) agonist that prompts the differentiation of monocytes/macrophages into a reduced pro-inflammatory phenotype, encapsulated in poly(lactic-*co*-glycolic acid) (PLGA) nanoparticles, significantly reduced MMP activity and pro-inflammatory cytokine expression, leading to increased lesion stability in the brachiocephalic arteries. It is also noteworthy to mention that the same dosage of pioglitazone, when administered orally, showed no effect, highlighting that targeted

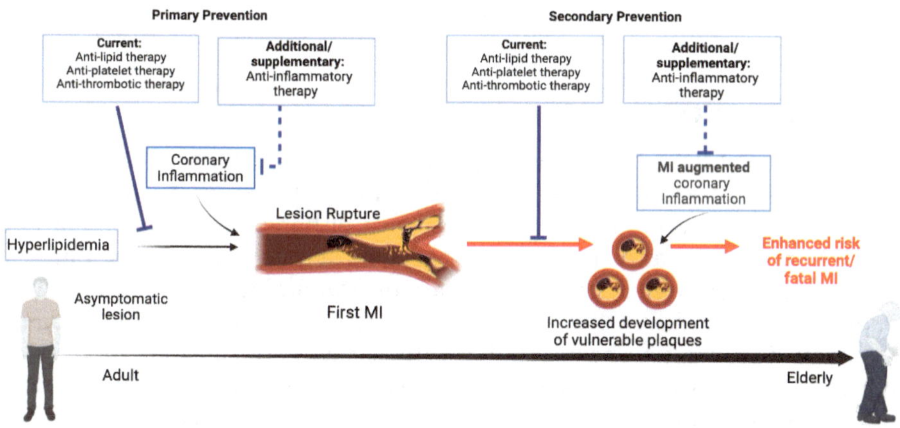

Figure 5.1 Points of therapeutic prevention in the pathogenesis of atherosclerosis. Atherosclerotic lesions often manifest asymptomatically. Patients are identified as being hyperlipidaemic, and clinicians often prescribe cholesterol-lowering agents as a primary prevention method. Post-MI, secondary prevention methods to reduce the instances of often fatal recurrent MI include continued use of cholesterol-lowering agents and anti-platelet and anti-thrombotic therapies, which are largely ineffective, hence contributing to the frequency of recurrent MI.

nanoparticles offer specific therapeutic delivery. It was also discovered that there was no reduction in the macrophage number within the lesion, meaning that these effects were purely the result of modulated MMP activity. In targeting the cells responsible for lesion destabilisation/rupture, there may be a significant reduction in both atherosclerosis and its associated cardiovascular manifestations.

After an initial MI, the risk of recurrent attacks increases greatly and remains elevated for many years,[45,46] likely due to MI-accelerated arterial inflammation.[47,48] Cholesterol-lowering statins are ineffective in preventing an increased risk of recurrent, often fatal events in MI survivors.[49] Additionally, chronic administration of systemic anti-inflammatory therapies appears to yield only marginal benefits, with an increased incidence of fatal infection.[3,50,51] Therefore, ideal therapeutic strategies aiming to prevent MI-related mortality should firstly focus on reducing new arterial injuries and, secondly, on decreasing arterial inflammation and halting the progression of pre-existing lesions into vulnerable plaques, without compromising the immune system (Figure 5.1).

5.3 Disease-modifying Risk Factors

Among many known risk factors in vulnerable atherosclerotic plaques and their associated myocardial ischaemic events, there are several underlying diseases that adversely impact atherogenesis. In this section, the role of diabetes mellitus (DM), hypertension, obesity and autoimmune diseases on atherosclerosis will be discussed and novel therapeutic targets will be explored.

5.3.1 Diabetes

It is well established that diabetes mellitus and atherosclerosis share a plethora of contributing factors, with diabetic patients at an increased risk of lesion development. Diabetics have enhanced dyslipidaemia and pro-inflammatory and oxidative stress burdens, which contribute to accelerated atherosclerosis in these patients. Dyslipidaemia and persistent oxidative stress drive the formation of oxLDL.[52] The role of oxidative stress in diabetes has been studied extensively, showing that there is an increase in reactive oxygen species (ROS) production and a subsequent downregulation of antioxidant activity.[53] Studies have linked ROS production to endothelial dysfunction, a key characteristic of atherosclerosis, and have shown that antioxidant treatment to restore this imbalance in homeostatic mechanisms aids in the restoration of vascular function.[54] Moreover, de Haan *et al.* showed that antioxidant deficiency combined with enhanced ROS production leads to diabetes-related vascular complications and that the preservation of antioxidant balance *via* the use of an antioxidant mimetic drastically reduced diabetic endothelial dysfunction.[55]

The preservation of oxidant/antioxidant homeostasis is crucial in reducing the accelerated LDL oxidation seen in patients with DM, ultimately reducing the risk of lesion development resulting from oxLDL and the loss of endothelial integrity.[56] Another key aspect of diabetic atherogenesis is the retention of oxLDL *via* proteoglycans, with a study showing enhanced proteoglycan (particularly biglycan) distribution at lesion sites, which anchor atherogenic lipoproteins, particularly oxLDL to the intima.[57] Although diabetic patients do not display increased proteoglycan-binding affinity to LDL when compared to non-diabetics, they do suffer from increased oxidative modification of LDL and have far more triglyceride-rich LDL, which contains substantially more cholesterol than that of non-diabetic patients.[58] This bound cholesterol-rich oxLDL–proteoglycan complex is then a target for infiltrating macrophages, causing rapid foam cell formation due to their enhanced lipid content. These findings present proteoglycans as a potential intervention for the treatment of diabetes-accelerated atherosclerosis.

Immune activation and chronic inflammation are key characteristics of both DM and atherosclerosis, with the inflammatory burden and necrotic core of lesions being significantly larger in diabetic patients when compared to non-diabetics.[59,60] Interestingly, an early study by Redgrave *et al.* showed that despite these diabetic patients having larger lesions, there were no differences in macrophage or lymphocyte cell number or local inflammation in both the plaque or fibrous cap.[60] More recent studies have proceeded to investigate this further, showing that immune cells can undergo differentiation to become pro-inflammatory, as well as promote enhanced effector functions. It was recently discovered that DM patients have enhanced levels of the pro-inflammatory mediator tumour necrosis factor alpha (TNFα) released from macrophages,[61] which can orchestrate lesion inflammation. Similarly, DM patients have elevated numbers of circulating activated leucocytes that are more granular and express elevated pro-inflammatory and apoptotic genes.[62] Moreover, T-cell function is altered in individuals with DM, with an increase in effector T cells and a decrease in regulatory T cells, which is believed to be the result of enhanced pro-inflammatory cytokine [interferon gamma (IFNγ) and TNFα] production in DM patients.[63] These now overactive immune cells may play a key role in the aggravated atherosclerosis observed in patients with DM, driving enhanced cell death and inflammation within the lesion.

5.3.2 Hypertension

Hypertension is a key contributing factor to atherogenesis and affects over one billion people worldwide.[64,65] This often-asymptomatic disease has been identified as a contributor to coronary plaque rupture[66] due to hypertension-induced plaque progression, increased inflammation and arterial stiffness/vascular remodelling, resulting in further endothelial damage and vascular dysfunction.[67,68] Both dyslipidaemia and hypertension are key risk factors associated with atherosclerosis, often acting synergistically to augment atherogenesis due to excessive ROS/oxLDL as a result of hyperlipidaemia. The loss of vessel homeostasis and enhanced sheer stress in hypertensive patients increase lipid permeability and fatty streak formation.[69] Similar to diabetes, hypertension also promotes antioxidant/oxidant imbalance, favouring ROS production, a known driver of atherogenesis. It has been well established that the renin–angiotensin system (RAS) is dysregulated in hypertension; however, it has also been directly linked to atherosclerosis by prompting endothelial damage, fibrosis, inflammation and reduced plaque stability due to dysregulated blood pressure.[70] RAS-mediated oxidative stress is driven by the interaction between angiotensin II (Ang-II) and the endothelial angiotensin-1 receptor (AT1R), driving an increase in intracellular calcium concentration and subsequent activation of NADPH oxidase enzymes, which are responsible for the generation of harmful ROS and the activation of the mitogen-activated protein kinase (MAPK) pathway. MAPK activation prompts the transactivation of nuclear factor kappa-light-chain-enhancer of activated B cells (NFκB) signalling, which drives the expression of a plethora of genes, including pro-inflammatory cytokines (TNFα and IL-6), chemokines [monocyte chemotactic protein 1 (MCP-1)] and various adhesion molecules, including VCAM/ICAM and selectins, all of which have been implicated in the development and progression of atherosclerosis.[71,72] Recently, it was demonstrated that Ang-II induces macrophage apoptosis within lesions *via* the activation of the AT1R–MAPK pathway and its associated endoplasmic reticulum (ER) stress/ROS, leading to the development of vulnerable, rupture-prone plaques.[73] The AT1R inhibitor losartan was then used, causing a drastic reduction in macrophage apoptosis and vulnerable plaque progression, highlighting the importance of Ang-II/hypertension in vulnerable plaque formation.

Interestingly, hypertension also drives an increase in vascular proteoglycan levels, contributing to atherosclerosis. It has been shown that attenuating Ang-II and/or inhibiting its interaction with AT1R drastically reduce the lesion size and progression.[64,74] It is understood that Apo-B, a major constituent of LDL cholesterol, is colocalised with proteoglycans, such as biglycan, within human atherosclerotic lesions.[75,76] An early study by Huang *et al.* provided proof-of-concept data into the mechanisms in which Ang-II stimulates the secretion of ECM proteins, particularly proteoglycans, and how they increase the retention of LDL cholesterol.[77] This study concluded that mice subject to Ang-II infusion on a Western diet showed a 3-fold increase in lesion size, lipid retention and a high level of colocalisation between Apo-B and biglycan.[77] Furthermore, a recent proteomics-based study has shown that hypertension alters the arterial proteome, causing a conformational change in mechanosensitive (emerin) and contractile (actin) proteins, leading to a reduction in the passage of macromolecules, smooth muscle hypertrophy, the accumulation of LDL cholesterol and a ~5-fold increase in coronary atheroma size.[78] These discoveries highlight the importance of hypertension as a key contributing factor of atherosclerosis, driving lesion progression and causing potential morbidity and mortality, as well as defining these common contributing factors as potential points of therapeutic intervention.

5.3.3 Obesity

It is well established that obesity is a key contributor to countless diseases, particularly diabetes, hypertension and atherosclerosis. Obesity affects ~30% of the global population and causes a staggering 3.4 million deaths annually.[79] Pathophysiologically, obesity drives systemic inflammation and atherogenesis *via* the release of pro-inflammatory cytokines and adipokines that promote the release of aldosterone; this is a hormone responsible for the maintenance of salt/water balance and blood pressure, which, upon excessive release, can drive an elevation in blood pressure/hypertension, aggravate platelets, contribute to cardiac fibrosis and drive endothelial dysfunction and vascular remodelling.[80] Moreover, the release of adipokines, such as leptin, adiponectin and MCP-1, drives the recruitment of monocytes into adipose tissue, which plays a pivotal role in atherogenesis driving tissue inflammation, endothelial dysfunction, ROS production and altered lipid metabolism.[81] The release of these inflammatory mediators and adipokines promotes the activation of the c-Jun N-terminal kinase (JNK) pathway and induces the formation of the NLR family pyrin domain-containing-3 (NLRP3) inflammasome. The NLRP3 inflammasome impairs glucose metabolism and insulin signalling and alters fibrin breakdown (fibrinolysis),[82] thus not only contributing to the onset of diabetes but also promoting local inflammation and vascular damage that manifest atherosclerosis. Moreover, leptin plays a crucial role in the regulation of food intake *via* its ability to signal the hypothalamus to limit hunger signalling; however, it has been shown that the concentration of leptin increases with adiposity, meaning that obese individuals should have reduced hunger signalling, although this is not the case. In obese individuals, there is an impairment in the transport of leptin across the blood–brain barrier, thus impairing leptin function as a result of oxidative modification[83] and causing a reduction in leptin signalling[84]; thus, these patients are resistant to leptin-induced appetite suppression. The key issue with hyperleptinaemia is the drastically increased MCP-1 signalling that orchestrates macrophage-/monocyte-mediated ROS and inflammatory mediator production, which increases the expression of adhesion molecules and downregulates cholesterol scavenging by adversely affecting high-density lipoprotein (HDL)-mediated clearance.[85] Leptin is also crucial in the accumulation of lipids/cholesterol esters *via* impaired cholesterol esterase and enhanced cholesterol acyltransferase activity, which was shown to prompt foam cell formation in a dose-dependent manner.[86]

There are countless ways in which leptin contributes to the manifestation of atherosclerotic disease, not only in the formation of the atheroma itself but also in leading to plaque destabilisation and rupture. The loss of plaque integrity *via* matrix remodelling has been demonstrated by Wågsäter *et al.*, who showed that both MMP-2 and MMP-9 isoforms are abundant in aortic lesions and correlated with macrophage accumulation, proteolytic activity and lesion progression towards a rupture-prone phenotype.[42] These findings not only highlight the importance of MMP activity and immune cell activation but also define obesity/hyperleptinaemia as a potential therapeutic target to treat atheroma progression and destabilisation.

5.3.4 Chronic Inflammation and Autoimmunity

The pathogenesis of atherosclerosis shares numerous contributing factors with several autoimmune diseases, which manifest as a result of chronic inflammation. Atherosclerosis is often seen in patients suffering from autoimmune diseases, such as rheumatoid arthritis (RA) and systemic lupus erythematosus (SLE). Autoimmunity is largely immune cell-mediated, as these cells have the potential to produce self-reactive autoantibodies that have been shown to heighten the risk of CVD.[87] Current treatments available for autoimmune diseases, such as SLE, include the administration of glucocorticoids, immunosuppressants and non-steroidal anti-inflammatory drugs, which all have poor efficacy and serious adverse side effects. Placing patients on these drugs often makes them extremely susceptible to potentially life-threatening infections. A recent clinical study by Katz *et al.* found a positive correlation between atherosclerotic CVD and SLE.[88] Findings from this study showed that, in hospitalised patients, SLE was associated with a significantly increased risk of atherosclerosis, coronary artery disease and cerebrovascular disease. By implicating nanomedicine into the treatment of such autoimmune diseases, several advantageous properties of these nanocarriers/particles can be exploited, including their ability to increase the accumulation, solubility, permeability and retention of their therapeutic payload to target tissues whilst possessing the ability to be targeted through surface modifications. Similarly, RA has been directly correlated to an increased cardiovascular risk, with atherosclerotic lesions being both more prevalent and more aggressive, *i.e.*, growing at an increased rate, showing a more inflammatory phenotype and displaying rupture-prone characteristics.[89] Moreover, endothelial dysfunction as a result of inflammation is also more prevalent in RA patients,[90] indicating that these complex pro-inflammatory autoimmune diseases are key contributors to CVD. A comparative study by Aubry *et al.* used histological techniques on samples taken post-mortem to evaluate the level of stenosis, plaque vulnerability and the number of lesions in RA *vs.* non-RA control patients.[91] Findings from this study showed no increase in lesion number; however, both plaque vulnerability and lesion inflammation were drastically enhanced in the tissue taken from RA patients. We have defined the pathogenic role of IgG antibodies in aggravating atherosclerotic lesions[92,93]; therefore, it is plausible that autoantibodies generated in response to diseases, such as SLE and RA, may act in a similar manner, accumulating within atherosclerotic lesions to further promote inflammation, lesion progression and cell death. This highlights the immense need for the development of novel therapies not only to combat autoimmune disease but also to aid in reducing the likelihood of cardiovascular complications in these high-risk populations.

5.4 Therapeutic Nanotechnology in Atherosclerosis Management

Current therapeutics for treating atherosclerosis include lipid-lowering statins, anti-platelet drugs and systemic anti-inflammatory drugs. Despite the use of these interventions either alone or in conjunction with one another, lesion progression and ischaemic attack are common; thus, the development of novel therapies is crucial to reducing the adverse impact of atherosclerosis and its associated cardiovascular complications, including MI. There are two key intervention points seen clinically, these being (1) the prevention of plaque progression into destabilised rupture-prone plaques and (2) post-ischaemic injury, as atherosclerosis is a disease that manifests in silence, often until the point of MI. Currently, the use of statins causes a drastic reduction in LDL cholesterol. Despite such a profound reduction in hyperlipidaemia, these patients often experience adverse cardiovascular events, including MI, as a result of atherosclerosis. A review by Falk *et al.* compiled data from various autopsy studies and concluded that the vast majority of unexpected coronary- and MI-associated deaths are the direct result of coronary thrombosis.[94] Further investigation into the efficacy of statins showed that their ability to reduce all-cause mortality, stroke and myocardial infarction is yet to be observed, with no conclusive association between reduced hyperlipidaemia and improved clinical outcomes observed in a meta-analysis consisting of 21 randomised clinical trials.[95] This is only one example of how current therapies are limited in efficacy. Therefore, there is an urgent need for the development of novel therapeutics to combat atherosclerosis. Since the emergence of nanotechnology, there have been some exciting research findings that may shape the development of such therapies, with some examples outlined below (Figure 5.2).

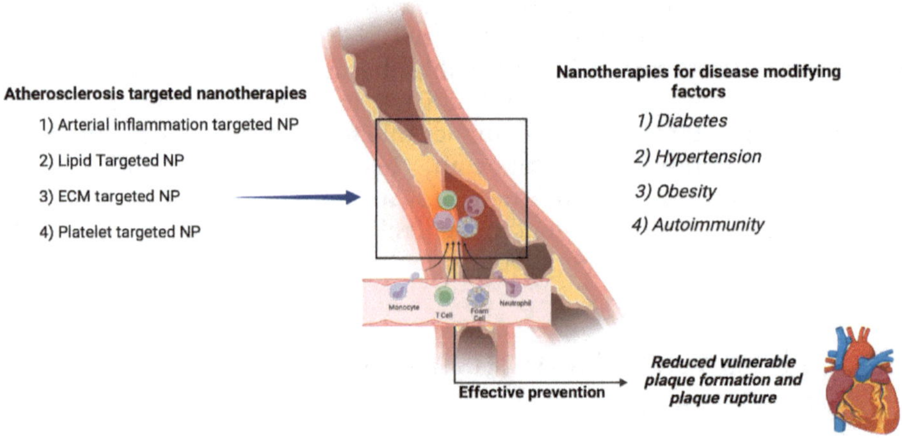

Figure 5.2 Nanotherapies in the prevention of vulnerable, rupture-prone plaques. Disease-modifying factors that play a crucial role in the development of atherosclerosis and potential nanotherapies that can aid in the treatment of atherosclerosis, MI and their underlying risk factors.

5.4.1 Immune-targeted Nanoparticle-based Therapies to Target Vulnerable Plaques

Due to atherosclerosis being largely asymptomatic, interventions are often initiated once fatty streaks have become more complex in nature, *i.e.*, atherosclerotic lesions. Once lesions have established, they become irreversible, and therefore, treatments to modify lipid accumulation display reduced efficacy, as the lesion becomes more multifaceted, implicating the immune system. The immune system drives vicious cycles of immune cell recruitment, ROS production, tissue injury and repair. The use of NP/nanomedicine has revolutionised targeted drug delivery, as they offer a higher level of controlled drug release as well as possess unique and modifiable chemical, biological and structural properties. These characteristics allow for reduced off-target degradation/liver uptake and a reduction in the drug concentration needed to elicit the same therapeutic effect due to targeted drug delivery. The ability to effectively target cell types or molecules has led to the development of numerous atherosclerosis-specific nanomedicines, including ones targeting macrophages, platelets, lipids and the lesion microenvironment.

Zhang *et al.* used a "two-pronged" approach to treat atherosclerosis, which involved the development of a pH-sensitive selenium NP (se-NP) with potent anti-inflammatory and anti-oxidant properties.[96] Resveratrol (Res), which is a natural phenol compound that inhibits cytokine secretion from inflammatory macrophages, as well as drives autophagy-mediated degradation of lipids, was encapsulated into the selenium-based NP. In using a nanotherapeutic approach, researchers effectively enhanced the anti-inflammatory/oxidative properties of their treatment, as selenium treatments have been shown to have potent ROS-scavenging properties. Findings from this study showed that se-NPs significantly reduced lesion oxidative stress, inflammation and macrophage differentiation. Furthermore, we have shown that selenium-based antioxidants can drastically reduce endothelial dysfunction due to reducing endothelial damage and the oxidant-dependent depletion of endothelial nitric oxide synthase, a nitric oxide-producing enzyme that plays a crucial role in vessel homeostasis and endothelial function.[54,97]

Modulating the inflammatory burden in atherosclerotic lesions using nanomedicine is gaining the attention of researchers due to lessened lesion immune cell recruitment and limited lesion cell death as a result of reduced immune cell effector function. Currently, the efficacy of corticosteroid-containing liposomes in inflammatory diseases, such as rheumatoid arthritis, cancers and atherosclerosis, is being investigated. A study by Bartneck *et al.* has utilised dexamethasone, a potent corticosteroid to mitigate the inflammatory burden in various cell types, particularly in fibroblasts and macrophages.[98] Under inflammatory conditions, treatment with dexamethasone liposomes caused a marked reduction in the expression of pro-inflammatory IL-6, TNFα and IL-1β whilst preventing the migration of both monocytes and macrophages, which was not observed in the unencapsulated native form of dexamethasone. Similarly, modulation of efferocytosis-driven lesion cell death has been investigated, targeting macrophage uptake of carbon nanotubules conjugated to an antiphagocytic CD47-SIRPα signal pathway inhibitor.[99] Findings from this study showed that nanotubes accumulate within lesions, reinstated phagocytic mechanisms and significantly reduced plaque progression whilst prompting a reduction in lesion-specific macrophage inflammatory gene expression. A similar study utilising siRNA to calmodulin-dependent protein kinase γ in lesion macrophages has also shown to be effective in improving lesion stability in mice.[100] Inhibition of this protein kinase prevents lesion necrosis and leads to the activation of efferocytotic signalling, prompting the clearance of apoptotic/necrotic cells within the lesion and increasing fibrous cap thickness, all of which are hallmark features of enhanced plaque stability. The use of nanocarriers to effectively increase drug efficacy and limit toxicity shows promise in the treatment of atherosclerosis and a subsequent reduction in MI/ischaemic injury.

5.4.2 Lipid-targeted Nanomedicine

The complex nature of atherosclerotic lesions involves both lipid- and immune-driven mechanisms, orchestrating atherogenesis in a synergistic manner. Lipid-dependent macrophages play a pivotal role in all stages of atherosclerosis and thus are an appealing therapeutic target. There have been numerous studies that have investigated the efficacy of statin-loaded nanoparticles, which have shown promising results both *in vivo* and *in vitro*.[101,102] A study by Katsuki *et al.* investigated the modulation of MCP-1-/CCR-2-dependent monocyte recruitment into atherosclerotic lesions using pitavastatin-loaded PLGA nanoparticles.[101] Findings from this study showed that pitvastatin administered alone, *i.e.*, not encapsulated in nanoparticle, had no effect on plaque stabilisation or lesion monocyte activity; however, pitvastatin PLGA nanoparticle treatment drastically decreased monocyte recruitment, gelatinase/protease activity, plaque destabilisation and rupture in atherogenic mice. Furthermore, the same group continued investigations *in vitro,* showing that NP treatment inhibited MCP-1 secretion and MCP-1-mediated chemotaxis from macrophages whilst significantly reducing MMP-9 release. A similar study investigating the use of statin-loaded reconstituted HDL (rHDL) to deliver low-dose statins to lesions has been completed.[102] This study analysed the anti-inflammatory effects of statins rather than their direct lipid-lowering effects, which are currently not utilised to their full potential due to reduced systemic drug bioavailability upon oral administration and excessive first-pass metabolism. By administering the statin-loaded rHDL intravenously, there was effective accumulation of the nanotherapeutic at the lesion, driving a significant reduction in both plaque area, monocyte recruitment and a reduction in the gene expression of pro-inflammatory mediators using both a 1-week high-dose and a 3-month low-dose treatment regime.

Moreover, a further study by Zhang *et al.* showed that modifying the se-NPs (as mentioned in Section 5.4.1[103]) also aided in the regulation of cholesterol metabolism. Their team developed Se-Res nanoparticles that interacted with the lectin-like oxidised low-density lipoprotein receptor 1 (LOX-1), known as LCS-Se/Res NPs.[96] Upon administration and binding to LOX-1 at the lesion site, there was a targeted release of Res and dissociation from LCS-Se, which acted synergistically to reduce plaque inflammation, promote macrophage differentiation to the anti-atherogenic M2 phenotype, reduce lesion ROS and cause an overall increase in plaque stability. Taken together, these studies highlight the significance of targeted administration of lipid-targeting therapies in the treatment of lesion progression, destabilisation and potential rupture in both a lipid-dependent and monocyte-/macrophage-dependent manner.

5.4.3 ECM-targeted Nanomedicine

It is well established that ECM integrity plays a crucial role in plaque stability,[42,104] and it is the loss of this integrity that leads to the generation of vulnerable, rupture-prone plaques. The shoulder region of the lesion is often the most vulnerable, as this is a site of high immune cell infiltration, MMP activity and ECM degradation. As previously described, the accumulation of immune cells within this region causes increased local inflammation and positive remodelling/stenosis. These immune cells, particularly macrophages, lead to enhanced MMP activity and fibrous cap degradation, which has been linked to cardiovascular mortality in patients suffering from coronary atherosclerosis.[43] The ECM is a complex network of proteins, with collagens being the most abundant; however, glycoproteins, proteoglycans, elastin and other structural proteins all play key roles in this highly structured tissue network. There are also numerous molecules that can prompt the recruitment of other cells to the ECM to aid in homeostasis, such as VCAM and integrins; however, under pathological conditions, this recruitment may become dysregulated, driving ECM damage. The fact that the lesion microenvironment can lead to differential expression of proteins and ligands means that it can be exploited as a therapeutic target, particularly in the case of nanomedicine. Fredman *et al.* used type IV collagen-targeted NPs to aid in increasing plaque stabilisation *via* targeted administration of a glucocorticoid-regulated annexin V A1 mimetic, Ac2-26, which acts on *via* formyl peptide receptor 2 (FPR2), mitigating inflammation.[105] In mice with pre-existing lesions, it was evident that the collagen IV-Ac2-26 NP localised within the lesion and led to an increase in protective lesion collagen, a decrease in collagenase activity, intraplaque necrosis and oxidative stress when compared to control mice, which lacked the FPR2 receptor. As ECM proteins such as collagen IV have enhanced activity in lesions and play a pivotal role in dys/regulation of both endothelial and smooth muscle integrity, the ECM is an appealing therapeutic target for drugs such as Ac2-26 as well as being a target in itself.

Adhesion molecules, such as VCAMs, are expressed by endothelial cells in response to injury and trigger the recruitment of immune cells. Integrins, transmembrane adhesion receptors, are also mediators of the local inflammatory response seen in atherosclerosis. Initial damage to the intima during atherogenesis causes increased integrin–ligand interactions that prompt leucocyte trafficking and immune cell adhesion to the lesion. These integrins elicit their effects by forming a mechanical link between the cell and the ECM. Under atherogenic conditions, endothelial cells can produce integrins in response to sheer stress (*i.e.*, turbulent blood flow or high blood pressure),[106] as well as in response to oxLDL to facilitate its removal.[107] Innovative nanoparticles/nanoemulsions to both VCAM-1 and integrins have been developed, respectively. Zhou *et al.* used microRNA-92a (miR-92a), a disturbed flow-induced miRNA that promotes endothelial dysfunction, inhibitor-loaded VCAM-1-targeted micelles,[108] to effectively target pathological vascular remodelling using a miR-92a over-expressing transgenic, atherogenic mouse model. The study provides proof-of-concept data about how miR-92a contributes to atherogenesis and defines VCAM-1-targeted anti-miR-92a micelles as a viable nano-therapeutic approach to reduce vascular remodelling and stenosis in atherosclerosis. Additionally, the efficacy of $\alpha_V\beta_3$ integrin-targeted fumagillin NPs loaded with atorvastatin has been investigated. The findings from this study showed that a single dose of $\alpha_V\beta_3$-fumagillin NPs in atherogenic rabbits led to a 3-week reduction in aortic angiogenesis, and it was shown that, when conjugated with atorvastatin, the effects were prolonged further when compared to control animals. This prolonged effect may be attributed to the specific accumulation of the nanoparticles in the aortic adventitia, which, under atherogenic conditions, express high levels of $\alpha_V\beta_3$ integrin. By limiting angiogenesis and targeting the direct release of statins to potential or developed lesion sites, this represents a novel approach to target either prophylactically (*i.e.*, prior to disease onset in high-risk patients) or in established disease states.

5.4.4 Platelet-targeted Nanomedicine

Platelets are key components of developing lesions, particularly following complications in eroded or ruptured plaques. Platelets adhere to the ECM *via* interaction with collagen and/or the damaged endothelium, driving the formation of the platelet–monocyte complex that promotes vulnerable plaque progression. The atherogenic environment exposes both non-collagenic adhesion molecules, such as von-Willebrand factor and fibrin, and collagens, to which platelets have a high affinity.[109] These properties present mimicking platelet function and adherence as an attractive method of atherosclerosis treatment. Rapamycin, an immunosuppressant used clinically to prevent tissue transplant rejection and inflammation, as well as stimulate autophagy and the clearance of damaged tissue, has been investigated as an anti-atherogenic drug. Administration of rapamycin in high concentrations has been shown to cause a reduction in lesion area *via* enhanced autophagy-mediated macrophage clearance; however, this led to immunosuppression and serious side effects resulting from low accumulation in the lesion and severe toxicity when administered systemically.[110] Recently, Song *et al.*[111] used this information and loaded rapamycin into platelet membrane-coated nanoparticles to effectively target the lesion microenvironment. Findings from this study showed that there was a ~5-fold increase in the accumulation of platelet NPs within the lesion than uncoated nanoparticles, further reinforcing that platelets bind to lesions in a highly effective manner. Furthermore, targeted delivery of rapamycin to the lesion *via* platelet NPs caused a significant reduction in the atherosclerotic lesion size and contributed to the stabilisation of lesions in atherogenic $ApoE^{-/-}$ mice. Taken together, these data suggest that rapamycin when administered directly to the lesion offers substantial anti-atherogenic effects *via* the promotion of autophagy in macrophages. One key limitation to rapamycin treatment is the adverse side effects upon systemic administration; however, this study showed that there is in fact an increase in long-term safety, with nanoparticle administration showing no adverse side effects whilst providing a greater therapeutic effect at a much lower drug concentration. These outcomes define platelet-coated NPs as an efficacious targeted delivery platform for the treatment of atherosclerosis.

5.5 Nanoparticle-based Therapies in Risk Management

As outlined in Section 5.3, there are various conditions that contribute to the manifestation of atherosclerosis, including diabetes, hypertension and obesity, with each being a potential target of NP-based treatment (Table 5.1). Recent interest in targeting these conditions using nanomedicine may prove to be beneficial in not only reducing the likelihood of direct complications but also mitigating the risk of adverse cardiovascular outcomes in these high-risk populations.

5.5.1 Diabetes-targeted Nanotherapies

Diabetes is described as the inability to regulate blood glucose levels, with modern medicine aiming to maintain glycaemic homeostasis through the regular administration of insulin. Currently, insulin is administered regularly *via* injection or an insulin pump; however, these methods can often be problematic due to them often being patient-regulated, and thus they are termed open-loop administration. Self-administration of insulin often contributes to poor glucose control and compliance, which accounts for a large portion of diabetic complications.

Recent advances in the field of nanomedicine have led to the development of insulin-loaded nanoparticles that release their cargo in response to a particular stimulus; for example, these can be coated with either a glucose-sensitive[122,141] or pH-sensitive[123] material, promoting a conformational change in the matrix structure and thus targeted drug release. Yu *et al.* developed glucose-sensitive nanovesicles that carry insulin and glucose oxidase, which undergo bioreduction leading to the dissociation of NP cargo in response to hyperglycaemic hypoxia. The use of these smart insulin-releasing NPs led to the effective regulation of blood glucose levels in diabetic mice, bringing them to normal levels whilst limiting the risk of hypoglycaemia for prolonged periods of time. This method of treating diabetes was shown to provide closed-loop insulin delivery in a manner that was safer and fast-acting upon glucose sensing and required a lower insulin dosage due to a much more targeted and self-regulated delivery approach. Alternatively, Wang *et al.* generated a "bioartificial pancreas" that used encapsulated islet cells, the cells responsible for insulin production, using nanoporous membranes to offer immunoprotection from degradation/transplant rejection.[142] One of the key limitations in islet cell transplantation is the ongoing shortage of donor islet cells and poor recipient transplant efficacy even in patients on immunosuppressants. The findings from this study showed that when coupled with the use of an insulin pump to provide insulin to patients during the initial stages following nano-encapsulated islet cell transplant allowed for the successful regulation of blood glucose levels and islet-cell-mediated insulin release. The use

Table 5.1 Novel nanoparticle therapies to treat atherosclerosis, diabetes, hypertension, obesity and autoimmunity.

Molecular/functional target		Therapeutic agent	Type of nanoparticles	*In vivo*/*in vitro*	Results
Atherosclerosis	Macrophage	Pitavastatin[112]	PLGA NP	*In vivo*	↓ Plaque rupture
		Dexamethasone[113]	Liposomal NP	*In vivo* + *in vitro*	↓ Size of necrotic core
		SHP1 inhibitor[114]	Single-walled carbon nanotubes (SWNTs—PEG—Cy5.5)	*In vivo* + *in vitro*	↓ Pro-inflammatory cytokines, especially IL-1β, TNFα and IL-6
		siRNA-targeting CaMKIIγ[115]	PEG—lipid—PLGA	*In vivo* + *in vitro*	↓ Migration of inflammatory cells ↓ Size of necrotic core ↓ Cell death ↑ Plaque stability ↑ Efferocytosis (lesion cell clearance)
	LDL	Statin[116]	rHDL	*In vivo* + *in vitro*	↑ Plaque stability ↓ Plaque inflammation ↓ Inflammatory cell recruitment
	VCAM-1	MicroRNA-92a inhibitor[117]	VCAM-1-internalising peptide-28 (VHPKQHR) nanoparticles (VINP28-NPs)	*In vivo*	↑ Plaque stability ↓ Atherosclerosis ↓ Neointimal hyperplasia ↓ Endothelial migration and adhesion to fibronectin
	Integrin	Fumagillin[118]	Nanoemulsions	*In vitro*	↓ Neovasculature

(*continued*)

Table 5.1 (*continued*)

Molecular/functional target		Therapeutic agent	Type of nanoparticles	*In vivo/in vitro*	Results
	Type IV collagen	Annexin A1 biomimetic peptide Ac2—26[119,120]	PLGA–PEG	*In vivo*	↑ Plaque stability ↓ Size of necrotic core ↓ Inflammation
	Lipid metabolism	Resveratrol[96]	pH-sensitive selenium NP	*In vivo* + *in vitro*	↑ Plaque stability ↑ Lipid degradation ↓ Endothelial injury ↓ Plaque oxidative stress
	Plaque	Rapamycin[111]	Platelet-coated PLGA	*In vivo* + *in vitro*	↓ Plaque development ↓ Size of necrotic core ↓ Inflammatory cell recruitment ↓ Atherosclerosis
Diabetes	Glucose	Insulin[121]	Hypoxia-sensitive hyaluronic acid nanovesicle	*In vivo* + *in vitro*	↑ Controlled release of insulin upon changes in the glucose level ↑ Glycaemic control
	Glucose	Insulin + glucose oxidase[122]	Glucose-sensitive dextran NP	*In vivo* + *in vitro*	↑ Conversion of glucose into gluconic acid ↑ Insulin release in hyperglycaemic conditions
	pH	Insulin[123]	pH-sensitive, chitosan-coated freeze-dried NP	*In vivo* + *in vitro*	↑ Insulin absorption in the small intestine ↑ Effective delivery and release of insulin

(*continued*)

Table 5.1 (*continued*)

Molecular/functional target		Therapeutic agent	Type of nanoparticles	*In vivo/in vitro*	Results
	T1D/T2D	Islet cells[124]	Sodium alginate cellulose sulphate beads	*In vivo*	↑ Diabetes management
Hypertension	RAAS	Aliskiren[125]	PLA–NP	*In vivo*	↓ Blood pressure
	ACE	YF4[126]	Lipid–PLGA	*In vivo* + *in vitro*	↓ Blood pressure
	Calcium channel	Felodipine (FD)[127,128]	PLGA	*In vitro*	↓ Systolic blood pressure ↑ Bioavailability
	Ang-II receptor	Olmesartan medoxomil (OM)[129]	Nanoemulsion	*In vitro*	↑ Anti-hypertensive activity ↓ Arterial blood pressure
Obesity	White adipose tissue (WAT)	Dibenzazepine[130]	PLGA	*In vivo* + *in vitro*	↑ Adipose browning ↑ Glucose homeostasis ↑ Anti-obesity effects
	Adipose endothelial cells	Cytochrome C[131]	Prohibitin-targeted NP	*In vivo* + *in vitro*	↓ Leptin levels ↓ Adipose tissues ↓ Obesity
	Adipose tissue vasculature	Peroxisome proliferator-activated receptor gamma (PPARγ) activator rosiglitazone (Rosi)[132]	PLGA-b-PEG	*In vivo* + *in vitro*	↑ Adipose browning ↑ Angiogenesis ↓ Weight gain
	Inflammatory cells	Celastrol[133]	BSA NP	*In vivo* + *in vitro*	↑ Insulin sensitivity ↓ Lipid accumulation ↓ Inflammation ↓ Body weight
Autoimmunity	CD71	CsA[134]	Gambogic acid-conjugated NP	*In vitro* + *ex vivo*	↓ Cytokine infiltration

(*continued*)

Table 5.1 (*continued*)

Molecular/functional target	Therapeutic agent	Type of nanoparticles	*In vivo*/*in vitro*	Results
MΦ	CsA[135]	RBCm-coated, CsA-loaded PLGA NP	*In vivo* + *in vitro*	↓ Inflammatory cell infiltration
MΦ	Mycophenolic acid (MPA)[136]	MPA-loaded dextran NP	*In vivo* + *in vitro*	↓ Macrophages ↓T-cell proliferation
Glomerulus	MAPK1 SiRNA[137]	Poly(ethylene glycol)-poly(L-lysine) conjugated MAPK1 SiRNA	*In vivo* + *in vitro*	↓ Glomerulonephritis ↑ Kidney function[137]
Inflammatory cytokines		Neutrophil-coated PLGA NP	*In vivo* + *in vitro*	↓ Synovial inflammation ↓ Arthritis
IL-6 receptor	Tocilizumab[138]	Hyaluronate Au–NP	*In vivo* + *in vitro*	↑ Anti-angiogenic effects ↑ IL-6 and CD68 expressions
Arthritic tissues	Tacrolimus (TAC)[139]	TAC-loaded human serum albumin NP	*In vivo* + *in vitro*	↑ Anti-arthritic effect
Dihydrofolic acid	Methotrexate (MTX)[140]	MTX-loaded PLGA NP	*In vitro*	↓ Arthritis

of these islet cells was not intended to completely reverse diabetes but to drive increased stability of blood glucose levels in a pancreas-like manner, offering improved insulin management whilst requiring significantly fewer islet donor cells per patient. It was also highlighted that this intervention can be used earlier during the pathogenesis of diabetes, allowing existing damaged islet cells to recover as well as the immune system to remain intact due to no immunosuppressive or anti-inflammatory drugs being required for transplantation, thereby limiting opportunistic infections in these patients.

5.5.2 Hypertension-targeted Nanotherapies

The use of nanovesicles and liposomes in the treatment of hypertension was initially studied over three decades ago[143,144]; however, these treatments were used broadly, and investigations into their mechanisms of action, therapeutic targets and biodistribution were limited. Recent advances in technology now allow for these once limitations to be explored in detail, leading to enhanced targeting and efficacy. One of the key limitations of current anti-hypertensive therapies when administered orally is extensive first-pass metabolism, thus requiring a higher drug concentration to be administered. Shah *et al.* recently investigated the effect of FD, a calcium channel antagonist to treat hypertension and angina using drug-loaded PLGA nanoparticles administered orally.[127] This study emphasised that encapsulating FD made this once poorly soluble and low-bioavailable drug more stable, allowed it to bypass extensive liver metabolism, and increased bioavailability and stability. Pharmacodynamic studies also provided proof-of-concept data into the efficacy of FD–PLGA when compared to FD-alone by drastically reducing blood pressure and normalising cardiac function (ST elevation) for a total of 3 days post-administration. A similar study investigated nanoemulsion efficacy in increasing the bioavailability and therapeutic effects of an Ang-II-AT1 receptor inhibitor OM.[145] When compared to non-emulsified OM, there was a 2.8-fold increase in OM concentration following oral administration, and a much more sustained drug release profile was observed. Nanoemulsion delivery led to a 3-fold decrease in the overall dose required to elicit the same anti-hypertensive effect. These two examples of anti-hypertensive treatments showcase the potential impacts of nanomedicine in increasing drug efficacy by simply altering their route of administration *via* encapsulation, either *via* PLGA coating or emulsification. In doing this, there were drastic reductions in first-pass metabolism, a more targeted delivery and a drastic reduction in the amount of drug required, defining nanomedicine as a likely candidate to aid in the treatment of hypertension and its associated comorbidities.

5.5.3 Obesity-targeted Nanotherapies

It is well established that obesity is associated with angiogenesis, with adipose tissue accumulation leading to vascular remodelling and altered adipose tissue metabolism.[146] WAT, whose accumulation leads to obesity, has been a target of anti-angiogenic therapy, as it has been shown in studies that modulators of angiogenesis alter the metabolic activity, inhibit WAT accumulation and drastically reduce obesity in mouse models.[147,148] By aiding the conversion of energy-storing WAT into the energy-expending brown adipose tissue (BAT), there should be severe reductions in obesity due to reduced energy (fatty acid) storage. Countless weight loss drugs have been abandoned due to their undesired side effects, which are largely associated with non-specific interactions between the therapeutic and multiple off-target tissues and organs. The introduction of nanoparticles to administer these drugs in a far more targeted manner has been advantageous and opened multiple new avenues for the administration of once ineffective therapeutics. Recently, there have been two key studies that investigated the ability of nanoparticle-mediated administration of drugs to promote the browning of adipose tissue, both of which yielded positive data to suggest that nanoparticles may be a viable option for the treatment of obesity.[130,149] A study by Xue *et al.* used this adipose tissue transformation approach along with exploiting the pro-angiogenic properties of developing WAT deposits as a therapeutic target to administer rosiglitazone (PPARγ activator) or prostaglandin E2 analogue (PGE2) to promote the transformation of WAT into BAT and ultimately reduce obesity.[149] Findings from this study concluded that generating atherogenic targeting NPs to administer rosiglitazone or PGE2 led to a significant change in adipose tissue browning, inhibited body weight gain and reduced serological cholesterol, insulin and triglyceride levels, suggesting that this NP-based treatment may be effective in not only the treatment of obesity but also its associated metabolic disorders, including diabetes.

Furthermore, NP-based treatments have been generated to limit lipid accumulation and inflammation[150] and to administer targeted pro-apoptotic protein for ablating primary adipose cells.[131] Fan *et al.* used a biodegradable albumin nanoparticle to administer the poorly bioavailable and low water-soluble drug known as celastrol. Once encapsulated in the nanocarrier, there was a significant increase in the bioavailability of the drug as well as a prolonged/sustained release that ameliorated lipid accumulation and inflammation, reducing obesity and its associated hypercholesterolemia in mice. A similar approach was taken to administer pro-apoptotic cytochrome C, which has a poor circulatory half-life (~4 minutes) and potentially catastrophic off-target side effects.[131] However, by using a targeted nanocarrier that interacts specifically with prohibitin, a WAT vessel-specific receptor, Hossen *et al.* were able to effectively administer cytochrome C directly to adipose tissue, resulting in a dose-dependent apoptotic response of "fat cells" as well as reduced serum leptin levels when compared to untreated controls.[131] Taken together, these findings highlight the benefits of using NP-based therapies to specifically target tissue/cell types, as well as allowing for the use of treatments that were once rendered useless due to poor metabolic properties or off-target side effects. By modulating either atherosclerosis or its associated risk factors with nanomedicine, there should be a significant reduction in both the instance and severity of adverse cardiovascular events, suggesting that nanomedicine may be the future of cardiovascular research.

5.5.4 Autoimmunity-targeted Nanotherapies

There is an emerging body of evidence to suggest that patients with underlying autoimmune diseases, such as SLE and RA, are at a substantially increased risk of CVD-associated complications. The current issue faced by clinicians is the severe lack of effective therapies that can effectively combat both cardiovascular disease and autoimmunity simultaneously. Moreover, targeted therapies are lacking, with current treatments having a plethora of detrimental off-target effects. The use of nanotherapies to treat these diseases is appealing and an area of growing investigation due to their countless benefits regarding drug delivery whilst limiting off-target effects. A recent study by Hao *et al.* investigated the therapeutic effects of red blood cell membrane-coated biomimetic nanoparticles in the treatment of SLE.[135] The investigators utilised cyclosporin A (CsA), an immunosuppressant commonly used in SLE patients, irrespective of its poor bioavailability and pharmacokinetics due to it being highly lipophilic, and CsA is still used clinically. To combat these adverse effects and enhance the pharmacokinetic profile, CsA was incorporated into a PLGA polymeric nanoparticle that was cloaked in a red blood cell (RBC) membrane. Findings from this study showed that there was a sustained drug release, enhanced biocompatibility and a 2.6-fold increase in drug half-life when compared to conventional CsA treatment.

Nanotherapies have gained interest among researchers focusing on inflammatory disease; for instance, Zhang *et al.* used a neutrophil membrane-coated nanoparticle to modulate the inflammatory response within joints of patients with RA.[151] The use of such membranes allowed for the nanoparticles to retain the membrane antigens and functions of the cells from which they were isolated. The use of these nano-decoys allowed them to effectively suppress joint inflammation, reduce pro-inflammatory cytokine expression and suppress overall RA. These findings show that targeted inhibition of inflammation can be effectively achieved using nanotherapies. The effects of these treatments on autoantibody production are yet to be investigated and may offer a novel means to the treatment of autoimmune diseases without the adverse side effects seen with current therapies. Alleviating the overexuberant inflammation and autoantibody production seen in autoimmune diseases, such as SLE and RA, may provide a novel means to concurrently treat the autoimmune disease itself and also reduce the instance and severity of deadly cardiovascular complications.

5.6 Prospects and Conclusions

The growing interest in the field of nanomedicine and a substantial number of emerging therapies approved for use clinically highlight the benefits of these targeted treatments. Nano-based therapies exhibit a plethora of advantageous qualities over conventional drug treatments, including their ability to be targeted to specific tissues and cells; they can be designed to control the release of their therapeutic payload as well as offer enhanced stability, bioavailability and efficacy. The evolvement of modern technology is allowing for more in-depth investigation into various pathological conditions, including atherosclerosis, enabling researchers to define new therapeutic targets that may ultimately reduce atherogenesis and limit life-threatening MI.

Acknowledgements

This work was supported by the Baker Heart and Diabetes Institute.

References

1. C. D. Mathers and D. Loncar, Projections of global mortality and burden of disease from 2002 to 2030, *PLoS Med.*, 2006, **3**(11), e442.
2. Y.-C. Chen, A. L. Huang, T. S. Kyaw, A. Bobik and K. Peter, Atherosclerotic Plaque Rupture, *Arterioscler., Thromb., Vasc. Biol.*, 2016, **36**(8), e63–e72.
3. P. M. Ridker, B. M. Everett, T. Thuren, J. G. MacFadyen, W. H. Chang, C. Ballantyne, F. Fonseca, J. Nicolau, W. Koenig, S. D. Anker, J. J. P. Kastelein, J. H. Cornel, P. Pais, D. Pella, J. Genest, R. Cifkova, A. Lorenzatti, T. Forster, Z. Kobalava, L. Vida-Simiti, M. Flather, H. Shimokawa, H. Ogawa, M. Dellborg, P. R. F. Rossi, R. P. T. Troquay, P. Libby and R. J. Glynn, Antiinflammatory Therapy with Canakinumab for Atherosclerotic Disease, *N. Engl. J. Med.*, 2017, **377**(12), 1119–1131.
4. J. K. Patra, G. Das, L. F. Fraceto, E. V. R. Campos, M. D. P. Rodriguez-Torres, L. S. Acosta-Torres, L. A. Diaz-Torres, R. Grillo, M. K. Swamy, S. Sharma, S. Habtemariam and H.-S. Shin, Nano based drug delivery systems: recent developments and future prospects, *J. Nanobiotechnol.*, 2018, **16**(1), 71.
5. A. C. Anselmo and S. Mitragotri, Nanoparticles in the clinic: An update, *Bioeng. Transl. Med.*, 2019, **4**(3), e10143.

6. R. Brusini, M. Varna and P. Couvreur, Advanced nanomedicines for the treatment of inflammatory diseases, *Adv. Drug Delivery Rev.*, 2020, **157**, 161–178.
7. H. Wang, Y. Zhou, Q. Sun, C. Zhou, S. Hu, C. Lenahan, W. Xu, Y. Deng, G. Li and S. Tao, Update on Nanoparticle-Based Drug Delivery System for Anti-inflammatory Treatment, *Front. Bioeng. Biotechnol.*, 2021, **9**, 630352.
8. S. Nakashiro, T. Matoba, R. Umezu, J. Koga, M. Tokutome, S. Katsuki, K. Nakano, K. Sunagawa and K. Egashira, Pioglitazone-Incorporated Nanoparticles Prevent Plaque Destabilization and Rupture by Regulating Monocyte/Macrophage Differentiation in ApoE-/- Mice, *Arterioscler., Thromb., Vasc. Biol.*, 2016, **36**(3), 491–500.
9. N. Tsukie, K. Nakano, T. Matoba, S. Masuda, E. Iwata, M. Miyagawa, G. Zhao, W. Meng, J. Kishimoto, K. Sunagawa and K. Egashira, Pitavastatin-incorporated nanoparticle-eluting stents attenuate in-stent stenosis without delayed endothelial healing effects in a porcine coronary artery model, *J. Atheroscler. Thromb.*, 2013, **20**(1), 32–45.
10. H. C. McGill Jr, C. A. McMahan, E. E. Herderick, G. T. Malcom, R. E. Tracy and J. P. Strong, Origin of atherosclerosis in childhood and adolescence, *Am. J. Clin. Nutr.*, 2000, **72**(5 Suppl), 1307S–1315S.
11. W. F. Enos, R. H. Holmes and J. Beyer, Coronary disease among United States soldiers killed in action in Korea; preliminary report, *J. Am. Med. Assoc.*, 1953, **152**(12), 1090–1093.
12. C. K. Glass and J. L. Witztum, Atherosclerosis. the road ahead, *Cell*, 2001, **104**(4), 503–516.
13. E. Galkina and K. Ley, Immune and inflammatory mechanisms of atherosclerosis (*), *Annu. Rev. Immunol.*, 2009, **27**, 165–197.
14. D. Wolf and K. Ley, Immunity and Inflammation in Atherosclerosis, *Circ. Res.*, 2019, **124**(2), 315–327.
15. V. Stoneman, D. Braganza, N. Figg, J. Mercer, R. Lang, M. Goddard and M. Bennett, Monocyte/macrophage suppression in CD11b diphtheria toxin receptor transgenic mice differentially affects atherogenesis and established plaques, *Circ. Res.*, 2007, **100**(6), 884–893.
16. T. Kyaw, C. Tay, A. Khan, V. Dumouchel, A. Cao, K. To, M. Kehry, R. Dunn, A. Agrotis, P. Tipping, A. Bobik and B. H. Toh, Conventional B2 B cell depletion ameliorates whereas its adoptive transfer aggravates atherosclerosis, *J. Immunol.*, 2010, **185**(7), 4410–4419.
17. A. Zernecke and C. Weber, Chemokines in atherosclerosis: proceedings resumed, *Arterioscler., Thromb., Vasc. Biol.*, 2014, **34**(4), 742–750.
18. A. Zlotnik, O. Yoshie and H. Nomiyama, The chemokine and chemokine receptor superfamilies and their molecular evolution, *Genome Biol.*, 2006, **7**(12), 243.
19. W. Wan, M. S. Lionakis, Q. Liu, E. Roffê and P. M. Murphy, Genetic deletion of chemokine receptor Ccr7 exacerbates atherogenesis in ApoE-deficient mice, *Cardiovasc. Res.*, 2013, **97**(3), 580–588.

20. M. Luchtefeld, C. Grothusen, A. Gagalick, K. Jagavelu, H. Schuett, U. J. F. Tietge, O. Pabst, K. Grote, H. Drexler, R. Förster and B. Schieffer, Chemokine receptor 7 knockout attenuates atherosclerotic plaque development, *Circulation*, 2010, **122**(16), 1621–1628.
21. E. Galkina and K. Ley, Vascular Adhesion Molecules in Atherosclerosis, *Arterioscler., Thromb., Vasc. Biol.*, 2007, **27**(11), 2292–2301.
22. K. Khodabandehlou, J. J. Masehi-Lano, C. Poon, J. Wang and E. J. Chung, Targeting cell adhesion molecules with nanoparticles using in vivo and flow-based in vitro models of atherosclerosis, *Exp. Biol. Med.*, 2017, **242**(8), 799–812.
23. K. Habas and L. Shang, Alterations in intercellular adhesion molecule 1 (ICAM-1) and vascular cell adhesion molecule 1 (VCAM-1) in human endothelial cells, *Tissue Cell*, 2018, **54**, 139–143.
24. M. Yin, C. Li, J. Jiang, J. Le, B. Luo, F. Yang, Y. Fang, M. Yang, Z. Deng, W. Ni and J. Shao, Cell adhesion molecule-mediated therapeutic strategies in atherosclerosis: From a biological basis and molecular mechanism to drug delivery nanosystems, *Biochem. Pharmacol.*, 2021, **186**, 114471.
25. J. Kzhyshkowska, C. Neyen and S. Gordon, Role of macrophage scavenger receptors in atherosclerosis, *Immunobiology*, 2012, **217**(5), 492–502.
26. K. J. Moore and M. W. Freeman, Scavenger receptors in atherosclerosis: beyond lipid uptake, *Arterioscler., Thromb., Vasc. Biol.*, 2006, **26**(8), 1702–1711.
27. E. A. Van Vré, H. Ait-Oufella, A. Tedgui and Z. Mallat, Apoptotic cell death and efferocytosis in atherosclerosis, *Arterioscler., Thromb., Vasc. Biol.*, 2012, **32**(4), 887–893.
28. I. V. Smirnova, M. Kajstura, T. Sawamura and M. S. Goligorsky, Asymmetric dimethylarginine upregulates LOX-1 in activated macrophages: role in foam cell formation, *Am. J. Physiol.: Heart Circ. Physiol.*, 2004, **287**(2), H782–H790.
29. M. Febbraio, E. A. Podrez, J. D. Smith, D. P. Hajjar, S. L. Hazen, H. F. Hoff, K. Sharma and R. L. Silverstein, Targeted disruption of the class B scavenger receptor CD36 protects against atherosclerotic lesion development in mice, *J. Clin. Invest.*, 2000, **105**(8), 1049–1056.
30. J. L. Mehta, N. Sanada, C. P. Hu, J. Chen, A. Dandapat, F. Sugawara, H. Satoh, K. Inoue, Y. Kawase, K. Jishage, H. Suzuki, M. Takeya, L. Schnackenberg, R. Beger, P. L. Hermonat, M. Thomas and T. Sawamura, Deletion of LOX-1 reduces atherogenesis in LDLR knockout mice fed high cholesterol diet, *Circ. Res.*, 2007, **100**(11), 1634–1642.
31. A. Javadifar, S. Rastgoo, M. Banach, T. Jamialahmadi, T. P. Johnston and A. Sahebkar, Foam Cells as Therapeutic Targets in Atherosclerosis with a Focus on the Regulatory Roles of Non-Coding RNAs, *Int. J. Mol. Sci.*, 2021, **22**(5), 2529.

32. S. L. Stephen, K. Freestone, S. Dunn, M. W. Twigg, S. Homer-Vanniasinkam, J. H. Walker, S. B. Wheatcroft and S. Ponnambalam, Scavenger receptors and their potential as therapeutic targets in the treatment of cardiovascular disease, *Int. J. Hypertens.*, 2010, **2010**, 646929.
33. N. Herijgers, M. P. de Winther, M. Van Eck, L. M. Havekes, M. H. Hofker, P. M. Hoogerbrugge and T. J. Van Berkel, Effect of human scavenger receptor class A overexpression in bone marrow-derived cells on lipoprotein metabolism and atherosclerosis in low density lipoprotein receptor knockout mice, *J. Lipid Res.*, 2000, **41**(9), 1402–1409.
34. M. Van Eck, M. P. De Winther, N. Herijgers, L. M. Havekes, M. H. Hofker, P. H. E. Groot and T. J. C. Van Berkel, Effect of Human Scavenger Receptor Class A Overexpression in Bone Marrow–Derived Cells on Cholesterol Levels and Atherosclerosis in ApoE-Deficient Mice, *Arterioscler., Thromb., Vasc. Biol.*, 2000, **20**(12), 2600–2606.
35. A. C. Li and C. K. Glass, The macrophage foam cell as a target for therapeutic intervention, *Nat. Med.*, 2002, **8**(11), 1235–1242.
36. Y. Gui, H. Zheng and R. Y. Cao, Foam Cells in Atherosclerosis: Novel Insights Into Its Origins, Consequences, and Molecular Mechanisms, *Front. Cardiovasc. Med.*, 2022, **9**, 845942.
37. E. Yamamoto, T. Yonetsu, T. Kakuta, T. Soeda, Y. Saito, B. P. Yan, O. Kurihara, M. Takano, G. Niccoli, T. Higuma, S. Kimura, Y. Minami, J. Ako, T. Adriaenssens, N. F. Boeder, H. M. Nef, F. Fracassi, T. Sugiyama, H. Lee, F. Crea, T. Kimura, J. G. Fujimoto, V. Fuster and I. K. Jang, Clinical and Laboratory Predictors for Plaque Erosion in Patients With Acute Coronary Syndromes, *J. Am. Heart Assoc.*, 2019, **8**(21), e012322.
38. T. Gerhardt, A. Haghikia, P. Stapmanns and D. M. Leistner, Immune Mechanisms of Plaque Instability, *Front. Cardiovasc. Med.*, 2021, **8**, 797046.
39. P. E. Rautou, A. C. Vion, N. Amabile, G. Chironi, A. Simon, A. Tedgui and C. M. Boulanger, Microparticles, Vascular Function, and Atherothrombosis, *Circ. Res.*, 2011, **109**(5), 593–606.
40. T. Kubo, T. Imanishi, S. Takarada, A. Kuroi, S. Ueno, T. Yamano, T. Tanimoto, Y. Matsuo, T. Masho, H. Kitabata, K. Tsuda, Y. Tomobuchi and T. Akasaka, Assessment of culprit lesion morphology in acute myocardial infarction: ability of optical coherence tomography compared with intravascular ultrasound and coronary angioscopy, *J. Am. Coll. Cardiol.*, 2007, **50**(10), 933–939.
41. J. L. Johnson, S. J. George, A. C. Newby and C. L. Jackson, Divergent effects of matrix metalloproteinases 3, 7, 9, and 12 on atherosclerotic plaque stability in mouse brachiocephalic arteries, *Proc. Natl. Acad. Sci.*, 2005, **102**(43), 15575–15580.
42. D. Wågsäter, C. Zhu, J. Björkegren, J. Skogsberg and P. Eriksson, MMP-2 and MMP-9 are prominent matrix metalloproteinases during

atherosclerosis development in the Ldlr(-/-)Apob(100/100) mouse, *Int. J. Mol. Med.*, 2011, **28**(2), 247–253.

43. A. Tanindi, A. Sahinarslan, S. Elbeg and M. Cemri, Relationship between MMP-1, MMP-9, TIMP-1, IL-6 and risk factors, clinical presentation, extent and severity of atherosclerotic coronary artery disease, *Open Cardiovasc. Med. J.*, 2011, **5**, 110.
44. P. T. Elkington, J. A. Green and J. S. Friedland, Analysis of Matrix Metalloproteinase Secretion by Macrophages, in *Macrophages and Dendritic Cells: Methods and Protocols*, ed. N. E. Reiner, Humana Press, Totowa, NJ, 2009, pp. 253–265.
45. A. A. Motivala, U. Tamhane, V. S. Ramanath, F. Saab, D. G. Montgomery, J. Fang, E. Kline-Rogers, N. May, G. Ng, J. Froehlich, H. Gurm and K. A. Eagle, A prior myocardial infarction: how does it affect management and outcomes in recurrent acute coronary syndromes?, *Clin. Cardiol.*, 2008, **31**(12), 590–596.
46. T. Jernberg, P. Hasvold, M. Henriksson, H. Hjelm, M. Thuresson and M. Janzon, Cardiovascular risk in post-myocardial infarction patients: nationwide real world data demonstrate the importance of a long-term perspective, *Eur. Heart J.*, 2015, **36**(19), 1163–1170.
47. N. V. Joshi, I. Toor, A. S. Shah, K. Carruthers, A. T. Vesey, S. R. Alam, A. Sills, T. Y. Hoo, A. J. Melville, S. P. Langlands, W. S. Jenkins, N. G. Uren, N. L. Mills, A. M. Fletcher, E. J. van Beek, J. H. Rudd, K. A. Fox, M. R. Dweck and D. E. Newby, Systemic Atherosclerotic Inflammation Following Acute Myocardial Infarction: Myocardial Infarction Begets Myocardial Infarction, *J. Am. Heart Assoc.*, 2015, **4**(9), e001956.
48. Y. Han, J. Jing, S. Tu, F. Tian, H. Xue, W. Chen, J. Chen, J. H. Reiber and Y. Chen, ST elevation acute myocardial infarction accelerates non-culprit coronary lesion atherosclerosis, *Int. J. Cardiovasc. Imaging*, 2014, **30**(2), 253–261.
49. F. M. Sacks, M. A. Pfeffer, L. A. Moye, J. L. Rouleau, J. D. Rutherford, T. G. Cole, L. Brown, J. W. Warnica, J. M. Arnold, C. C. Wun, B. R. Davis and E. Braunwald, The effect of pravastatin on coronary events after myocardial infarction in patients with average cholesterol levels, *N. Engl. J. Med.*, 1996, **335**(14), 1001–1009.
50. P. M. Ridker, B. M. Everett, A. Pradhan, J. G. MacFadyen, D. H. Solomon, E. Zaharris, V. Mam, A. Hasan, Y. Rosenberg, E. Iturriaga, M. Gupta, M. Tsigoulis, S. Verma, M. Clearfield, P. Libby, S. Z. Goldhaber, R. Seagle, C. Ofori, M. Saklayen, S. Butman, N. Singh, M. Le May, O. Bertrand, J. Johnston, N. P. Paynter, R. J. Glynn, and CIRT Investigators, Low-Dose Methotrexate for the Prevention of Atherosclerotic Events, *N. Engl. J. Med.*, 2019, **380**(8), 752–762.
51. S. Verma, J. W. Eikelboom, S. M. Nidorf, M. Al-Omran, N. Gupta, H. Teoh and J. O. Friedrich, Colchicine in cardiac disease: a systematic review and meta-analysis of randomized controlled trials, *BMC Cardiovasc. Disord.*, 2015, **15**, 96.

52. A. Poznyak, A. V. Grechko, P. Poggio, V. A. Myasoedova, V. Alfieri and A. N. Orekhov, The Diabetes Mellitus-Atherosclerosis Connection: The Role of Lipid and Glucose Metabolism and Chronic Inflammation, *Int. J. Mol. Sci.*, 2020, **21**(5), 1835.
53. E. D. Marco, J. C. Jha, A. Sharma, J. L. Wilkinson-Berka, K. A. Jandeleit-Dahm and J. B. de Haan, Are reactive oxygen species still the basis for diabetic complications?, *Clin. Sci.*, 2015, **129**(2), 199–216.
54. K. Brassington, S. M. H. Chan, H. J. Seow, A. Dobric, S. Bozinovski, S. Selemidis and R. Vlahos, Ebselen reduces cigarette smoke-induced endothelial dysfunction in mice, *Br. J. Pharmacol.*, 2021, **178**(8), 1805–1818.
55. J. B. de Haan and M. E. Cooper, Targeted antioxidant therapies in hyperglycemia-mediated endothelial dysfunction, *Front. Biosci., Scholar Ed.*, 2011, **3**(2), 709–729.
56. Y.-J. Xu, P. S. Tappia, N. S. Neki and N. S. Dhalla, Prevention of diabetes-induced cardiovascular complications upon treatment with antioxidants, *Heart Failure Rev.*, 2014, **19**(1), 113–121.
57. E. B. Neufeld, L. M. Zadrozny, D. Phillips, A. Aponte, Z.-X. Yu and R. S. Balaban, Decorin and biglycan retain LDL in disease-prone valvular and aortic subendothelial intimal matrix, *Atherosclerosis*, 2014, **233**(1), 113–121.
58. L. R. Tannock and V. L. King, Proteoglycan mediated lipoprotein retention: a mechanism of diabetic atherosclerosis, *Rev. Endocr. Metab. Disord.*, 2008, **9**(4), 289–300.
59. A. P. Burke, F. D. Kolodgie, A. Zieske, D. R. Fowler, D. K. Weber, P. J. Varghese, A. Farb and R. Virmani, Morphologic Findings of Coronary Atherosclerotic Plaques in Diabetics, *Arterioscler., Thromb., Vasc. Biol.*, 2004, **24**(7), 1266–1271.
60. J. N. Redgrave, J. K. Lovett, A. B. Syed and P. M. Rothwell, Histological Features of Symptomatic Carotid Plaques in Patients with Impaired Glucose Tolerance and Diabetes (Oxford Plaque Study), *Cerebrovasc. Dis.*, 2008, **26**(1), 79–86.
61. N. G. Nikiforov, K. O. Galstyan, L. V. Nedosugova, N. V. Elizova, K. I. Kolmychkova and E. A. Ivanova, Proinflammatory monocyte polarization in type 2 diabetes mellitus and coronary heart disease, *Vessel Plus*, 2017, **1**, 192–195.
62. X. Fang, B. Dorcely, X.-p. Ding, S. Yin, N.-H. Son, S.-L. Hu and I. J. Goldberg, Glycemic reduction alters white blood cell counts and inflammatory gene expression in diabetes, *J. Diabetes Its Complications*, 2018, **32**(11), 1027–1034.
63. J. Moura, J. Rodrigues, M. Gonçalves, C. Amaral, M. Lima and E. Carvalho, Impaired T-cell differentiation in diabetic foot ulceration, *Cell. Mol. Immunol.*, 2017, **14**(9), 758–769.
64. B. Zhou, R. M. Carrillo-Larco and G. Danaei, *et al.*, Worldwide trends in hypertension prevalence and progress in treatment and control

from 1990 to 2019: a pooled analysis of 1201 population-representative studies with 104 million participants, *Lancet*, 2021, **398**(10304), 957–980.
65. A. Al-Sharea, M. K. Lee, A. Whillas, D. L. Michell, W. A. Shihata, A. J. Nicholls, O. D. Cooney, M. J. Kraakman, C. B. Veiga and A. M. Jefferis, Chronic sympathetic driven hypertension promotes atherosclerosis by enhancing hematopoiesis, *Haematologica*, 2019, **104**(3), 456.
66. E. J. Benjamin, P. Muntner, A. Alonso, M. S. Bittencourt and C. W. Callaway, *et al.*, Heart Disease and Stroke Statistics—2019 Update: A Report From the American Heart Association, *Circulation*, 2019, **139**(10), e56–e528.
67. R. Del Giorno, L. Balestra, P. S. Heiniger and L. Gabutti, Blood pressure variability with different measurement methods: Reliability and predictors. A proof of concept cross sectional study in elderly hypertensive hospitalized patients, *Medicine*, 2019, **98**(28), e16347.
68. G. Parati, J. E. Ochoa, C. Lombardi and G. Bilo, Assessment and management of blood-pressure variability, *Nat. Rev. Cardiol.*, 2013, **10**(3), 143–155.
69. J. Hurtubise, K. McLellan, K. Durr, O. Onasanya, D. Nwabuko and J. F. Ndisang, The Different Facets of Dyslipidemia and Hypertension in Atherosclerosis, *Curr. Atheroscler. Rep.*, 2016, **18**(12), 82.
70. A. C. Montezano, A. Nguyen Dinh Cat, F. J. Rios and R. M. Touyz, Angiotensin II and Vascular Injury, *Curr. Hypertens. Rep.*, 2014, **16**(6), 431.
71. A. Durante, G. Peretto, A. Laricchia, F. Ancona, M. Spartera, A. Mangieri and D. Cianflone, Role of the Renin-Angiotensin-Aldosterone System in the Pathogenesis of Atherosclerosis, *Curr. Pharm. Des.*, 2012, **18**(7), 981–1004.
72. L. Piqueras and M.-J. Sanz, Angiotensin II and leukocyte trafficking: New insights for an old vascular mediator. Role of redox-signaling pathways, *Free Radicals Biol. Med.*, 2020, **157**, 38–54.
73. S. Shu, Y. Zhang, W. Li, L. Wang, Y. Wu, Z. Yuan and J. Zhou, The role of monocyte chemotactic protein-induced protein 1 (MCPIP1) in angiotensin II-induced macrophage apoptosis and vulnerable plaque formation, *Biochem. Biophys. Res. Commun.*, 2019, **515**(2), 378–385.
74. G. M. Silva, M. S. França-Falcão, N. T. M. Calzerra, M. S. Luz, D. D. A. Gadelha, C. M. Balarini and T. M. Queiroz, Role of renin-angiotensin system components in atherosclerosis: focus on Ang-II, ACE2, and Ang-1–7, *Front. physiol.*, 2020, **11**, 1067.
75. G. Camejo, E. Hurt-Camejo, O. Wiklund and G. Bondjers, Association of apo B lipoproteins with arterial proteoglycans: Pathological significance and molecular basis, *Atherosclerosis*, 1998, **139**(2), 205–222.
76. E. Hurt-Camejo and G. Camejo, ApoB-100 Lipoprotein Complex Formation with Intima Proteoglycans as a Cause of Atherosclerosis

and Its Possible Ex Vivo Evaluation as a Disease Biomarker, *J. Cardiovasc. Dev. Dis.*, 2018, **5**(3), 36.

77. F. Huang, J. C. Thompson, P. G. Wilson, H. H. Aung, J. C. Rutledge and L. R. Tannock, Angiotensin II increases vascular proteoglycan content preceding and contributing to atherosclerosis development, *J. Lipid Res.*, 2008, **49**(3), 521–530.
78. R. H. Al-Mashhadi, A. L. Al-Mashhadi, Z. P. Nasr, M. B. Mortensen, E. A. Lewis, E. Camafeita, K. Ravlo, Z. Al-Mashhadi, D. W. Kjaer, J. Palmfeldt, P. Bie, J. M. Jensen, B. L. Nørgaard, E. Falk, J. Vázquez and J. F. Bentzon, Local Pressure Drives Low-Density Lipoprotein Accumulation and Coronary Atherosclerosis in Hypertensive Minipigs, *J. Am. Coll. Cardiol.*, 2021, **77**(5), 575–589.
79. R. J. Henning, Obesity and obesity-induced inflammatory disease contribute to atherosclerosis: a review of the pathophysiology and treatment of obesity, *Am. J. Cardiovasc. Dis.*, 2021, **11**(4), 504–529.
80. A. M. Thøgersen, S. Söderberg, J. H. Jansson, G. Dahlén, K. Boman, T. K. Nilsson, B. Lindahl, L. Weinehall, H. Stenlund, V. Lundberg, O. Johnson, B. Ahrén and G. Hallmans, Interactions between fibrinolysis, lipoproteins and leptin related to a first myocardial infarction, *Eur. J. Cardiovasc. Prev. Rehabil.*, 2004, **11**(1), 33–40.
81. R. Adya, B. K. Tan and H. S. Randeva, Differential Effects of Leptin and Adiponectin in Endothelial Angiogenesis, *J. Diabetes Res.*, 2015, **2015**, 648239.
82. H. Kwon and J. E. Pessin, Adipokines mediate inflammation and insulin resistance, *Front. Endocrinol.*, 2013, **4**, 71.
83. L. Ozcan, A. S. Ergin, A. Lu, J. Chung, S. Sarkar, D. Nie, M. G. Myers Jr and U. Ozcan, Endoplasmic reticulum stress plays a central role in development of leptin resistance, *Cell Metab.*, 2009, **9**(1), 35–51.
84. P. Kievit, J. K. Howard, M. K. Badman, N. Balthasar, R. Coppari, H. Mori, C. E. Lee, J. K. Elmquist, A. Yoshimura and J. S. Flier, Enhanced leptin sensitivity and improved glucose homeostasis in mice lacking suppressor of cytokine signaling-3 in POMC-expressing cells, *Cell Metab.*, 2006, **4**(2), 123–132.
85. T. Lundåsen, W. Liao, B. Angelin and M. Rudling, Leptin Induces the Hepatic High Density Lipoprotein Receptor Scavenger Receptor B Type I (SR-BI) but Not Cholesterol 7α-Hydroxylase (Cyp7a1) in Leptin-deficient (ob/ob) Mice, *J. Biol. Chem.*, 2003, **278**(44), 43224–43228.
86. R.-S. Li, S.-H. Zhou, Z.-P. Wang, Y.-Y. Du and H.-W. Pan, Effects of leptin on acyl-CoA: cholesterol O-acyltransferase-1 expression and activity during foam cell differentiation in human mononuclear THP-1 cells, *Int. J. Clin. Exp. Pathol.*, 2016, **9**(3), 3417–3423.
87. S. Ajeganova, I. Hafström and J. Frostegård, Patients with SLE have higher risk of cardiovascular events and mortality in comparison with controls with the same levels of traditional risk factors and intima-media measures, which is related to accumulated disease damage and

antiphospholipid syndrome: a case-control study over 10 years, *Lupus Sci. Med.*, 2021, **8**(1), e000454.
88. G. Katz, N. R. Smilowitz, A. Blazer, R. Clancy, J. P. Buyon and J. S. Berger, Systemic Lupus Erythematosus and Increased Prevalence of Atherosclerotic Cardiovascular Disease in Hospitalized Patients, *Mayo Clin. Proc.*, 2019, **94**(8), 1436–1443.
89. S. Skeoch, P. L. H. Cristinacce, H. Williams, P. Pemberton, D. Xu, J. Sun, J. James, C. Yuan, T. Hatsukami, P. D. Hockings, M. Y. Alexander, J. C. Waterton and I. N. Bruce, Imaging atherosclerosis in rheumatoid arthritis: evidence for increased prevalence, altered phenotype and a link between systemic and localised plaque inflammation, *Sci. Rep.*, 2017, **7**(1), 827.
90. S. Skeoch and I. N. Bruce, Atherosclerosis in rheumatoid arthritis: is it all about inflammation?, *Nat. Rev. Rheumatol.*, 2015, **11**(7), 390–400.
91. M. C. Aubry, H. Maradit-Kremers, M. S. Reinalda, C. S. Crowson, W. D. Edwards and S. E. Gabriel, Differences in atherosclerotic coronary heart disease between subjects with and without rheumatoid arthritis, *J. Rheumatol.*, 2007, **34**(5), 937–942.
92. T. Kyaw, P. Loveland, P. Kanellakis, A. Cao, A. Kallies, A. L. Huang, K. Peter, B.-H. Toh and A. Bobik, Alarmin-activated B cells accelerate murine atherosclerosis after myocardial infarction via plasma cell-immunoglobulin-dependent mechanisms, *Eur. Heart J.*, 2021, **42**(9), 938–947.
93. C. Tay, Y. H. Liu, P. Kanellakis, A. Kallies, Y. Li, A. Cao, H. Hosseini, P. Tipping, B. H. Toh, A. Bobik and T. Kyaw, Follicular B Cells Promote Atherosclerosis via T Cell-Mediated Differentiation Into Plasma Cells and Secreting Pathogenic Immunoglobulin G, *Arterioscler., Thromb., Vasc. Biol.*, 2018, **38**(5), e71–e84.
94. E. Falk, M. Nakano, J. F. Bentzon, A. V. Finn and R. Virmani, Update on acute coronary syndromes: the pathologists' view, *Eur. Heart J.*, 2013, **34**(10), 719–728.
95. P. Byrne, M. Demasi, M. Jones, S. M. Smith, K. K. O'Brien and R. DuBroff, Evaluating the Association Between Low-Density Lipoprotein Cholesterol Reduction and Relative and Absolute Effects of Statin Treatment: A Systematic Review and Meta-analysis, *JAMA Intern. Med.*, 2022, **182**(5), 474–481.
96. H. Zhang, L. Gao, M. Sui, J. Wang, Y. Wang, X. Xuan, Z. Zhang, L. Zhu and L. Hou, The "two-pronged" nanosystem to precisely improve lipid metabolism and inflammatory microenvironment for atherosclerotic plaque stabilization, *Nano Res.*, 2023, **16**(2), 2706–2718.
97. K. Brassington, S. M. H. Chan, S. N. De Luca, A. Dobric, S. A. Almerdasi, K. Mou, H. J. Seow, O. Oseghale, S. Bozinovski, S. Selemidis and R. Vlahos, Ebselen abolishes vascular dysfunction in influenza A virus-induced exacerbations of cigarette smoke-induced lung inflammation in mice, *Clin. Sci.*, 2022, **136**(8), 537–555.

98. M. Bartneck, F. M. Peters, K. T. Warzecha, M. Bienert, L. van Bloois, C. Trautwein, T. Lammers and F. Tacke, Liposomal encapsulation of dexamethasone modulates cytotoxicity, inflammatory cytokine response, and migratory properties of primary human macrophages, *Nanomedicine*, 2014, **10**(6), 1209–1220.
99. A. M. Flores, N. Hosseini-Nassab, K.-U. Jarr, J. Ye, X. Zhu, R. Wirka, A. L. Koh, P. Tsantilas, Y. Wang, V. Nanda, Y. Kojima, Y. Zeng, M. Lotfi, R. Sinclair, I. L. Weissman, E. Ingelsson, B. R. Smith and N. J. Leeper, Pro-efferocytic nanoparticles are specifically taken up by lesional macrophages and prevent atherosclerosis, *Nat. Nanotechnol.*, 2020, **15**(2), 154–161.
100. W. Tao, A. Yurdagul, N. Kong, W. Li, X. Wang, A. C. Doran, C. Feng, J. Wang, M. A. Islam, O. C. Farokhzad, I. Tabas and J. Shi, siRNA nanoparticles targeting CaMKIIγ in lesional macrophages improve atherosclerotic plaque stability in mice, *Sci. Transl. Med.*, 2020, **12**(553), eaay1063.
101. S. Katsuki, T. Matoba, S. Nakashiro, K. Sato, J.-i. Koga, K. Nakano, Y. Nakano, S. Egusa, K. Sunagawa and K. Egashira, Nanoparticle-Mediated Delivery of Pitavastatin Inhibits Atherosclerotic Plaque Destabilization/Rupture in Mice by Regulating the Recruitment of Inflammatory Monocytes, *Circulation*, 2014, **129**(8), 896–906.
102. R. Duivenvoorden, J. Tang, D. P. Cormode, A. J. Mieszawska, D. Izquierdo-Garcia, C. Ozcan, M. J. Otten, N. Zaidi, M. E. Lobatto, S. M. van Rijs, B. Priem, E. L. Kuan, C. Martel, B. Hewing, H. Sager, M. Nahrendorf, G. J. Randolph, E. S. G. Stroes, V. Fuster, E. A. Fisher, Z. A. Fayad and W. J. M. Mulder, A statin-loaded reconstituted high-density lipoprotein nanoparticle inhibits atherosclerotic plaque inflammation, *Nat. Commun.*, 2014, **5**(1), 3065.
103. L. Guo, J. Xiao, H. Liu and H. Liu, Selenium nanoparticles alleviate hyperlipidemia and vascular injury in ApoE-deficient mice by regulating cholesterol metabolism and reducing oxidative stress, *Metallomics*, 2020, **12**(2), 204–217.
104. C. Reimann, J. Brangsch, F. Colletini, T. Walter, B. Hamm, R. M. Botnar and M. R. Makowski, Molecular imaging of the extracellular matrix in the context of atherosclerosis, *Adv. Drug Delivery Rev.*, 2017, **113**, 49–60.
105. G. Fredman, N. Kamaly, S. Spolitu, J. Milton, D. Ghorpade, R. Chiasson, G. Kuriakose, M. Perretti, O. Farokzhad and I. Tabas, Targeted nanoparticles containing the proresolving peptide Ac2-26 protect against advanced atherosclerosis in hypercholesterolemic mice, *Sci. Transl. Med.*, 2015, **7**(275), 275ra20. Erratum in: *Sci. Transl. Med.*, 2015, **7**(277), 277er2.

106. E. Tzima, M. Irani-Tehrani, W. B. Kiosses, E. Dejana, D. A. Schultz, B. Engelhardt, G. Cao, H. DeLisser and M. A. Schwartz, A mechanosensory complex that mediates the endothelial cell response to fluid shear stress, *Nature*, 2005, **437**(7057), 426–431.
107. A. Yurdagul Jr, J. Green, P. Albert, M. C. McInnis, A. P. Mazar and A. W. Orr, α5β1 integrin signaling mediates oxidized low-density lipoprotein–induced inflammation and early atherosclerosis, *Arterioscler., Thromb., Vasc. Biol.*, 2014, **34**(7), 1362–1373.
108. Z. Zhou, C.-F. Yeh, M. Mellas, M.-J. Oh, J. Zhu, J. Li, R.-T. Huang, D. L. Harrison, T.-P. Shentu, D. Wu, M. Lueckheide, L. Carver, E. J. Chung, L. Leon, K.-C. Yang, M. V. Tirrell and Y. Fang, Targeted polyelectrolyte complex micelles treat vascular complications invivo, *Proc. Natl. Acad. Sci.*, 2021, **118**(50), e2114842118.
109. L. Badimon, T. Padró and G. Vilahur, Atherosclerosis, platelets and thrombosis in acute ischaemic heart disease, *Eur. Heart J. Acute Cardiovasc. Care*, 2012, **1**(1), 60–74.
110. W. Martinet, H. De Loof and G. R. Y. De Meyer, mTOR inhibition: a promising strategy for stabilization of atherosclerotic plaques, *Atherosclerosis*, 2014, **233**(2), 601–607.
111. Y. Song, Z. Huang, X. Liu, Z. Pang, J. Chen, H. Yang, N. Zhang, Z. Cao, M. Liu, J. Cao, C. Li, X. Yang, H. Gong, J. Qian and J. Ge, Platelet membrane-coated nanoparticle-mediated targeting delivery of Rapamycin blocks atherosclerotic plaque development and stabilizes plaque in apolipoprotein E-deficient (ApoE−/−) mice, *Nanomedicine*, 2019, **15**(1), 13–24.
112. S. Katsuki, T. Matoba, S. Nakashiro, K. Sato, J. Koga, K. Nakano, Y. Nakano, S. Egusa, K. Sunagawa and K. Egashira, Nanoparticle-mediated delivery of pitavastatin inhibits atherosclerotic plaque destabilization/rupture in mice by regulating the recruitment of inflammatory monocytes, *Circulation*, 2014, **129**(8), 896–906.
113. M. Bartneck, F. M. Peters, K. T. Warzecha, M. Bienert, L. van Bloois, C. Trautwein, T. Lammers and F. Tacke, Liposomal encapsulation of dexamethasone modulates cytotoxicity, inflammatory cytokine response, and migratory properties of primary human macrophages, *Nanomedicine*, 2014, **10**(6), 1209–1220.
114. A. M. Flores, N. Hosseini-Nassab, K. U. Jarr, J. Ye, X. Zhu, R. Wirka, A. L. Koh, P. Tsantilas, Y. Wang, V. Nanda, Y. Kojima, Y. Zeng, M. Lotfi, R. Sinclair, I. L. Weissman, E. Ingelsson, B. R. Smith and N. J. Leeper, Pro-efferocytic nanoparticles are specifically taken up by lesional macrophages and prevent atherosclerosis, *Nat. Nanotechnol.*, 2020, **15**(2), 154–161.

115. W. Tao, A. Yurdagul, N. Kong, W. Li, X. Wang, A. C. Doran, C. Feng, J. Wang, M. A. Islam, O. C. Farokhzad, I. Tabas and J. Shi, siRNA nanoparticles targeting CaMKIIγ in lesional macrophages improve atherosclerotic plaque stability in mice, *Sci. Transl. Med.*, 2020, **12**(553), eaay1063.
116. R. Duivenvoorden, J. Tang, D. P. Cormode, A. J. Mieszawska, D. Izquierdo-Garcia, C. Ozcan, M. J. Otten, N. Zaidi, M. E. Lobatto, S. M. van Rijs, B. Priem, E. L. Kuan, C. Martel, B. Hewing, H. Sager, M. Nahrendorf, G. J. Randolph, E. S. G. Stroes, V. Fuster, E. A. Fisher, Z. A. Fayad and W. J.Mulder, A statin-loaded reconstituted high-density lipoprotein nanoparticle inhibits atherosclerotic plaque inflammation, *Nat. Commun.*, 2014, **5**, 3065.
117. Z. Zhou, C. F. Yeh, M. Mellas, M. J. Oh, J. Zhu, J. Li, R. T. Huang, D. L. Harrison, T. P. Shentu, D. Wu, M. Lueckheide, L. Carver, E. J. Chung, L. Leon, K. C. Yang, M. V. Tirrell and Y. Fang, Targeted polyelectrolyte complex micelles treat vascular complications invivo, *Proc. Natl. Acad. Sci. U.S.A.*, 2021, **118**(50), e2114842118.
118. P. M. Winter, S. D. Caruthers, H. Zhang, T. A. Williams, S. A. Wickline and G. M. Lanza, Antiangiogenic synergism of integrin-targeted fumagillin nanoparticles and atorvastatin in atherosclerosis, *JACC Cardiovasc. Imaging*, 2008, **1**(5), 624–634.
119. G. Fredman, N. Kamaly, S. Spolitu, J. Milton, D. Ghorpade, R. Chiasson, G. Kuriakose, M. Perretti, O. Farokzhad and I. Tabas, Targeted nanoparticles containing the proresolving peptide Ac2-26 protect against advanced atherosclerosis in hypercholesterolemic mice, *Sci. Transl. Med.*, 2015, **7**(275), 275ra20.
120. N. Kamaly, G. Fredman, M. Subramanian, S. Gadde, A. Pesic, L. Cheung, Z. A. Fayad, R. Langer, I. Tabas and O. C. Farokhzad, Development and in vivo efficacy of targeted polymeric inflammation-resolving nanoparticles, *Proc. Natl. Acad. Sci. U. S. A.*, 2013, **110**(16), 6506–6511.
121. J. Yu, Y. Zhang, Y. Ye, R. DiSanto, W. Sun, D. Ranson, F. S. Ligler, J. B. Buse and Z. Gu, Microneedle-array patches loaded with hypoxia-sensitive vesicles provide fast glucose-responsive insulin delivery, *Proc. Natl. Acad. Sci. U. S. A.*, 2015, **112**(27), 8260–8265.
122. Z. Gu, A. A. Aimetti, Q. Wang, T. T. Dang, Y. Zhang, O. Veiseh, H. Cheng, R. S. Langer and D. G. Anderson, Injectable nano-network for glucose-mediated insulin delivery, *ACS Nano*, 2013, **7**(5), 4194–4201.
123. K. Sonaje, Y. J. Chen, H. L. Chen, S. P. Wey, J. H. Juang, H. N. Nguyen, C. W. Hsu, K. J. Lin and H. W. Sung, Enteric-coated capsules filled with freeze-dried chitosan/poly(γ-glutamic acid) nanoparticles for oral insulin delivery, *Biomaterials*, 2010, **31**(12), 3384–3394.
124. T. Wang, Successful diabetes management without immunosuppressivedrugs in NHP model has been demonstrated. Encapsulation system with taperednanopore conduits achieved normal glycaemia with

regulated insulin release, *Artif. Cells, Nanomed., Biotechnol.*, 2018, **46**(sup3), 1162–1168.
125. O. Pechanova, A. Barta, M. Koneracka, V. Zavisova, M. Kubovcikova, J. Klimentova, J. Török, A. Zemancikova and M. Cebova, Protective Effects of Nanoparticle-Loaded Aliskiren on Cardiovascular System in Spontaneously Hypertensive Rats, *Molecules*, 2019, **24**(15), 2710.
126. S. Zhao, J. Li, Y. Zhou, L. Huang, Y. Li, J. Xu, C. Fu, X. Guo and J. Yang, Lipid Nanoparticles-Encapsulated YF4: A Potential Therapeutic Oral Peptide Delivery System for Hypertension Treatment, *Front. Pharmacol.*, 2019, **10**, 102.
127. U. Shah, G. Joshi and K. Sawant, Improvement in antihypertensive and antianginal effects of felodipine by enhanced absorption from PLGA nanoparticles optimized by factorial design, *Mater. Sci. Eng., C*, 2014, **35**, 153–163.
128. I. S. Fancher, I. Rubinstein and I. Levitan, Potential Strategies to Reduce Blood Pressure in Treatment-Resistant Hypertension Using Food and Drug Administration–Approved Nanodrug Delivery Platforms, *Hypertension*, 2019, **73**(2), 250–257.
129. B. Gorain, H. Choudhury, A. Kundu, L. Sarkar, S. Karmakar, P. Jaisankar and T. K. Pal, Nanoemulsion strategy for olmesartan medoxomil improves oral absorption and extended antihypertensive activity in hypertensive rats, *Colloids Surf., B*, 2014, **115**, 286–294.
130. C. Jiang, M. A. Cano-Vega, F. Yue, L. Kuang, N. Narayanan, G. Uzunalli, M. P. Merkel, S. Kuang and M. Deng, Dibenzazepine-Loaded Nanoparticles Induce Local Browning of White Adipose Tissue to Counteract Obesity, *Mol. Ther.*, 2017, **25**(7), 1718–1729.
131. M. N. Hossen, K. Kajimoto, H. Akita, M. Hyodo, T. Ishitsuka and H. Harashima, Therapeutic assessment of cytochrome C for the prevention of obesity through endothelial cell-targeted nanoparticulate system, *Mol. Ther.*, 2013, **21**(3), 533–541.
132. Y. Xue, X. Xu, X. Q. Zhang, O. C. Farokhzad and R. Langer, Preventing diet-induced obesity in mice by adipose tissue transformation and angiogenesis using targeted nanoparticles, *Proc. Natl. Acad. Sci. U. S. A*, 2016, **113**(20), 5552–5557.
133. N. Fan, J. Zhao, W. Zhao, Y. Shen, Q. Song, H. C. Shum, Y. Wang and J. Rong, Biodegradable celastrol-loaded albumin nanoparticles ameliorate inflammation and lipid accumulation in diet-induced obese mice, *Biomater. Sci.*, 2022, **10**(4), 984–996.
134. R. Ganugula, M. Arora, D. Zou, S. K. Agarwal, C. Mohan and M. Kumar, A highly potent lymphatic system–targeting nanoparticle cyclosporine prevents glomerulonephritis in mouse model of lupus, *Sci. Adv.*, 2020, **6**(24), eabb3900.
135. X. Hao, H. Zhang, R. Liu, J. Che, D. Zhang, J. Liang and L. Sun, Red blood cell membrane functionalized biomimetic nanoparticles for

systemic lupus erythematosus treatment, *Mater. Today Adv.*, 2022, **16**, 100294.
136. B. Jiang, Y. Zhang, Y. Li, Y. Chen, S. Sha, L. Zhao, D. Li, J. Wen, J. Lan, Y. Lou, H. Su, C. Zhang, J. Zhu and J. Tao, A Tissue-Tended Mycophenolate-Modified Nanoparticle Alleviates Systemic Lupus Erythematosus in MRL/Lpr Mouse Model Mainly by Promoting Local M2-Like Macrophagocytes Polarization, *Int. J. Nanomed.*, 2022, **17**, 3251–3267.
137. H. Shimizu, Y. Hori, S. Kaname, K. Yamada, N. Nishiyama, S. Matsumoto, K. Miyata, M. Oba, A. Yamada, K. Kataoka and T. Fujita, siRNA-based therapy ameliorates glomerulonephritis, *J. Am. Soc. Nephrol.*, 2010, **21**(4), 622–633.
138. K. Janakiraman, V. Krishnaswami, V. Rajendran, S. Natesan and R. Kandasamy, Novel nano therapeutic materials for the effective treatment of rheumatoid arthritis-recent insights, *Mater. Today Commun.*, 2018, **17**, 200–213.
139. L. Q. Thao, H. J. Byeon, C. Lee, S. Lee, E. S. Lee, H.-G. Choi, E. S. Park and Y.-S. Youn, Pharmaceutical potential of tacrolimus-loaded albumin nanoparticles having targetability to rheumatoid arthritis tissues, *Int. J. Pharm.*, 2016, **497**(1), 268–276.
140. Y.-J. Ha, S.-M. Lee, C.-H. Mun, H. J. Kim, Y. Bae, J.-H. Lim, K.-H. Park, S.-K. Lee, K.-H. Yoo and Y.-B. Park, Methotrexate-loaded multifunctional nanoparticles with near-infrared irradiation for the treatment of rheumatoid arthritis, *Arthritis Res. Ther.*, 2020, **22**(1), 146.
141. J. Yu, Y. Zhang, Y. Ye, R. DiSanto, W. Sun, D. Ranson, F. S. Ligler, J. B. Buse and Z. Gu, Microneedle-array patches loaded with hypoxia-sensitive vesicles provide fast glucose-responsive insulin delivery, *Proc. Natl. Acad. Sci.*, 2015, **112**(27), 8260–8265.
142. T. Wang, Successful diabetes management without immunosuppressivedrugs in NHP model has been demonstrated. Encapsulation system with taperednanopore conduits achieved normal glycaemia with regulated insulin release, *Artif. Cells, Nanomed., Biotechnol.*, 2018, **46**(sup3), 1162–1168.
143. H. N. Hodis, J. K. Amartey, D. W. Crawford, E. Wickham and D. H. Blankenhorn, In vivo hypertensive arterial wall uptake of radiolabeled liposomes, *Hypertension*, 1990, **15**(6_pt_1), 600–605.
144. J. B. Laursen, S. Rajagopalan, Z. Galis, M. Tarpey, B. A. Freeman and D. G. Harrison, Role of superoxide in angiotensin II–induced but not catecholamine-induced hypertension, *Circulation*, 1997, **95**(3), 588–593.
145. B. Gorain, H. Choudhury, A. Kundu, L. Sarkar, S. Karmakar, P. Jaisankar and T. K. Pal, Nanoemulsion strategy for olmesartan medoxomil improves oral absorption and extended antihypertensive activity in hypertensive rats, *Colloids Surf., B*, 2014, **115**, 286–294.
146. Y. Cao, Adipose tissue angiogenesis as a therapeutic target for obesity and metabolic diseases, *Nat. Rev. Drug Discovery*, 2010, **9**(2), 107–115.

147. M. A. Rupnick, D. Panigrahy, C. Y. Zhang, S. M. Dallabrida, B. B. Lowell, R. Langer and M. J. Folkman, Adipose tissue mass can be regulated through the vasculature, *Proc. Natl. Acad. Sci. U. S. A.*, 2002, **99**(16), 10730–10735.
148. K. F. Barnhart, D. R. Christianson, P. W. Hanley, W. H. P. Driessen, B. J. Bernacky, W. B. Baze, S. Wen, M. Tian, J. Ma, M. G. Kolonin, P. K. Saha, K. A. Do, J. F. Hulvat, J. G. Gelovani, L. Chan, W. Arap and R. Pasqualini, A peptidomimetic targeting white fat causes weight loss and improved insulin resistance in obese monkeys, *Sci. Transl. Med.*, 2011, **3**(108), 108ra112.
149. Y. Xue, X. Xu, X.-Q. Zhang, O. C. Farokhzad and R. Langer, Preventing diet-induced obesity in mice by adipose tissue transformation and angiogenesis using targeted nanoparticles, *Proc. Natl. Acad. Sci.*, 2016, **113**(20), 5552–5557.
150. N. Fan, J. Zhao, W. Zhao, Y. Shen, Q. Song, H. C. Shum, Y. Wang and J. Rong, Biodegradable celastrol-loaded albumin nanoparticles ameliorate inflammation and lipid accumulation in diet-induced obese mice, *Biomater. Sci.*, 2022, **10**(4), 984–996.
151. Q. Zhang, D. Dehaini, Y. Zhang, J. Zhou, X. Chen, L. Zhang, R. H. Fang, W. Gao and L. Zhang, Neutrophil membrane-coated nanoparticles inhibit synovial inflammation and alleviate joint damage in inflammatory arthritis, *Nat. Nanotechnol.*, 2018, **13**(12), 1182–1190.

CHAPTER 6

Advanced Nanomaterials for Cardiovascular Implants

I. CICHA*[a], C. ALEXIOU[a] AND T. CIACH[b,c]

[a]Department of Otolaryngology, Head and Neck Surgery, Section of Experimental Oncology and Nanomedicine (SEON), Else Kröner-Fresenius-Stiftung Professorship, Universitätsklinikum Erlangen, Friedrich-Alexander-Universität Erlangen-Nürnberg, Erlangen, Germany; [b]Warsaw University of Technology, Faculty of Chemical and Process Engineering, Division of Biotechnology and Bioprocess Engineering, BioMedical Engineering Laboratory, Waryńskiego 1, 00-645 Warsaw, Poland; [c]Centre for Advanced Materials and Technologies CEZAMAT, Warsaw University of Technology, Poleczki 19, 02-822, Warsaw, Poland
*E-mail: Iwona.Cicha@uk-erlangen.de; Iwona.Cicha@yahoo.com

6.1 Introduction

The global burden of atherothrombotic disorders continues to grow.[1] Cardiovascular diseases (CVDs) cause about 3.9 million deaths in Europe and above 800 thousand deaths in the USA each year. Among the different types of CVD, coronary heart disease (CHD) is the most common and most lethal form, contributing to about 40% of cardiovascular mortality. According to the 2016 Heart Disease and Stroke Statistics update of the American Heart Association (AHA), about 15.5 million persons ≥20 years of age in the USA have CHD,[2] and it is estimated that approximately every 42 seconds, an American suffers from myocardial infarction (MI). To treat patients affected by acute MI and its chronic consequences, invasive

Nanoscience & Nanotechnology Series No. 64
Cardiovascular Nanomedicine
Edited by Roger Narayan and Tanveer Tabish
© The Royal Society of Chemistry 2024
Published by the Royal Society of Chemistry, www.rsc.org

treatment approaches or surgical procedures combined with implant placement are often required. These include percutaneous coronary intervention (PCI), which commonly involves stent placement, coronary artery bypass grafting (CABG), cardiac patch implantation, and heart transplantation in patients with congestive heart failure. The procedures are recommended depending on patients' overall health, age, and other disease-related factors.

This chapter addresses the types of cardiovascular implants for which advanced nanomaterials are of particular relevance.

6.1.1 Types of Cardiovascular Implants

Cardiovascular implants comprise several types of biological or artificial devices, including *stents*, which are implanted during endovascular intervention and play a key role in the revascularization of patients with CHD; *vascular grafts,* which are used for surgical bypassing of atherosclerosis-affected coronary arteries or peripheral vessel replacement; *cardiac patches*, which should provide mechanical support and regeneration of infarcted cardiac muscle; and prosthetic *heart valves*, which are used in patients with heart valve stenosis.

The most common method of coronary revascularization is an endovascular procedure called PCI, which is also one of the most frequently performed therapeutic procedures in medicine.[3] During PCI, stents, most commonly made of stainless steel or different alloys, *e.g.* cobalt–chromium or nickel–titanium, are implanted to prevent the lumen collapse after ballooning. With over 600 000 stent implantations recorded in the National Cardiovascular Data Registry (NCDR) CathPCI Registry (2014 data), stents represent the most commonly implanted medical device.[4] Nonetheless, several safety issues have been identified in the follow-up of stent implantations, including the risk of restenosis resulting from vascular remodeling and neointimal hyperplasia [in particular in bare metal stents (BMS)],[5] increased risk of in-stent thrombosis associated with drug-eluting stents (DESs),[6] and increased risk of thrombosis and MI associated with bioresorbable vascular stents.[7]

Vascular prostheses or vascular grafts are implanted surgically and comprise *biological grafts*, including, *e.g.* autologous (patient's own) blood vessels; allografts (vessels from other human donors); xenografts (decellularized vessels of bovine or porcine origin); and artificial *synthetic grafts* made from polymer-based, durable, or biodegradable materials. Among the latter, expanded polytetrafluoroethylene (ePTFE), polyethylene terephthalate (PET/Dacron), or polyurethane (PU) are used as low-immunogenic and low-thrombogenic materials suitable for large-diameter vascular grafts. Vascular graft implantation requires open surgery, during which the prosthesis is sutured to the healthy parts of an artery, and autologous graft implantation is associated with the necessity of a second surgical procedure to harvest a vessel. Besides the general risks related to surgical interventions (*e.g.* infections), implantation of non-autologous grafts bears the risk of foreign body reaction, thrombosis, loss of graft integrity, and graft failure. The artificial grafts, which maintain satisfactory patency over years when implanted in large peripheral arteries, are not suitable as small-diameter vascular implants in coronary positions. Apart from required mechanical stability, suturability, biocompatibility, and thromboresistance, an ideal vascular graft should be leakage-free, compliant and enable rapid endothelialization. Technical factors, including facile sterilization, off-shelf availability, and cost-effectiveness, are alike important.

Recently, the combination of metal stents and polymer grafts has been introduced in the clinical practice for the treatment of aneurysms. The implantation of a dense polymer mesh placed on a balloon-expandable or self-expanding metal stent enables blocking the entrance to an aneurysmal sack and thus prevents blood from entering it. This minimally invasive procedure has been shown to reduce the risk of morbidity and mortality associated with open surgery.[8]

Cardiac patches can be implanted surgically or injected in the form of hydrogels. They comprise *synthetic patches*, made of durable ePTFE or PET,[9] and *biological patches*, including, *e.g.* decellularized pericardium of xenogeneic origin, or scaffolds fabricated from naturally occurring materials,[10] such as extracellular matrix components (*e.g.* collagen and hyaluronic acid) or marine polysaccharides (*e.g.* alginate). Pure material-based patches, made of either synthetic or natural materials, are bioinert and provide mechanical support to the failing heart muscle, but they lack biological cues and may prevent cell–cell contacts and electrical coupling between cardiomyocytes. Cell-based patches support the integration into native microenvironment, but their mechanical strength and *in vivo* stability are unknown. The potential immunogenicity of cell-based or decellularized xenogeneic patches remains an important concern.

Heart valve prostheses comprise mechanical or biological valves.[11,12] Mechanical valves, made of durable materials, usually combine stainless steel and bioinert polymers. The valves are available in different designs, including bileaflet, monoleaflet, or caged-ball type, and their advantages are durability and low risk of re-operations. However, their implantation requires surgical procedures and subsequent life-long antiplatelet medication. Bioprosthetic valves, based on decellularized, glutaraldehyde-fixed porcine tissue, include stentless bioprostheses manufactured from whole porcine aortic valves and stented bioprostheses containing porcine aortic valve leaflets mounted on a polymer-supporting stent, both of which are implanted surgically. A further type of bioprosthetic valve contains porcine leaflets mounted on expendable metallic stents for percutaneous implantation. The advantage of the bioprosthetic valves is their structural similarity to the native tissue, but they are not free of thrombotic risk. They are also prone to leaflet calcification and may cause immune reactions, if the material is not sufficiently decellularized. Compared to mechanical valves, their durability is poorer, but they may also be easier to replace using the percutaneous implantation method.

In general, all the cardiovascular implants listed earlier enable the effective therapy of patients suffering from CVD, but all of them have considerable limitations. Consequently, the development of advanced structure-mimicking and function-supporting materials is expected to improve the outcomes of implanted devices.

6.1.2 Nanomaterials: Definition and Classification

The definition of "nanomaterial" differs among organizations and countries, even though standardized size limits are necessary for regulatory purposes.[13] Nanomaterials are differently defined in the USA and the EU, whereby the Food and Drug Administration's (FDA) definition implemented in the USA describes a nanomaterial as any material with at least one dimension smaller than 1000 nm and a nanoparticle as an object with all three external dimensions in the 1–100 nm size range. The EU regulation states that a nanomaterial is a "natural, incidental or manufactured material containing particles, in an unbound state, or as an aggregate, or as an agglomerate and where, for 50% or more of the particles in the number size distribution, one or more external dimensions is in the size range 1–100 nm. In specific cases and where warranted by concerns for the environment, health, safety, or competitiveness, the number size distribution threshold of 50% may be replaced by a threshold between 1% and 50%".[14]

Scientists have not unanimously agreed on a precise definition of nanomaterials as yet, but the cutoff size limit for nanoparticulate matter implemented by some scientific journals is 400 nm, while in biomaterial sciences, even submicron fibers and structures sized up to 600–700 nm are commonly termed "nano". The ISO norm TS80004 (ISO/TS 80004) defined nanomaterial as the "material with any external dimension in the nanoscale, or having an internal structure or surface structure in the nanoscale". *Nanoscale* is, in turn, defined as the "length range from approximately 1 nm to 100 nm". This definition includes both *nanoobjects* or *nanoparticles*, *i.e.* small fragments of materials, and *nanostructured materials*, the surface or internal structure of which is in the nanoscale. According to this definition, a nanomaterial may be a member of both these categories,[15] and both of them have relevant applications in medicine. Depending on the number of dimensions that fall in the nanoscale, nanomaterials can be further grouped into *nanoparticles*, which are nanoobjects with all three external dimensions in the nanoscale and with similar lengths of their longest and shortest axes; *nanorods*, with all three external dimensions in the nanoscale but a large aspect ratio; and *nanofibers*, with two external dimensions in the nanoscale and length in μm or mm scale, whereas *nanotubes* are defined

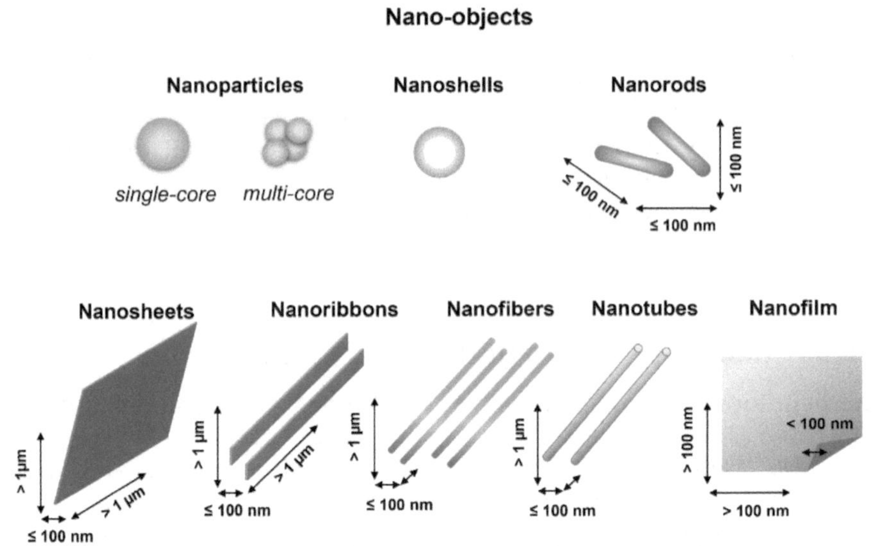

Figure 6.1 Classification of nanoscale objects.

as hollow nanofibers. Materials with only one external dimension in the nanoscale are termed *nanosheets* or *nanoribbons* (if two other directions are significantly different). The term *nanofilm* is sometimes used alternately with *nanosheet*, but more commonly, it refers to the layer of surface coating with nanoscale thickness. The types of nanoobjects are summarized in Figure 6.1.

Depending on their structure and composition, nanostructured materials can be classified into nanoporous materials, nanofoams, nanocrystalline materials, or nanocomposites. Nanoporous materials contain nanopores of nanolength sizes, and nanofoam is a material filled with nanobubbles or nanoholes containing the gaseous phase (aerogels). A nanocomposite material is a solid matrix containing physically or chemically distinct nanoscale objects, while a nanocrystalline material contains a significant fraction of nano-sized crystal grains.[16]

While nano-sized particles exist in nature (*e.g.* plasma proteins, lipids and lipoproteins, histones, ribosomes, and some viruses), biomedical nanotechnology commonly utilizes materials that are engineered to create nanoscale objects or structures. Due to their small size and very large surface-to-volume ratio, they possess unique optical, magnetic, and electrical properties. Nanomaterials can also bind with and neutralize toxins, detect biomarkers in human blood and urine, act as targeted or untargeted drug nanocarriers to increase drug bioavailability and enhance the efficacy of therapy, or control the drug release. Importantly, different forms of nanomaterials, *e.g.* nanoparticles and nanofibers, can be combined to create multifunctional tools with drug-release or disease-sensing abilities, as well as adjustable mechanical, biological, and structural properties.

6.2 Cardiovascular Stents

The development of interventional methods of revascularization was marked by the introduction of balloon angioplasty by Andreas Grüntzig in Zurich in the late 1970s. This method enabled the opening of the arterial lumen narrowed by atherosclerotic plaque, but it showed high rates of arterial closure and considerable risk of vessel dissection.[17] In 1986, Jacques Puel in Toulouse, France, and Ulrich Sigwart in Lausanne, Switzerland, implanted the first self-expanding stents into patients undergoing coronary balloon angioplasty to prevent the vessel from collapsing and to avoid restenosis.[18] Soon afterward, in 1987, the first balloon-expandable iliac stent was implanted in Germany by Julio Palmaz and Richard Schatz, and in 1988, they performed the first balloon-expandable coronary stent placement in Brazil. The FDA approved coronary stents in 1994. The use of these BMS, usually fabricated from corrosion-resistant materials, such as stainless steel (316L), cobalt–chromium (Co–Cr), platinum–iridium (Pt–Ir) alloys, or nitinol (Ni–Ti), improved the interventional treatment of atherosclerosis by achieving immediate restoration of blood flow in obstructed arteries and providing mechanical support to the disease-affected vessels. However, vascular wall injury resulting from stent implantation was observed to induce excessive smooth muscle cell (SMC) proliferation, leading to restenosis. In 2002, the first successful clinical trials of DES were performed, marking another important step in interventional cardiology.[19] The first-generation DESs, based on a stainless steel platform coated with durable polymer-eluting antiproliferative drugs, enabled further reduction of in-stent restenosis and were subsequently developed toward biodegradable coating and bioresorbable struts. However, stent implantation leads to disruption of the endothelial monolayer on the plaque surface, which may initiate acute thrombotic events and/or contribute to the development of neoatherosclerosis.[20] As the antiproliferative drugs eluted by DES significantly inhibit vascular healing and endothelialization of struts, prolonged dual antiplatelet therapy is necessary in the recipients of these implants,[21] which may pose serious risks to patients with comorbidities. Consequently, efforts focusing on the development of advanced stent materials with improved micro-/nanostructure and regenerative properties have been intensively pursued in recent years. The improvement of stent materials aims at reducing restenosis and thrombosis without preventing endothelialization and commonly involves the incorporation of growth factors or endothelial cell adhesion-promoting peptides to induce stent endothelialization, anticoagulants to prevent thrombosis, or a combination of both. There is also an increasing interest in materials capable of release, or *in situ* generation, of nitric oxide (NO)[22] due to its crucial role in vascular homeostasis and its broad-ranging vasoprotective, antithrombotic, and anti-hyperplastic activities.[23] The examples of recently reported advanced nanomaterials with promising perspectives for stent implants are highlighted in Sections 6.2.1–6.2.5.

6.2.1 Nanofibrous Materials for Stents

One of the approaches developed toward improving vascular healing involves the fabrication of stents containing nanofibrous polymeric mesh, whereby the nanofibers are often loaded with drugs that promote endothelialization and/or prevent thrombosis. Examples of such stents have been recently described by Cheng *et al.*, who developed poly(lactic-*co*-glycolic acid) (PLGA) nanofibers eluting platelet inhibitor ticagrelor, which blocks adenosine diphosphate receptors of subtype P2Y12.[24] The nanofibrous mesh-covered stents implanted in rabbit aorta were shown to improve vascular healing, enhance the strut endothelialization rate, and limit restenosis *in vivo*. Using a similar approach, Lee *et al.* demonstrated improved strut endothelialization and reduced intimal hyperplasia upon implantation of sub-micron PLGA tubes eluting telmisartan (angiotensin II receptor type 1 antagonist and antihypertensive drug with pleiotropic anti-inflammatory activity) in a rabbit model.[25] Electrospun sub-micron core–shell fibers with 500–700 nm fiber thickness and approximately 50 nm shell thickness were developed as a stent cover in a study by Wang *et al.*[26] A polylactide (PLA)-based shell enclosed a core containing antiproliferative drug rapamycin and reactive oxygen species scavenger TEMPOL in polyvinyl alcohol (PVA) solution. Upon implantation in minipig coronary arteries, the covered stents enabled rapid endothelialization while reducing inflammation and intimal hyperplasia compared with control DES.[26] In a study by Chen *et al.*, stent coating made of amine group-functionalized polysiloxane nanofibers, and serving as a reservoir of immobilized collagen type III, was developed.[27] Upon implantation in the abdominal aorta of rabbits, improved endothelialization, reduced platelet adhesion and intimal hyperplasia, and a mild inflammatory response were observed. Biodegradable cellulose acetate fibers with diameters of 200–800 nm were also reported by Janjic *et al.*[28] The fibers loaded with rosuvastatin were electrospun on different commercially available stents (Co–Cr, Ni–Ti, and stainless steel stent). In the second step, the solution of heparin was spun directly onto rosuvastatin-loaded nanofibers, producing an anticoagulant nanofibrous layer. The authors concluded that sustained release of heparin and rosuvastatin from the fibrous coating is expected to prevent thrombosis, improve endothelialization, and reduce vascular wall inflammation upon stent implantation.[28]

Another interesting approach was reported by Cheng *et al.*, who developed coated stents using a layer-by-layer microfluidic electrospinning method.[29] The inner-layer nanofibers, deposited on the inner surface of 316L stainless steel stents, contained methylacrylated gelatin–polyethylene glycol diacrylate–polydopamine loaded with heparin and vascular endothelial growth factor (VEGF), and the outer layer nanofibers were based on polycaprolactone (PCL)–polydopamine and contained rapamycin. The developed multifunctional coated stent released anticoagulant heparin into the artery lumen to inhibit thrombosis and VEGF to facilitate endothelial healing. The outer, arterial wall-facing layer released rapamycin to prevent proliferation and migration of VSMCs, effectively preventing restenosis in a rabbit carotid artery.[29]

Recruitment of endothelial cells to the surface of implanted stents can also be realized without drug loading, utilizing magnetic forces for targeting cells labeled with cytocompatible superparamagnetic iron oxide nanoparticles. In a pilot *in vitro* study by Choe *et al.*, a nanofibrous composite material was developed as stent coating, combining PU nanofibers with ferromagnetic 2205 stainless steel (2205-SS) particles.[30] Stents combining ferromagnetic struts and the developed ferromagnetic particle-spiked nanofibrous coating demonstrated good mechanical properties, hydrophobicity, and uniform cell capture on the surface. Further studies will be necessary, however, to confirm the efficacy of this approach *in vivo*.

6.2.2 Surface Modifications

Several recent approaches utilized stent surface modifications or nanocoating techniques to reduce the risk of stent thrombosis and restenosis or to improve strut endothelialization. An example of this approach is a commercially available COBRA Polyzene-F NanoCoated Coronary Stent System (CeloNova BioSciences).[31] This cobalt–chromium coronary stent coated with a nanothin layer of Polyzene-F (poly-bis[trifluoroethoxy]phosphazene) was shown in preclinical studies to inhibit inflammation, thrombogenicity, and neointimal hyperplasia compared with a stainless steel stent or uncoated COBRA stents.[32] Compared with published results for currently approved BMS, COBRA PzF stent showed a reduced occurrence of target vessel failure (defined as a composite of cardiac death, myocardial infarction, or clinically driven target vessel revascularization), restenosis, and no occurrence of stent thrombosis at 9 months after the implantation in 287 patients.[31] Additionally, an *in vivo* rabbit study showed a rapid endothelialization of the struts, with above 50% of coverage on day 3, nearly 80% on day 5, and complete endothelialization on day 7 in the absence of thrombosis.[33]

As another example of this method, a multifunctional lysozyme nanofilm on cardiovascular stent was reported by Xu *et al.*[34] The authors used this nanoplatform to immobilize heparin and copper ions, with the dual aim of reducing thrombogenicity and *in situ* catalyzing NO generation, respectively. The *in vitro* study showed that the developed nanofilm reduced adhesion and activation of platelets, enhanced NO synthesis, and induced endothelial cell growth, while in parallel inhibiting SMC proliferation.[34]

Another approach to supporting endothelial healing has been tested by Ma *et al.*, who modified the surface of titanium using an anodic oxidation process to create titanium dioxide (TiO_2) nanotube arrays.[35] Subsequently, the authors immobilized sodium alginate/carboxymethyl chitosan complex on the nanotube's surface, followed by grafting the carbon monoxide (CO)-releasing molecule, termed CORM-401, onto the surface. The modifications improved hydrophilicity and biocompatibility of the material, and the slow release of CO molecules could be enhanced in the presence of cysteine, which resulted in reduced platelet adhesion and activation, decreased hemolysis rate, and improved endothelial cell proliferation, in parallel with enhanced expression of NO and VEGF. The authors suggested that the release of CO from TiO_2 nanotubes can improve the hemocompatibility of implants and simultaneously enhance surface endothelialization.[35]

6.2.3 Nanoparticulate Materials for Drug-eluting Stents

The therapeutic properties and biological function of drug-eluting stents can be improved not only by utilizing biodegradable materials to elute drugs but also by incorporating drugs into nanoparticles to tackle delayed re-endothelialization, inflammatory response, or in-stent thrombosis. In a recent study by Li *et al.*, the nanoparticle-coated stent was developed by combining PLGA nanoparticles, loaded with rapamycin and coated with the platelet membrane (PNP, diameter of 91 ± 17 nm), with a bioresorbable PLA stent.[36] The produced nanoparticles were spray-coated on the PLA surface, forming a homogeneous layer. *In vitro*, the presence of particles coated with platelet membranes on the stent struts enhanced the adhesion of endothelial cells while reducing SMC attachment and proliferation. The implantation in rabbit abdominal aorta showed that the PNP-coated stents promote endothelialization, suppress thrombosis and restenosis, and reduce macrophage infiltration compared with control PLA stents.[36]

Nanoparticles based on PLA were also used as carriers of tauroursodeoxycholic acid (TUDCA), a natural hydrophilic bile acid known as an endoplasmic reticulum (ER) stress inhibitor.[37] Following *in vitro* studies, BMS coated with the produced particles using soaking-and-drying treatment were implanted in rabbit carotid arteries for a period of 4 weeks. Compared to uncoated control BMS, TUDCA-eluting stents promoted endothelialization by inhibiting ER stress and pro-apoptotic pathways, and the drug was released slowly during the experimental period.[37]

Electrostatically self-assembled particles (330 nm), composed of heparin and NO donor (DETA-NONOate), were also developed by Zhu *et al.* and immobilized on 316L SS coated with polyglycidyl methacrylate by dipping in nanoparticle suspension.[38] Upon implantation in the iliac artery of rabbits exposed to a high-fat diet, reduced intimal hyperplasia was observed. Furthermore, the struts of nanoparticle-coated stent were fully covered by the layer of endothelium, while the uncoated BMS were not completely endothelialized 4 weeks after the implantation.[38]

A similar approach, based on nanoparticles releasing heparin and generating NO from nitrosothiols, was tested by Luo *et al.*[39] Multifunctional coating of stainless steel stent pre-coated with polydopamine was realized by immobilization of cystamine-modified heparin/polyethyleneimine (Hep–Cys/PEI) particles (400 nm) on the stent surface. The resulting stent inhibited the adhesion and activation of platelets, enhanced the proliferation of endothelial cells, and reduced SMC proliferation *in vitro*. In a rat model of subcutaneous implantation, the nanoparticle-coated stents showed less inflammatory cell infiltration and reduced granulation tissue development compared with the uncoated steel surface.[39]

Dopamine coating of 316L steel stent has also been tested as a method to immobilize VEGF-loaded heparin/poly-L-lysine (Hep–PLL) nanoparticles (140–170 nm) created by electrostatic interactions.[40] *In vitro* studies showed the promising potential of this nanoparticle coating in terms of biocompatibility, anti-thrombogenic properties, and inhibition of SMC proliferation. The coated surfaces promoted endothelial cell adhesion, proliferation, migration, and biological activity compared with uncoated samples.[40] In a study by Liu *et al.*, 316L stainless steel coated with polydopamine was functionalized with nanoparticles containing laminin incorporated into the inner core and stromal cell-derived factor-1α (SDF-1α) immobilized on the external surface.[41] Following the promising *in vitro* studies, *in vivo* implantation demonstrated confluent endothelial monolayer, reduced thrombus formation, and neointimal hyperplasia at 3 months post-implantation.[41]

To support endothelial healing, stent surfaces can be further improved by the incorporation of chitosan-modified PLGA nanoparticles containing miR-126 dsRNA, which enhances the proliferation and migration of human umbilical vein endothelial cells (HUVECs) while reducing the proliferation and migration of SMCs.[42] The coating was realized by repeated immersion and drying of 316L stainless steel in nanoparticle solution. In a rabbit model of iliac stenting, miR-126-eluting stents significantly inhibited neointimal hyperplasia 4 weeks after implantation compared to stents coated with nanoparticles without dsRNA (Figure 6.2).[42]

6.2.4 Nanofibrous Stents for Aneurysms

Stents have further been developed as a tool for minimally invasive treatment of aortic aneurysms. For this purpose, stent struts are combined with a very dense polymeric mesh, usually electrospun in the form of nanofibers. Such covered stents enable the effective endovascular management of complicated aneurysms, which could not be treated with endovascular coiling and surgical clipping. The polyurethane-covered cobalt–chromium papyrus stent has been in clinical use since 2015,[43,44] but the reduction of in-stent restenosis and delayed endothelialization remain challenging. Recent examples of the advances in the covered stent approach include the studies by Zhang *et al.* (2019), who reported a stent covered with dual drug-loaded electrospun fibers aiming at rapid endothelialization and long-term inhibition of stenosis.[45] To achieve this

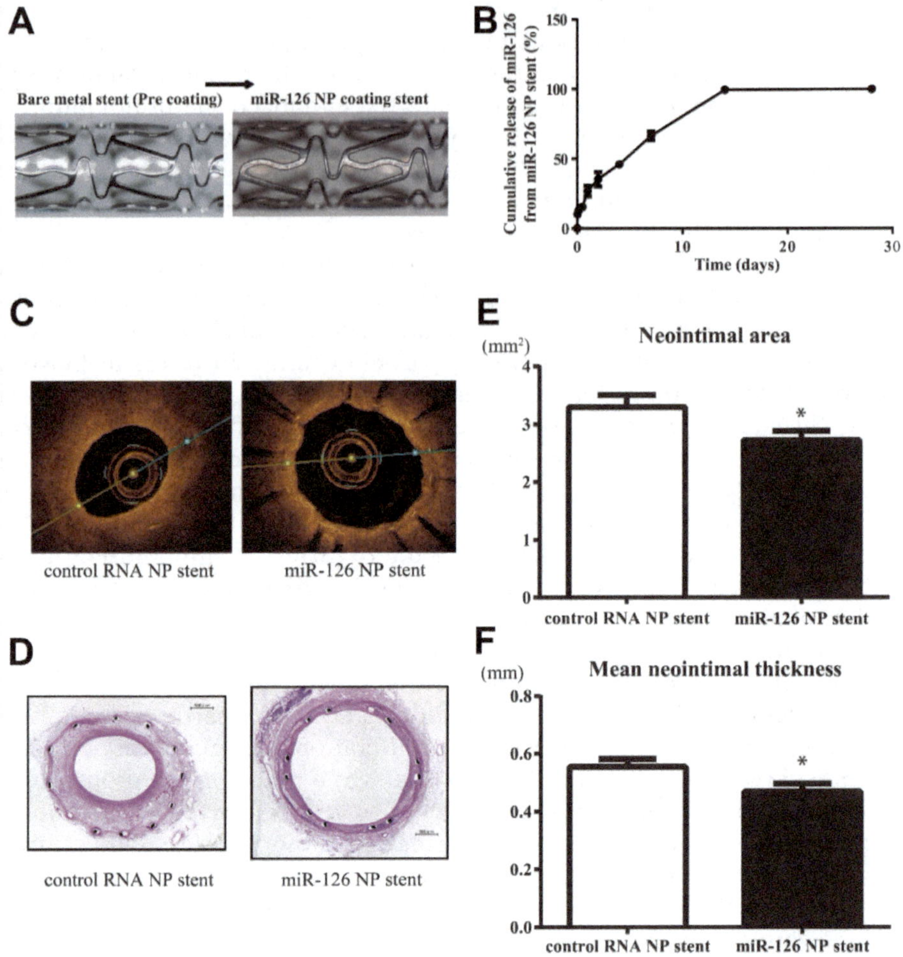

Figure 6.2 Evaluation of miR-126 nanoparticle (NP)-conjugated stents in a rabbit model of iliac atherosclerosis.[42] Adapted from ref. 42, https://doi.org/10.1371/journal.pone.0172798, under the terms of the CC BY 4.0 license, https://creativecommons.org/licenses/by/4.0/. (A) Images of pre-coated BMS and a stent coated with miR-126 NPs. (B) Cumulative release of miR-126 from an miR-126 NP-conjugated stent. (C) Optical coherence tomography (OCT) images of a rabbit iliac artery after the implantation of a control RNA NP-conjugated stent or miR-126 NP-conjugated stent. (D) Hematoxylin and eosin (HE) staining of a rabbit iliac artery after the implantation of a control RNA NP-conjugated or miR-126 NP-conjugated stent. (E) Changes in the neointimal area of rabbit iliac arteries after the implantation of a control RNA NP-conjugated or miR-126 NP-conjugated stent. (F) Changes in the mean neointimal thickness of rabbit iliac arteries after the implantation of a control RNA NP-conjugated or miR-126 NP-conjugated stent ($n = 8$ per group; $^*P < 0.05$).

aim, mesoporous silica nanoparticles (MSNs) loaded with paclitaxel (PTX) were encapsulated in electrospun PLA fibers, while VEGF was conjugated onto the PLA surface using polydopamine coating. Implantation in vein pouch aneurysm model in dogs showed that the drug-loaded covered stents improved the complete aneurysm occlusion rates and endothelialization, as well as reduced lumen restenosis over 12 weeks compared to control stents.[45]

Incorporation of rosuvastatin and heparin into the electrospun nanofibrous poly(L-lactide-*co*-caprolactone) (PLCL)-covered stents has also been proposed for treating aneurysms.[46] Covered stents were implanted in a rabbit model of carotid aneurysm for 4 months and monitored using digital subtraction angiography, which demonstrated that stents with the highest concentration of the rosuvastatin group had better outcomes than the control group treated with phosphate-buffered saline (PBS). Furthermore, the nanofibrous drug-releasing stents improved endothelialization and showed favorable antithrombotic properties *in vivo*.[46] Another approach to incorporate statins was reported by Chu *et al.*, who used emulsion electrospinning to produce PLCL-based, slow-degrading core–shell nanofibrous (400–700 nm) mesh containing atorvastatin in the inner fiber cores.[47] *In vivo*, the drug-releasing covered stents enabled effective aneurysm occlusion and promoted endothelialization in a rabbit model of carotid artery aneurysm. Covered stents with a core–shell fiber structure of polymer mesh have also been used to control inflammatory responses in aortic dissection.[48] For this purpose, methylprednisolone was encapsulated in the inner core of PLCL nanofibers using coaxial electrospinning. The implantation in a porcine aortic dissection model demonstrated the ability of a methylprednisolone-loaded nanofiber-covered stent to reduce aortic inflammation and promote aortic wall remodeling.[48]

6.2.5 Smart Stents

The last decade brought about further developments in material technologies that enabled the fabrication of advanced "smart" stents, which allow biological process monitoring, condition sensing, and controlled drug release. In 2015, Son *et al.* reported a bioresorbable electronic stent (BES) fabricated from bioresorbable/bioinert materials (ZM21 alloy, Mg 97%, Zn 2%, and Mn 1%) covered by a PLA layer and containing a suite of sensors and actuators on their surface,[49] including flow/temperature sensors, as well as a memory module for pattern analysis and diagnosis. The system was furthermore equipped with dual drug delivery systems, *i.e.* reactive oxygen species-scavenging ceria nanoparticles and rapamycin-loaded gold nanorod core/mesoporous silica shell nanoparticles that were activated by either a local hyperthermia or an external optical stimulus (NIR laser). The implantation of BES in a canine artery showed an inhibition of inflammatory responses and reduced macrophage infiltration.

A more recent study by Herbert *et al.* reported the development of an implantable soft electronic stent system containing nanomembrane strain sensors that enable continuous wireless monitoring of restenosis in real time without batteries or circuits.[50] The wireless stent based on stainless steel with microscale polyimide (PI) connectors to enable a solenoid-like flow of current was electroplated with gold and insulated in parylene. The soft, arterial strain sensor was aerosol jet-printed to create a bottom PI layer, a middle layer of silver nanoparticles (AgNP), and a top PI layer between two plates, forming a parallel-plate capacitor enclosed in poly(styrene-isoprene-styrene) elastomer. The sliding of the top and bottom plates relative to each other allowed the two PI base layers to be pulled in opposite directions when the elastomer was stretched and enabled the capacitive sensing of strain. The catheter-implantable sensor was inductively coupled with an external antenna, which allowed monitoring of changes in the arterial wall strain. The efficacy of the system has been demonstrated in silicone models of arteries with different stiffness. The integrated stent and sensor were also implanted *ex vivo* in the ovine coronary artery, showing lower changes in resonant frequency in an artery stiffened by silicone compared to the normal artery.[50] However, the *in vivo* demonstration of efficacy in terms of restenosis detection and wireless coupling with the device remains to be carried out.

Nonetheless, these innovative technologies, combining mechanical, electronic, sensing, and therapeutic functionalities with advanced nanomaterials, open promising opportunities for improved stent designs.

6.3 Vascular Prostheses

CABG is a surgical operation, during which the coronary arteries narrowed or occluded by atherosclerotic plaques are bypassed with venous or arterial grafts to restore blood flow to the ischemic myocardium and relieve the symptoms of angina.[51] According to different estimates, the yearly numbers of CABG procedures in the USA vary between 260 000 and 400 000.[52] In the EU, more than 180 000 patients per year undergo CABG,[53] making it the most frequently performed type of open-heart revascularization surgery.[54] CABG commonly involves an additional surgery to harvest a length of blood vessel from the leg, chest, or another part of the body. Apart from surgical risks and possible complications, about 20% to 30% of bypass patients need a second procedure within 10 years if the grafted arteries or veins become blocked. While arterial bypasses usually have a very good 10 year patency rate of 80–90%,[55] up to 40% of vein grafts may be closed at just 1 year after the surgery.[56] Longer-term studies indicate that up to 40–50% of vein grafts require reoperation 10 years after the bypass.[57] Given the number of yearly procedures, this corresponds to a considerable number of patients. Another problem related to CABG is the limited availability of autologous graft material with quality suitable for coronary vessel bypass due to the patient's age, status of the vascular tree, pre-surgeries, and/or comorbidities. One of the most addressed, but thus far unanswered, needs in cardiovascular tissue engineering is the development of small-diameter vascular grafts, which would mimic the functional complexity of an autologous vessel.

Large-diameter durable synthetic grafts (*e.g.* PET, ePTFE, or PU) are routinely implanted in bypass surgery in peripheral arteries (reviewed in ref. 58), with acceptable long-term patency despite poor endothelialization and the presence of a mild foreign body reaction. However, these prostheses are unsuitable in coronary localizations, where grafts with diameters of 2–4 mm are needed. Among the main disadvantages of durable grafts are the innate incapability to mimic the characteristic features of arterial walls, *i.e.* its high tensile strength, elasticity and resistance to fatigue, and biological functionality, including contractility, compliance, thromboresistance, and responsiveness to stimuli.

As an alternative to durable materials, vascular tissue engineering with natural and/or biodegradable polymers was proposed.[59] However, similar to durable grafts, clinical studies on large-diameter tissue-engineered PGA/PCL prostheses showed that the central issue limiting their long-term patency has been graft stenosis due to intimal hyperplasia.[60] This factor is expected to greatly affect also the patency of small-diameter prostheses as a result of the specific geometrical challenges in coronary localizations. Due to the noncompliance of the tissue-engineered grafts implanted *in vivo*, small-caliber prostheses have proven prone to multiple complications, including vessel thrombosis, aneurysm formation, and intimal hyperplasia leading to stenosis, which resulted in low patency rates and graft failure.[61,62] Thus, despite the scope and intensity of experimental work during the last four decades, the clinical impact of tissue-engineered small-diameter vascular implants has been negligible so far. As highlighted in Sections 6.3.1–6.3.2, recent advances in material design may pave the road for the development of improved small-diameter vascular grafts in the future.

6.3.1 Advanced Vascular Grafts: Nanocomposites, Nanocoatings, and Nanostructures

There are multiple requirements a vascular graft should fulfill, such as appropriate mechanical properties and suturability, compliance, thromboresistance, and biocompatibility. Ideally, biological cues enabling rapid endothelialization and improving patency are of advantage too. While mechanical properties are relatively easy to tailor using different polymeric scaffolds and a number of state-of-the-art fabrication techniques, improving biological function – thromboresistance and rapid endothelialization – remains challenging. Previous studies have shown that graft endothelialization depends not only on the chemical composition of a material but also on its surface structure. An interesting example of structural nanotopography modifications, which are expected to inhibit the adhesion of platelets and induce oriented growth of endothelial cells, was reported by Wang *et al.*[63] They developed a vascular graft that consisted of a freeze-cast inner layer with nanolamellar structure, made of cross-linked silk fibroin (SF) and gelatin, and the outer sheath of PCL fibers that improved mechanical properties of the graft (elastic modulus and tensile strength). The vascular grafts maintained structural integrity upon subcutaneous implantation in rats for one month. Notably, the patency rate of lamellar grafts was maintained at 66% three months post-implantation, whereas all control non-lamellar grafts were obstructed.[63] The endothelialization-enhancing effect of surface structure modification has also been investigated by Kianpour *et al.* using *in situ* synthesis of TiO_2 nanoparticles on the PU matrix.[64] The nanocomposite film on the surface of scaffold improved cell adhesion, migration, and extracellular matrix production *in vitro*, leading to enhanced endothelialization compared with a pure PU surface.

Furthermore, in our previous studies, nanofibrous scaffolds were shown to promote endothelial cell attachment and monolayer formation, while endothelialization of flat or microfibrous surfaces was more difficult to achieve.[65] This is one of the reasons why the researchers who develop vascular scaffolds commonly use (electro)spinning techniques (electrospinning, solution blow spinning, and melt electrowriting), which enable good control over fiber diameter and, to various degrees, their alignment. Although the nanofibrous surface commonly supports endothelial cell growth on 2D scaffolds, it is often insufficient to enable efficient 3D endothelialization of materials that do not contain any biological cues. For this reason, multiple types of nanocoatings, fiber modifications, and drug-release systems have been recently investigated. As also in the case of stent coatings, one of the promising approaches is coating of graft nanofibers with polydopamine, a mussel-inspired adhesive. Apart from hydrophilizing the surfaces and increasing cell attachment, the nanolayer of polydopamine enables the immobilization of further peptides, nanoparticles, or drugs that promote endothelial healing and prevent thrombosis on the surface of grafts.[66] Several of the many interesting examples of nanomaterials used to advance the development of vascular grafts are highlighted in the following paragraphs of this section.

In a study by Emechebe *et al.*, a dual-layer nanofiber vascular graft was developed, combining a faster-degrading luminal layer of polydioxanone fibers (PDO, fiber diameter 250 to 300 nm) and a slow-degrading outer layer composed of PCL/PDO with antithrombotic drug dipyridamole and reinforced with 3D-printed PCL spiral filaments.[67] The PDO/PCL nanofibers were coated with polydopamine, which enabled functionalization with VEGF. Good patency of the graft was demonstrated in the porcine model of carotid artery grafting during two week post-implantation period.[67] Dual-layer micro- or nanofibrous scaffolds, offering the possibility of prolonged drug release, have been utilized by other groups too. Matsuzaki *et al.* reported the development of grafts composed of an inner PLCL layer containing heparin and an outer PCL nanofibrous layer, which were implanted in sheep carotid arteries for 8 weeks.[68] In comparison to control grafts without heparin, which had early thrombotic occlusion at <1 week in 3 of 5 implanted conduits, all drug-releasing grafts remained patent for 8 weeks, demonstrating native-like cellularity and lack of calcification.[68]

Another example of this approach was reported by Zhang *et al.*, who produced bilayer grafts containing an inner layer composed of freeze-thawed nanofibrillated cellulose/PVA and incorporating heparin/poly-L-lysine nanoparticles (320 nm) and an outer layer of electrospun PCL.[69] Minor tissue response to those scaffolds was demonstrated upon subcutaneous implantation in rats, indicating their biocompatibility. The grafts implanted for 5 days in rabbit carotid arteries remained patent, without thrombosis, and showed good compliance similar to native arteries.[69] Grafts fabricated from composite core–shell nanofibers containing synthetic PLCL core (542 nm diameter) coated with a natural material shell layer of silk or heparin/silk crosslinked gel were also tested.[70] The Hep/silk–PLCL grafts had a layer of coating of approximately 140 nm thickness, whereas the silk–PLCL layer was 154 nm. The implantation of Hep/silk–PLCL grafts in rabbit carotid arteries showed a good patency, without lumen narrowing even after eight months *in vivo*, and a progressive autologous tissue remodeling to replace the degrading scaffold. By contrast, PLCL grafts were occluded 2 weeks post-implantation, and significant stenosis was observed in silk–PLCL grafts at 6 weeks *in vivo*.[70]

Collagen nanofibers modified with hyaluronic acid oligosaccharides (183 nm fiber thickness) were also tested as a carrier platform for antiplatelet enzymes apyrase and 5′-nucleotidase (5′-NT).[71] Covalent grafting of these enzymes enabled the conversion of adenosine diphosphate (ADP) into adenosine monophosphate (AMP) and adenosine, leading to decreased platelet activation. The enzyme-laden nanofibers were further combined with PCL to improve the mechanical properties before implantation *in vivo*. Their good biocompatibility was confirmed upon subcutaneous implantation in rats, followed by carotid artery grafting in rabbits, which demonstrated reduced platelet adhesion, enhanced endothelialization, and improved patency at 1 month post-implantation as compared with control grafts without enzyme-loading.[71]

Apart from the above-highlighted reports, some *in vitro* studies demonstrated that nanomaterials with drug-release functionality may have promising potential in the development of improved vascular grafts. As an example, functionalization of electrospun PCL fibers (420 nm) using layer-by-layer self-assembly of gelatin, polylysine, and heparin nanoparticles, followed by genipin crosslinking to form a multiphase nanocoating, enabled both an improved endothelial cell attachment and a reduced platelet adhesion to the surface of the developed grafts.[72]

6.3.2 Advanced Vascular Grafts: Nanoparticles

Another approach to improving graft patency and functionality involves the incorporation of bioactive or drug-releasing nanoparticles into the graft materials, which can be either synthetic or natural. In one of the recent studies, decellularized and photo-oxidized bovine internal mammary arteries were loaded with heparin- and VEGF-containing nanoparticles (diameter of 61.8 ± 18.3 nm) composed of copolymer of chitosan and Arg–Glu–Asp–Val (REDV) peptide.[73] The bioprostheses (see Figure 6.3) containing the multifunctional nanoparticles, capable of sustained drug release over 1 month *in vitro*, were subsequently implanted in the abdominal aorta of rabbits for up to 3 months, showing 100% patency after 1 month and 83.3% patency after 3 months. The grafts had very low thrombogenicity, and a rapid lumen coverage with endothelial cells was achieved.[73]

Bilayered grafts, composed of a sponge-like inner layer of gelatin and an outer layer of submicron poly(ester-urethane)urea fibers (700 ± 58 nm), were also used as a carrier of MNPs loaded with salvianic acid (210 ± 45 nm), which were incorporated into the inner layer of the graft.[74] Drug encapsulation in nanoparticles and in the gelatin matrix ensured a sustained release profile. Upon implantation in a rabbit carotid artery for 20 days, the grafts showed good performance with no intimal hyperplasia or thrombosis.[74] An interesting approach to improving implant functionality and patency was also reported by Zhang *et al.*, who developed PCL/poly-D-lysine (PDL) tubes containing erythrocyte membrane-derived nanovesicles (152.8 nm diameter and −25.9 mV Zeta potential), which are rich in natural bioactive components.[75] The stable incorporation of negatively charged erythrocyte membrane-derived nanovesicles was facilitated by the presence of PDL, which, as cationic poly-amino acid, has a positive charge and promotes the adhesion of cells. Upon implantation in the carotid artery of rabbits, the developed grafts showed good hemocompatibility and patency, as well as anti-inflammatory properties and rapid remodeling within 21 days.[75]

Overall, the described nanomaterial-based vascular grafts demonstrate good capacity for rapid endothelialization and antithrombotic activity, which indicates their promising potential in the development of improved, clinically applicable small-diameter vascular grafts.

6.4 Cardiac Patches

Pathological left ventricle (LV) remodeling, the aftermath of MI, results in mostly irreversible structural changes, including ventricular wall thinning and dilatation, which lead to progressive decline in contractile function and reduced stroke volume. Cardiac patch implantation or "patch"-plasty was first described by Dor *et al.*, who used endoventricular circular patch to treat MI patients with LV scarring.[76] This therapy is aimed not only

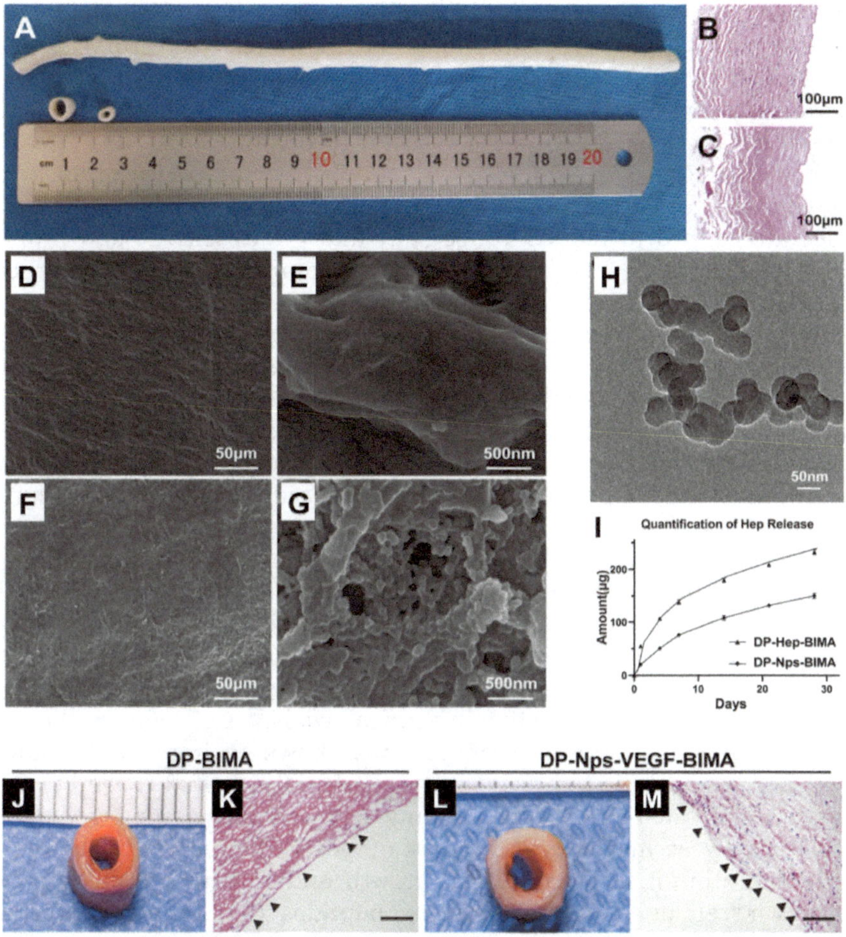

Figure 6.3 Modification of decellularized bovine internal mammary arteries (BIMA) with drug-eluting NPs.[73] Adapted from ref. 73, https://doi.org/10.3389/fbioe.2023.1109058, under the terms of the CC BY 4.0 license, https://creativecommons.org/licenses/by/4.0/. (A) Fresh BIMA. (B, C) HE staining of native BIMA cross-section and decellularized and photo-oxidized BIMA (DP–BIMA). (D, E) Scanning electron microscopy (SEM) images of native BIMA. (F, G) SEM images of CS–REDV–NPs–VEGF-modified DP–BIMA. (H) Transmission electron microscopy (TEM) images of the CS–REDV–Hep nanoparticles. (I) Quantification of heparin release in PBS (n = 3). (J, K) Macroscopic and HE-stained images of the middle segments of the implanted non-modified DP–BIMA and of DP–NPs–VEGF–BIMA (L, M).

at supporting the injured ventricle wall mechanically but potentially also at promoting the regeneration of the myocardium. Further applications of cardiovascular patches involve the repairs of septal detects and other vascular defects. Commercially available patches include *e.g.* durable GORE-TEX® Cardiovascular Patch, which has been on the market for nearly

4 decades, as well as a decellularized bovine pericardial patch (St. Jude Medical Inc.). GORE-TEX® Patch made of multidirectional-strength ePTFE is designed for optimal tissue ingrowth and helps to prevent aneurysmal dilation. Its application is not limited to the myocardium, being also suitable for aortic and peripheral vascular reconstructions. The Pericardial Patch with EnCap AC Technology is made of bovine pericardium, which was exposed to calcification-reducing treatment, in which residual aldehydes are capped to reduce antigenicity and cytotoxicity[77,78] and promote endothelialization.[78,79] The bovine patch is recommended for reconstructive repairs of septum, leaflets, and vessels, rather than myocardial patching.

In the last two decades, multiple material types, including synthetic polymers, such as PLA, PVA, PLGA, poly(glycerol sebacate) (PGS), and PU,[80] have been utilized to fabricate cardiac patches. The advantages of cardiac patches made of synthetic materials include the possibility of long-term storage and a lower-cost, easy, and reproducible manufacturing process. Apart from synthetic polymers and decellularized extracellular matrix from xenogenic sources,[81,82] natural materials and hydrogels have been previously used in the engineering of the cardiac patch, some of them with promising results.[83] Those materials, characterized by good biocompatibility, mimic thecellular microenvironment, enabling tissue ingrowth and preventing LV remodeling. Alginate, hyaluronic acid, fibrin, and collagen are examples of natural materials widely used as cardiac patches. Injectable alginate hydrogels were tested in clinical trials.[84] Collagen in the form of electrospun nanofibrous scaffolds was shown to improve the attachment and survival of stem cells and endothelial cells.[85,86] However, it has been demonstrated that patches aiming at myocardial tissue regeneration, and not only at mechanical support of post-ischemic ventricle, require additional stimuli to promote cell growth and function. Besides growth factors or cell attachment-promoting peptides that mimic the cellular microenvironment, physical stimulation has been shown to promote cell–cell contacts, contractility, and signal propagation between the cardiomyocytes. Electrical signals,[87,88] mechanical stretching or compression systems, perfusion-induced shear stress alone,[89] or in combination with electrical stimuli,[90] as well as magnetic stimuli,[91,92] contribute to tissue maturation and functional coupling between the cells and can be enhanced by the incorporation of nanoscale objects in the patch material to improve conductivity.

6.4.1 Hydrogel-based Cardiac Patches

Commercially available cardiac patches require surgical implantation, which is associated with considerable risk. The injectable scaffolds based on crosslinkable hydrogels have multiple benefits in this context, but their mechanical properties upon implantation and the lack of electrical conductivity may reduce the benefits of these materials and prevent tissue repair. Incorporating nanomaterials into the natural hydrogels has been therefore proposed already in 2011[93] as a method to improve the conductivity and mechanical stability of the hydrogel-based patches. Embedding of conductive nanoscale objects, such as carbon nanotubes, metal nanoparticles, or graphene oxide, thus promotes the regenerative function of the novel cardiac patches.

In line with this, Navaei *et al.* developed gelatin methacrylate (GelMA)-based hybrid hydrogels containing gold nanorods (AuNRs) to improve both electrical conductivity and mechanical stiffness of the hydrogels.[94] The scaffolds enabled the attachment and retention of seeded cardiomyocytes, the formation of a tissue layer, and synchronous contraction. The expression of cardiac markers (*i.e.* sarcomeric α-actinin and connexin 43) was observed in cells grown on these hydrogels as a function of AuNR concentration.[94] Injectable hydrogels based on methacrylated elastin and gelatin-containing carbon nanotubes (CNTs) were also reported by Wang *et al.*[95] The hydrogel patches, characterized by shape-memory behavior and a hierarchical porous structure, were implanted in a rat model of MI as cell-free patches or patches pre-seeded with rat cardiomyocytes. Functional recovery and a decrease in the infarcted area were observed 4 weeks post-implantation, and the results were further confirmed in the minipig model of MI using cell-free CNT-containing patches or patches seeded with cardiomyocytes differentiated from human pluripotent stem cells.[95]

An interesting approach to improving material conductivity was recently reported by Lee *et al.*, who combined 10% gelatin and 5% dextran aldehyde hydrogel with nanoflakes of ($Ti_3C_2T_x$) MXene, consisting of a two-dimensional, atomically thin layer of titanium carbide (3.0 mg mL^{-1}).[96] The resulting adhesive hydrogel was resistant to deformation and demonstrated suitable characteristics for cardiac patch applications, including a high electrical conductivity (18.3 mS cm^{-1}), cardiac tissue-like elasticity (30.4 kPa), and good cytocompatibility with cultured cardiomyocytes. Upon *in vivo* application in rats, the adhesive nanoflake-containing patches prevented adverse remodeling and improved cardiac function two weeks after MI.[96]

Another approach to improving patch functionality and survival of cells colonizing 3D-printed GelMA hydrogel patches was developed by Wang *et al.*[97] The authors first incorporated the hydrogel nanodroplets that underwent ultrasound-induced phase transition, leading to increased porosity and permeability of hydrogel. Subsequently, hemoglobin was encapsulated into nanodroplets (ND–Hb) to fabricate acoustically sensitive nano-oxygen carriers. *In vitro*, the highest cell survival was observed in the patches containing ND–Hb irradiated by the low-intensity pulsed ultrasound. Using this experimental approach in a rat model of MI enabled ultrasound-controlled oxygen release, which improved the survival of implanted cells and increased revascularization, resulting in improved cardiac function after MI.[97]

Notably, also non-conductive epicardial patches have the potential to support the regeneration of cardiac muscle. As an example, a viscoelastic gel-point adhesive patch made of ionically crosslinked self-adhesive starch hydrogel with a low dynamic modulus, which was developed by Lin *et al.*, prevented LV remodeling in rat models of both acute and subacute MI.[98]

6.4.2 Nanofibrous Cardiac Patches

Biological cell-based cardiac patches have shown promising potential for MI therapy, but the size restrictions and the integration into the host heart tissue remain challenging. To address this problem, Chen *et al.* modified electrospun cellulose nanofibers with chitosan (CS) and SF (CS/SF) nanolayers using layer-by-layer coating technology driven by electrostatic interactions between the oppositely charged CS and SF.[99] The presence of increasing numbers of nanolayers gradually increased the diameter of cellulose fibers from 220 up to 400 nm. The resulting patches were seeded with adipose tissue-derived mesenchymal stem cells (AD-MSCs) and subsequently positioned on the infarcted region in rat hearts. Compared with animals who received intra-myocardial injection of MSCs, the patch-based stem cell delivery improved cell survival, reduced ventricular remodeling, and alleviated ventricular ejection fraction and fractional shortening, which were accompanied by reduced cardiac fibrosis and apoptosis and increased angiogenic response.[99] Cardiac patches based on cell sheet fabrication have also been proposed as a biocompatible approach to heart regeneration, but their therapeutic effects are restricted by low retention and poor mechanical strength. To overcome this problem, Wei *et al.* developed patches composed of electrospun PCL/silk fibroin nanofibers into which multi-walled CNTs were incorporated.[100] The resulting nanofibrous meshes were combined with cell sheets of brown adipose tissue-derived stem cells stacked layer by layer, leading to myocardium-like structures. Implantation of these patches in a rat model of MI resulted in restored myocardial function, accelerated angiogenesis, and enhanced M2 phenotype in macrophages compared with animals that received stem cell injection.[100] Single-walled CNTs in combination with gelatin were also tested *in vitro* as a method to modulate the physicochemical and biological properties of polyurethane nanofibers.[101] Incorporation of both gelatin and CNTs into the electrospun PU resulted in a reduced mean diameter of nanofibrous scaffolds from 210 to 140 nm, improved degradation profile due to the presence of gelatin, and enhanced electrical conductivity due to the incorporation of CNTs. The increased hydrophilicity of the scaffolds enabled effective colonization with myocardial myoblast layers forming after 7 days of culture.[101]

To enable the anisotropic electrical signal propagation after MI, functionalization of electrospun silk fibroin fibers with a 100 nm layer of reduced graphene oxide (rGO) was reported by Zhao *et al.* and tested in a rat model of MI.[102] Compared to the non-conductive silk patch and isotropic conductive rGO/silk, the anisotropic conductive rGO/silk patches improved LV ejection fraction, LV thickness, and survival of myocytes, in parallel with improved angiogenesis in the infarcted region.[102] In a study by Hosoyama *et al.*, collagen fibers were combined with different nano-gold- or nano-silver particles (AuNPs or AgNPs) to fabricate an electroconductive cardiac patch.[103] The most promising *in vitro* results in neonatal rat cardiomyocytes, as well as *in vivo* results in a rat model of MI, were obtained using patches containing spherical AuNPs, which enabled the recovery of cardiac function at 28 days post-implantation, in parallel with increased connexin-43 levels and angiogenic response, as well as scar size reduction.[103]

Gold nanorods (AuNRs, 60 × 20 nm) have been also used to improve conductivity and facilitate suture-free attachment of cardiac patches based on albumin electrospun fibers.[104] The nanocomposite material was seeded with rat neonatal cardiomyocytes, which after 7 days of culture showed pronounced actinin striation and Cx43 expression, indicating the ability to create a functional tissue patch with appropriate synchronous contractility. The nanocomposite patch was then placed on the surface of healthy rat myocardium and irradiated with a near-infrared laser (808 nm) to test the efficiency of suture-free adhesion. The presence of AuNRs enabled the conversion of light into thermal energy, which locally changed the molecular structure of the patch, leading to its strong attachment to the wall of the heart.[104]

Collectively, the use of nanomaterial-based approaches in cardiac patch engineering holds a promise of not only providing mechanical support to prevent the ventricular remodeling post-MI but also creating a 3D microenvironment capable of promoting vascularization and tissue regeneration.

6.5 Heart Valve Prostheses

Degenerative valvular disease affects >100 million people worldwide.[105] Due to the increasing disease burden, the yearly numbers of implanted valve prostheses exceed 250 000 globally, with 90 000 implantations in the USA alone.[11] Over the last 50 years, nearly 4 million prosthetic valves have been implanted in patients with severe valvular disease.[12] The field was revolutionized by the introduction of percutaneous transcatheter aortic valve implantation (TAVI) as an alternative to surgical valve replacement, which dramatically improved the quality of life and survival of patients with aortic stenosis. The first TAVI procedure in the world was performed in April 2002 by Alain Cribier in Rouen, France, using a polymeric valve placed on a balloon-expandable stent. This approach paved the way for the first-in-man implantation of self-expanding CoreValve in 2004, which featured a novel porcine pericardium-based valve and enabled reducing catheter size.[106,107] Before the introduction of TAVI, mechanical valves represented the vast majority of the implanted valves due to their durability and low risk of re-operation, but their disadvantages included a permanent need for anticoagulation, apart from the innate risks of surgical procedure. Recently, the number of biological valve prostheses started to surpass those of mechanical valve implants,[108,109] mostly owing to their increasing durability. While constituting an ultimate therapeutic solution for patients with valve stenosis,[110] the implants are not risk-free. Biological valve prostheses still present an increased risk of structural deterioration and the risk of calcification or thrombosis.[111]

The striated matrix of the aortic valve leaflet is organized into three layers: ventricularis, with radially aligned collagen and elastin fibers; glycosaminoglycan-rich spongiosa; and the fibrosa consisting of large circumferentially aligned bundles of collagen fibers.[112] The leaflets contain two main cell types: valvular endothelial cells (VECs) on the surface and valvular interstitial cells (VICs), which are localized throughout each layer and maintain the valvular structure and function. Due to this complex structure of valve leaflets, which is difficult to reproduce *in vitro*, engineering artificial valves is extremely challenging. The aim of the work in this field focuses on increasing durability, mimicking structure and function, and reducing the thrombogenicity of the engineered prostheses.

6.5.1 Nanomaterials for Improving Bioprosthetic Valve Function

An example of the current nanomaterial-based approaches to improving current valve prostheses was reported by Zhou *et al.*, who focused on enhancing endothelialization of bioprosthetic valves.[113] For this purpose, the authors prepared PCL nanoparticles loaded with VEGF, which were subsequently grafted on the surface of the decellularized aortic valve using the Michael addition reaction. Due to the high encapsulation efficacy and slow release of VEGF, the nanofunctionalized valve strongly improved the adhesion and proliferation of endothelial cells. In parallel, very good hemocompatibility was observed *in vitro*, as well as good cell infiltration and capillary formation upon subcutaneous implantation in rats.[113]

An interesting approach to improving the antithrombogenic properties and mechanical strength of prosthetic valves was reported by Seyrek *et al.*, who incorporated –COOH-modified multi-walled CNTs into a decellularized bovine pericardium.[114] The mechanical stability of the resulting composite pericardial scaffold was improved compared to non-modified pericardium. While the modification of the pericardial matrix with CNTs did not affect its biocompatibility, it has effectively prevented platelet adhesion, showing overall superb hemocompatibility.[114]

Since leaflet calcification is one of the drawbacks of bioprosthetic valves, nanomaterial-based approaches are currently being investigated to reduce the calcification rate. Rapid endothelialization may prevent the immune response to the decellularized xenogeneic matrix and inhibit calcification. In this context, Yu *et al.* developed MNPs (50 nm) serving as dual drug carriers of RunX2-siRNA and VEGF, which were immobilized on the decellularized porcine aortic valve by a layer self-assembly method (Figure 6.4).[115] *In vitro*, the hybrid decellularized valve showed enhanced endothelialization, reduced RunX2 gene expression in VICs, and a low hemolysis rate. Upon subcutaneous implantation in rats, the endothelialization of drug carrier-functionalized valves was enhanced compared with control valves without drug release.[115] In the study by Li *et al.*, anti-osteogenic PCL nanoparticles (133 nm) loaded with osteoprotegerin (OPG, osteogenesis-inhibiting glycoprotein) were modified with MAL–PEG–PCL and immobilized onto a thiolated decellularized porcine aortic valve.[116] The particles effectively inhibited the differentiation of rat bone marrow stem cells into osteoblasts *in vitro*, and their immobilization on the valve did not affect the structure of the matrix. Upon subcutaneous implantation in rats, the morphological structure of the valve was almost intact for 8 weeks. In comparison to the non-modified valve, the composite valve showed smaller inflammatory infiltrates and reduced osteocalcin expression.[116] Reducing calcification by surface modification of decellularized valves was also reported by Guo *et al.* and Li *et al.*, who developed a nanocomposite hydrogel by embedding polyhedral oligomeric silsesquioxane (POSS) nanoparticles into a poly(ethylene glycol) (PEG) hydrogel.[117,118] The resulting hydrogel was used as a coating layer on the surface of the decellularized porcine valve. Compared to the non-modified porcine valve, the composite prosthesis reduced both platelet adhesion and calcification processes while promoting endothelial cell adhesion and proliferation both *in vitro* and *in vivo*.[117]

Finally, a bio-inspired, nanoparticle-based approach to preventing thrombosis, inflammation, and calcification of the prosthetic valves in parallel to enhancing endothelialization was reported by Hu *et al.*[119] The authors developed biomimetic PLGA nanoparticles loaded with atorvastatin and rapamycin and encapsulated in erythrocyte membrane vesicles (103 nm). The nanoparticles were effectively crosslinked with the decellularized valves without affecting their structural integrity, stability, or mechanical properties. Importantly, the modification of decellularized bioprostheses with drug-loaded nanoparticles dramatically reduced platelet adhesion and improved endothelialization compared with unmodified, glutaraldehyde-treated valves. Apart from their anti-inflammatory effect, the nanoparticle-modified valves showed a strong resistance to calcification upon implantation in rats for 120 days, indicating the good potential of this approach toward engineering safer, more risk-free bioprosthetic valve implants.[119]

The highlighted studies collectively indicate that nanoparticle-based controlled drug delivery systems represent a promising approach for the

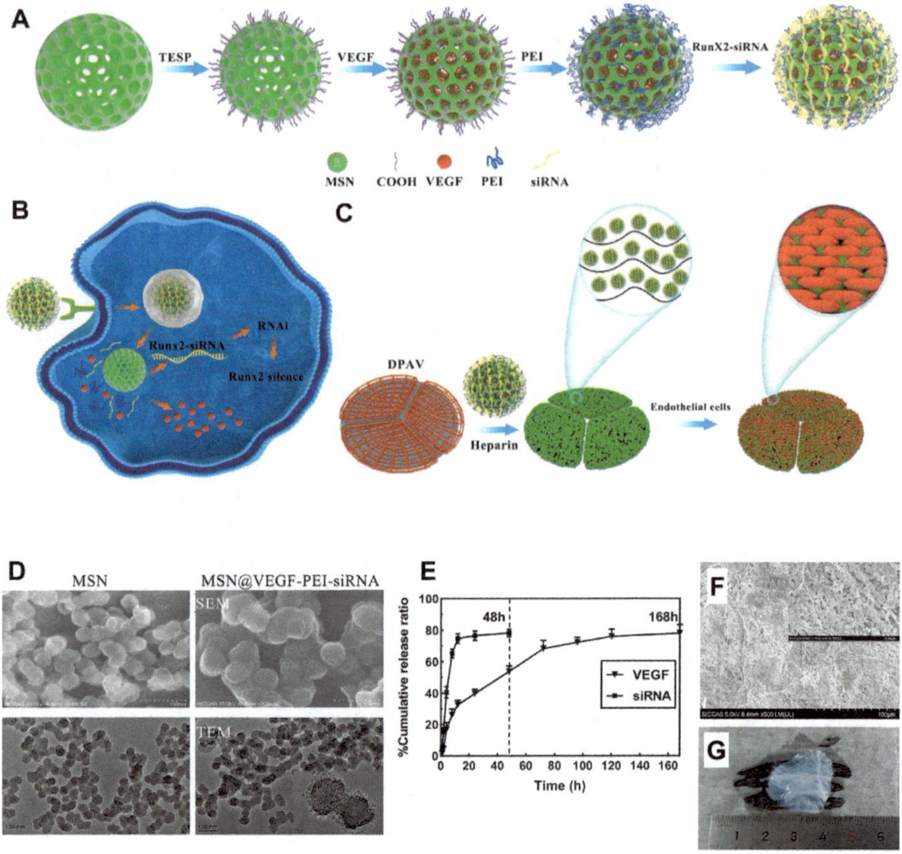

Figure 6.4 Hybrid heart valves with drug and siRNA delivery functions based on decellularized porcine aortic valves.[115] Adapted from ref. 115, https://doi.org/10.3390/bioengineering10070859, under the terms of the CC BY 4.0 license, https://creativecommons.org/licenses/by/4.0/. (A) Schematic depiction of MSN@VEGF–PEI–siRNA synthesis. (B) Cell uptake and mode of action of nanoparticles. (C) Schematic presentation of decellularized porcine aortic valve modification. (D) Representative SEM and TEM images of MNS and MSN@VEGF–PEI–siRNA. (E) Cumulative release curve of siRNA and VEGF from a hybrid valve. (F) Representative SEM image of the hybrid valve. (G) Macroscopic image of a modified hybrid valve.

biofunctionalization of the bioprosthetic scaffolds to accelerate endothelialization, limit calcification, and prevent thrombosis.

6.5.2 Nanomaterials Mimicking Valve Structure

Heart valve prostheses can also be produced *de novo* using polymer spinning methods that enable the fabrication of constructs with defined nanoarchitectures and anisotropic mechanical properties.[120,121] Recent approaches that aim at mimicking the three-layered structure of valve leaflets and providing a 3D microenvironment, which supports attachment and proliferation of human aortic valve interstitial cells, are briefly highlighted in the following paragraphs of this section.

In a study by Wu *et al.,* tri-layered gel-like nanofibrous scaffolds, consisting of electrospun PLGA and poly(aspartic acid) (PASP), were developed.[122] The tri-layered structure was achieved by modification of electrospinning parameters and provided the valve with necessary anisotropy, while the gel-like properties resulted from subsequent PASP hydrogel formation by means of polysuccinimide crosslinking. Compared to PLGA scaffolds, PLGA–PASP scaffolds with a high PASP content significantly promoted the production of collagen and glycosaminoglycans and increased the proliferation of human aortic VICs.[122]

Another electrospinning-based approach to the development of nanofibrous valve prostheses, mimicking the structure of native heart valve leaflets, was described by Jana *et al.*, who developed a tri-layered, oriented nanofibrous scaffold of PCL, with an average fiber diameter of 370 nm.[123] Following the mechanical characterization, an *in vivo* tissue engineering approach was used to test the feasibility of creating cellularized prosthetic valve leaflets. For this, the tri-layered nanofibrous constructs were implanted subcutaneously in a rat model for 2 months. Subsequent histological analyses demonstrated that infiltrated cells and collagen fibrils were aligned both in the circumferentially oriented and in the radially oriented layer, whereas the layer with randomly oriented nanofibers showed a random distribution of cells and collagen fibrils. Apart from collagen, the deposited extracellular matrix contained elastin and glycosaminoglycans, resembling the native composition of leaflets. No calcification was observed in the cellularized tri-layered constructs, suggesting that this approach could support the development of functional, native-like valve leaflets.[123]

Taken together, nanomaterial-based approaches can greatly support both the tissue engineering of heart valves and the modification of currently existing prosthetic valves. The proposed composite scaffolds enable the fabrication of heart valve implants with improved anisotropic structure and biomechanics, which is important for supporting valve cell phenotype and maintaining long-term leaflet integrity.

6.6 Conclusions

Nanomaterials have the potential not only to improve multiple physicochemical and structural features of implantable devices but also to enhance their biological or regenerative properties. However, despite multiple promising reports and considerable progress in (nano)material science, there are not enough data that would confirm the long-term stability of nanoengineered materials exposed to mechanical loading that are typical for aortic or intracardiac positions. With relatively scarce and short-term *in vivo* studies in animal models, predicting the behavior of these scaffolds in terms of durability, degradation, and tissue remodeling remains difficult. Furthermore, although some promising preclinical *in vivo* studies have been published in recent years, the clinical translation of nanomaterial-based implants remains challenging. Medical devices composed of nanofibrous materials, such as prostheses or covered stents, can be expected to find their way into the clinical routine much faster than implants that contain nanomedicinal products, such as drug-loaded nanoparticles that act by pharmacological, immunological, or metabolic means.[124] Incorporating cells or biologically active nanoobjects into the implants raises concerns of safety and potential immunogenicity,[125] besides the obvious challenges related to storage, upscaling, and quality control. Nonetheless, nanomaterials in combination with advanced fabrication techniques will certainly help to improve cardiovascular implants and to overcome the different clinical problems related to stents, cardiac patches, prosthetic vascular grafts, and heart valves. Advanced nanomaterials, enabling the precise engineering of scaffold topography and complex geometries, in combination with surface modifications mimicking the native tissue, should in the future promote the development and successful translation of long-lasting, biocompatible, and functional cardiovascular implants.

Acknowledgements

The support from Deutsche Forschungsgemeinschaft (DFG, German Research Foundation), Project number 326998133 – TRR 225 (subproject B02), and the DFG-NCN project “RESTORE” (grant No. CI 162/4–1, Germany and grant No. UMO-2020/39/I/ST5/01131, Poland) is gratefully acknowledged.

References

1. H. D. Wang, M. Naghavi and C. Allen, *et al.*, *Lancet*, 2016, **388**, 1459–1544.
2. Writing Group Members, D. Mozaffarian, E. J. Benjamin, A. S. Go, D. K. Arnett, M. J. Blaha, M. Cushman, S. R. Das, S. de Ferranti, J. P. Després, H. J. Fullerton, V. J. Howard, M. D. Huffman, C. R. Isasi, M. C. Jiménez, S. E. Judd, B. M. Kissela, J. H. Lichtman, L. D. Lisabeth, S. Liu, R. H. Mackey, D. J. Magid, D. K. McGuire, E. R. Mohler 3rd, C. S. Moy, P. Muntner, M. E. Mussolino, K. Nasir, R. W. Neumar, G. Nichol, L. Palaniappan, D. K. Pandey, M. J. Reeves, C. J. Rodriguez, W. Rosamond, P. D. Sorlie, J. Stein, A. Towfighi, T. N. Turan, S. S. Virani, D. Woo, R. W. Yeh, M. B. Turner, American Heart Association Statistics

Committee, and Stroke Statistics Subcommittee, *Circulation*, 2016, **133**, 447–454.

3. K. W. McDermott, W. J. Freeman and A. Elixhauser, in *Healthcare Cost and Utilization Project (HCUP) Statistical Briefs*, Rockville, MD, 2006.
4. F. A. Masoudi, A. Ponirakis, J. A. de Lemos, J. G. Jollis, M. Kremers, J. C. Messenger, J. W. M. Moore, I. Moussa, W. J. Oetgen, P. D. Varosy, R. N. Vincent, J. Wei, J. P. Curtis, M. T. Roe and J. A. Spertus, *J. Am. Coll. Cardiol.*, 2017, **69**, 1427–1450.
5. F. Alfonso, R. A. Byrne, F. Rivero and A. Kastrati, *J. Am. Coll. Cardiol.*, 2014, **63**, 2659–2673.
6. W. H. Maisel, *N. Engl. J. Med.*, 2007, **356**, 981–984.
7. S. S. Dhruva and J. P. Curtis, *Ann. Intern. Med.*, 2017, **167**, 675–676.
8. D. E. Kandzari, R. C. Sarao and R. Waksman, *Cardiovasc. Revasc. Med.*, 2022, **43**, 97–101.
9. V. Dor, M. Saab, P. Coste, M. Kornaszewska and F. Montiglio, *Thorac. Cardiovasc. Surg.*, 1989, **37**, 11–19.
10. M. Domenech, L. Polo-Corrales, J. E. Ramirez-Vick and D. O. Freytes, *Tissue Eng., Part B*, 2016, **22**, 438–458.
11. P. Pibarot and J. G. Dumesnil, *Circulation*, 2009, **119**, 1034–1048.
12. J. Dunning, H. Gao, J. Chambers, N. Moat, G. Murphy, D. Pagano, S. Ray, J. Roxburgh and B. Bridgewater, *J. Thorac. Cardiovasc. Surg.*, 2011, **142**, 776–782 e773.
13. M. Miernicki, T. Hofmann, I. Eisenberger, F. von der Kammer and A. Praetorius, *Nat. Nanotechnol.*, 2019, **14**, 208–216.
14. E. A. Bleeker, W. H. de Jong, R. E. Geertsma, M. Groenewold, E. H. Heugens, M. Koers-Jacquemijns, D. van de Meent, J. R. Popma, A. G. Rietveld, S. W. Wijnhoven, F. R. Cassee and A. G. Oomen, *Regul. Toxicol. Pharmacol.*, 2013, **65**, 119–125.
15. I. S. O. Online Browsing Platform, ISO/TS 80004-1:2010(en), 2010, Available from: https://www.iso.org/obp/ui/#iso:std:iso:ts:80004:-1:ed-1:v1:en:fn:2.
16. I. S. O. Online Browsing Platform, ISO/TS 80004-4:2011(en), 2011, Available from: https://www.iso.org/obp/ui/#iso:std:iso:ts:80004:-4:ed-1:v1:en.
17. J. Canfield and H. Totary-Jain, *J. Pers. Med.*, 2018, **8**, 33.
18. U. Sigwart, *Eur. Heart J.*, 2017, **38**, 2171–2172.
19. E. Grube, U. Gerckens and L. Buellesfeld, *Minerva Cardioangiol.*, 2002, **50**, 469–473.
20. F. Otsuka, R. A. Byrne, K. Yahagi, H. Mori, E. Ladich, D. R. Fowler, R. Kutys, E. Xhepa, A. Kastrati, R. Virmani and M. Joner, *Eur. Heart J.*, 2015, **36**, 2147–2159.
21. R. A. Byrne, M. Joner and A. Kastrati, *Eur. Heart J.*, 2015, **36**, 3320–3331.
22. T. A. Tabish, M. J. Crabtree, H. E. Townley, P. G. Winyard and C. A. Lygate, *JACC Basic Transl. Sci.*, 2024, **9**, 691–709.

23. M. J. Alef, R. Vallabhaneni, E. Carchman, S. M. Morris Jr, S. Shiva, Y. Wang, E. E. Kelley, M. M. Tarpey, M. T. Gladwin, E. Tzeng and B. S. Zuckerbraun, *J. Clin. Invest.*, 2011, **121**, 1646–1656.
24. C. H. Lee, M. J. Hsieh, K. S. Liu, C. W. Cheng, S. H. Chang, S. J. Liu, C. J. Wang, M. Y. Hsu, K. C. Hung, Y. H. Yeh, W. J. Chen, I. C. Hsieh, J. H. Juang and M. S. Wen, *Int. J. Nanomed.*, 2018, **13**, 6039–6048.
25. C. H. Lee, K. S. Liu, J. G. Roth, K. C. Hung, Y. W. Liu, S. H. Wang, C. C. Kuo and S. J. Liu, *Pharmaceutics*, 2021, **13**, 1756.
26. R. Wang, J. Lu, J. Yin, H. Chen, H. Liu, F. Xu, T. Zang, R. Xu, C. Li, Y. Wu, Q. Wu, X. Fei, M. Zhu, L. Shen and J. Ge, *Bioact. Mater.*, 2023, **19**, 666–677.
27. N. Chen, M. Li, H. Wu, Y. Qin, J. Wang, K. Xu, R. Luo, L. Yang, Y. Wang and X. Zhang, *Regen. Biomater.*, 2023, **10**, rbad055.
28. M. Janjic, F. Pappa, V. Karagkiozaki, C. Gitas, K. Ktenidis and S. Logothetidis, *Int. J. Nanomed.*, 2017, **12**, 6343–6355.
29. Y. Cheng, X. Zhang, R. Liu, Y. Li, J. Zeng, M. Zhou and Y. Zhao, *Adv. Healthcare Mater.*, 2022, **11**, e2200965.
30. J. A. Choe, S. Uthamaraj, D. Dragomir-Daescu, G. S. Sandhu and B. J. Tefft, *Tissue Eng., Part A*, 2023, **29**, 413–423.
31. D. E. Cutlip, K. N. Garratt, V. Novack, M. Barakat, P. Meraj, L. Maillard, A. Erglis, R. Jauhar, J. J. Popma, R. Stoler, S. Silber and F. S. T. I. Pz, *JACC*, 2017, **10**, 160–167.
32. T. Koppara, K. Sakakura, E. Pacheco, Q. Cheng, X. Zhao, E. Acampado, A. V. Finn, M. Barakat, L. Maillard, J. Ren, M. Deshpande, F. D. Kolodgie, M. Joner and R. Virmani, *Int. J. Cardiol.*, 2016, **222**, 217–225.
33. L. Maillard, D. Corseaux, A. Altié, A. Ung, J. Courageot, M. Barakat, E. Teiger and E. Van Belle, *Cardiovasc. Revasc. Med.*, 2020, **21**, 195–199.
34. L. Xu, C. He, S. Yang, Y. Zhu, P. Wang, S. Wu, F. Guo and Y. Wang, *Colloids Surf., B*, 2023, **230**, 113530.
35. W. Ma, X. Liu, M. Yang, Q. Hong, L. Meng, Q. Zhang, J. Chen and C. Pan, *Biomater. Adv.*, 2023, **149**, 213393.
36. Y. Li, B. Zhang, X. Liu, H. Wan, Y. Qin, H. Yan, Y. Wang, Y. An, Y. Yang, Y. Dai, L. Yang and Y. Wang, *Biomaterials*, 2023, **302**, 122288.
37. J. Zhou, J. Weng, X. Huang, S. Sun, Q. Yang, H. Lin, J. Yang, H. Guo and J. Chi, *Front. Cardiovasc. Med.*, 2022, **9**, 1025558.
38. T. Zhu, M. Zhou, W. Gao, D. Fang, Z. Liu, G. Wu, M. Wan, C. Mao and J. Shen, *Langmuir*, 2020, **36**, 2901–2910.
39. R. Luo, J. Zhang, W. Zhuang, L. Deng, L. Li, H. Yu, J. Wang, N. Huang and Y. Wang, *J. Mater. Chem. B*, 2018, **6**, 5582–5595.
40. J. Tan, Y. Cui, Z. Zeng, L. Wei, L. Li, H. Wang, H. Hu, T. Liu, N. Huang, J. Chen and Y. Weng, *J. Biomed. Mater. Res., Part A*, 2020, **108**, 1295–1304.
41. T. Liu, X. Wang, X. Tang, T. Gong, W. Ye, C. Pan, H. Ding, X. Luo, X. Li and Q. M. Wang, *ACS Appl. Mater. Interfaces*, 2017, **9**, 30373–30386.

42. M. Izuhara, Y. Kuwabara, N. Saito, E. Yamamoto, D. Hakuno, Y. Nakashima, T. Horie, O. Baba, M. Nishiga, T. Nakao, T. Nishino, F. Nakazeki, Y. Ide, M. Kimura, T. Kimura and K. Ono, *PLoS One*, 2017, **12**, e0172798.
43. B. Lattuca, L. Schmutz, L. Cornillet, B. Ledermann, V. Fernandez, P. Messner, F. Leclercq and G. Cayla, *Int. J. Cardiol.*, 2015, **201**, 208–209.
44. G. C. Inaraja Pérez, A. Rodríguez Morata, J. P. Reyes Ortega, R. Gómez Medialdea and P. Cabezudo García, *Ann. Vasc. Surg.*, 2015, **29**, 1018.e5–1018.e8.
45. Y. Zhang, J. Wang, J. Xiao, T. Fang, N. Hu, M. Li, L. Deng, Y. Cheng, Y. Zhu and W. Cui, *Acta Biomater.*, 2019, **94**, 295–305.
46. P. Liu, Y. Liu, P. Li, Y. Zhou, Y. Song, Y. Shi, W. Feng, X. Mo, H. Gao, Q. An and W. Zhu, *ACS Appl. Mater. Interfaces*, 2018, **10**, 41012–41018.
47. J. Chu, L. Chen, Z. Mo, G. L. Bowlin, B. A. Minden-Birkenmaier, Y. Morsi, A. Aldalbahi, M. El-Newehy, W. Wang and X. Mo, *Acta Biomater.*, 2020, **111**, 102–117.
48. J. Liu, H. Zhu, Y. Pei, H. Zhang, J. Zhou and Z. Jing, *Biomater. Res.*, 2022, **26**, 15.
49. D. Son, J. Lee, D. J. Lee, R. Ghaffari, S. Yun, S. J. Kim, J. E. Lee, H. R. Cho, S. Yoon, S. X. Yang, S. Lee, S. T. Qiao, D. S. Ling, S. Shin, J. K. Song, J. Kim, T. Kim, H. Lee, J. Kim, M. Soh, N. Lee, C. S. Hwang, S. Nam, N. S. Lu, T. Hyeon, S. H. Choi and D. H. Kim, *ACS Nano*, 2015, **9**, 5937–5946.
50. R. Herbert, M. Elsisy, B. Rigo, H. R. Lim, H. Kim, C. Choi, S. Kim, S. H. Ye, W. R. Wagner, Y. Chun and W. H. Yeo, *Nano Today*, 2022, **46**, 101557.
51. G. C. Kaiser, *Circulation*, 1985, **72**, V46–58.
52. B. J. Bachar and B. Manna, in *Coronary Artery Bypass Graft*, StatPearls, Treasure Island, FL, 2023.
53. EUROSTAT, in *Cardiovascular diseases statistics: Statistics Explained*, 2021, http://ec.europa.eu/eurostat/statistics-explained/index.php/Cardiovascular_diseases_statistics.
54. S. J. Head, T. M. Kieser, V. Falk, H. A. Huysmans and A. P. Kappetein, *Eur. Heart J.*, 2013, **34**, 2862–2872.
55. M. J. Mack, J. A. Osborne and H. Shennib, *Ann. Thorac. Surg.*, 1998, **66**, 1055–1059.
56. M. S. Lee, S. J. Park, D. E. Kandzari, A. J. Kirtane, W. F. Fearon, E. S. Brilakis, P. Vermeersch, Y. H. Kim, R. Waksman, J. Mehilli, L. Mauri and G. W. Stone, *JACC*, 2011, **4**, 831–843.
57. S. Goldman, K. Zadina, T. Moritz, T. Ovitt, G. Sethi, J. G. Copeland, L. Thottapurathu, B. Krasnicka, N. Ellis, R. J. Anderson, W. Henderson and VA Cooperative Study Group, *J. Am. Coll. Cardiol.*, 2004, **44**, 2149–2156.
58. S. Ravi and E. L. Chaikof, *Regener. Med.*, 2010, **5**, 107–120.
59. C. B. Weinberg and E. Bell, *Science*, 1986, **231**, 397–400.

60. N. Hibino, E. McGillicuddy, G. Matsumura, Y. Ichihara, Y. Naito, C. Breuer and T. Shinoka, *J. Thorac. Cardiovasc. Surg.*, 2010, **139**, 431–U233.
61. P. Zilla, D. Bezuidenhout and P. Human, *Biomaterials*, 2007, **28**, 5009–5027.
62. J. Chlupac, E. Filova and L. Bacakova, *Physiol. Res.*, 2009, **58 Suppl 2**, S119–S139.
63. Z. Wang, C. Liu, Y. Xiao, X. Gu, Y. Xu, N. Dong, S. Zhang, Q. Qin and J. Wang, *ACS Nano*, 2019, **13**, 10576–10586.
64. G. Kianpour, R. Bagheri, A. Pourjavadi and H. Ghanbari, *Mater. Sci. Eng., C*, 2020, **114**, 111043.
65. I. Łopianiak, W. Rzempołuch, M. Civelek, I. Cicha, T. Ciach and B. A. Butruk-Raszeja, *J. Biol. Eng.*, 2023, **17**, 20.
66. K. Kopeć, M. Wojasiński, M. Eichler, H. Genç, R. P. Friedrich, R. Stein, R. Singh, C. Alexiou, H. Hlawaty, T. Ciach and I. Cicha, *Biomater. Adv.*, 2022, **134**, 112544.
67. G. A. Emechebe, F. O. Obiweluozor, I. S. Jeong, J. K. Park, C. H. Park and C. S. Kim, *Nanomedicine*, 2020, **30**, 102306.
68. Y. Matsuzaki, S. Miyamoto, H. Miyachi, R. Iwaki, T. Shoji, K. Blum, Y. C. Chang, J. Kelly, J. W. Reinhardt, H. Nakayama, C. K. Breuer and T. Shinoka, *Ann. Thorac. Surg.*, 2021, **111**, 1234–1241.
69. C. Zhang, Q. Xie, R. Cha, L. Ding, L. Jia, L. Mou, S. Cheng, N. Wang, Z. Li, Y. Sun, C. Cui, Y. Zhang, Y. Zhang, F. Zhou and X. Jiang, *Adv. Healthcare Mater.*, 2021, **10**, e2100839.
70. H. Kuang, Y. Wang, Y. Shi, W. Yao, X. He, X. Liu, X. Mo, S. Lu and P. Zhang, *Biomaterials*, 2020, **259**, 120288.
71. W. Jia, L. Liu, M. Li, Y. Zhou, H. Zhou, H. Weng, G. Gu, M. Xiao and Z. Chen, *Acta Biomater.*, 2022, **153**, 287–298.
72. D. Wang, X. Wang, X. Li, L. Jiang, Z. Chang and Q. Li, *Mater. Sci. Eng., C*, 2020, **107**, 110212.
73. Y. Liu, H. Yuan, Y. Liu, C. Chen, Z. Tang, C. Huang, Z. Ning, T. Lu and Z. Wu, *Front. Bioeng. Biotechnol.*, 2023, **11**, 1109058.
74. X. Guo, J. Zhu, H. Zhang, Z. You, Y. Morsi, X. Mo and T. Zhu, *J. Colloid Interface Sci.*, 2019, **539**, 351–360.
75. C. Zhang, R. Cha, C. Wang, X. Chen, Z. Li, Q. Xie, L. Jia, Y. Sun, Z. Hu, L. Zhang, F. Zhou, Y. Zhang and X. Jiang, *Biomaterials*, 2023, **297**, 122124.
76. V. Dor, *J. Cardiovasc. Surg.*, 2002, **17**, 180-187.
77. G. Gong, Z. Ling, E. Seifter, S. M. Factor and R. W. Frater, *Eur. J. Cardiothorac. Surg.*, 1991, **5**, 288–299.
78. R. W. M. Frater, E. Seifter, K. Liao and F. Wasserman, in *Advances in Anticalcification and Antidegenerative Treatments of Biologocal Heart Valves and Coronary Conduits*, ed. S. Gabbay and D. J. Wheatley, Silent Partners Inc, Austin, TX, 1997, pp. 105–113.

79. D. Hoffman, G. Gong, K. Liao, F. Macaluso, S. D. Nikolic and R. W. Frater, *Circulation*, 1992, **86**, II75–II79.
80. G. Chan and D. J. Mooney, *Trends Biotechnol.*, 2008, **26**, 382-392.
81. T. J. Keane, I. T. Swinehart and S. F. Badylak, *Methods*, 2015, **84**, 25–34.
82. L. T. Saldin, M. C. Cramer, S. S. Velankar, L. J. White and S. F. Badylak, *Acta Biomater.*, 2017, **49**, 1–15.
83. B. Follin, A. A. Ghotbi, A. E. Clemmensen, S. Bentsen, M. Juhl, R. H. Sondergaard, L. D. Lund, M. Haack-Sorensen, P. Hasbak, S. Cohen, R. S. Ripa, J. Kastrup, A. Ekblond and A. Kjaer, *Stem Cells Int.*, 2018, **2018**, 7821461.
84. E. Ruvinov and S. Cohen, *Adv. Drug Delivery Rev.*, 2016, **96**, 54–76.
85. S. Das and J. Jang, *J. 3D Print. Med.*, 2018, **2**, 69–87.
86. F. Zhang, Y. Xie, H. Celik, O. Akkus, S. H. Bernacki and M. W. King, *Biofabrication*, 2019, **11**, 035020.
87. M. Yamada, K. Tanemura, S. Okada, A. Iwanami, M. Nakamura, H. Mizuno, M. Ozawa, R. Ohyama-Goto, N. Kitamura, M. Kawano, K. Tan-Takeuchi, C. Ohtsuka, A. Miyawaki, A. Takashima, M. Ogawa, Y. Toyama, H. Okano and T. Kondo, *Stem Cells*, 2007, **25**, 562–570.
88. M. Radisic, H. Park, H. Shing, T. Consi, F. J. Schoen, R. Langer, L. E. Freed and G. Vunjak-Novakovic, *Proc. Natl. Acad. Sci. U. S. A.*, 2004, **101**, 18129–18134.
89. T. Dvir, O. Levy, M. Shachar, Y. Granot and S. Cohen, *Tissue Eng.*, 2007, **13**, 2185–2193.
90. Y. Barash, T. Dvir, P. Tandeitnik, E. Ruvinov, H. Guterman and S. Cohen, *Tissue Eng., Part C*, 2010, **16**, 1417–1426.
91. Y. Sapir, B. Polyak and S. Cohen, *Nanotechnol.*, 2014, **25**, 014009.
92. Y. Sapir, E. Ruvinov, B. Polyak and S. Cohen, *Methods Mol. Biol.*, 2014, **1181**, 83–95.
93. T. Dvir, B. P. Timko, M. D. Brigham, S. R. Naik, S. S. Karajanagi, O. Levy, H. Jin, K. K. Parker, R. Langer and D. S. Kohane, *Nat. Nanotechnol.*, 2011, **6**, 720–725.
94. A. Navaei, H. Saini, W. Christenson, R. T. Sullivan, R. Ros and M. Nikkhah, *Acta Biomater.*, 2016, **41**, 133–146.
95. L. Wang, Y. Liu, G. Ye, Y. He, B. Li, Y. Guan, B. Gong, K. Mequanint, M. M. Q. Xing and X. Qiu, *Nat. Biomed. Eng.*, 2021, **5**, 1157–1173.
96. M. Lee, J. Park, G. Choe, S. Lee, B. G. Kang, J. H. Jun, Y. Shin, M. C. Kim, Y. S. Kim, Y. Ahn and J. Y. Lee, *ACS Nano*, 2023, **17**, 12290–12304.
97. H. Wang, Y. Guo, Y. Hu, Y. Zhou, Y. Chen, X. Huang, J. Chen, Q. Deng, S. Cao, B. Hu, R. Jiang, J. Pan, T. Tan, Y. Wang, Y. Chen, Q. Dong, P. Chen and Q. Zhou, *Int. J. Biol. Macromol.*, 2023, **244**, 125139.
98. X. Lin, Y. Liu, A. Bai, H. Cai, Y. Bai, W. Jiang, H. Yang, X. Wang, L. Yang, N. Sun and H. Gao, *Nat. Biomed. Eng.*, 2019, **3**, 632–643.
99. J. Chen, Y. Zhan, Y. Wang, D. Han, B. Tao, Z. Luo, S. Ma, Q. Wang, X. Li, L. Fan, C. Li, H. Deng and F. Cao, *Acta Biomater.*, 2018, **80**, 154–168.

100. X. Wei, L. Wang, C. Duan, K. Chen, X. Li, X. Guo, P. Chen, H. Liu and Y. Fan, *Bioact. Mater.*, 2023, **27**, 271–287.
101. F. Tondnevis, H. Keshvari and J. A. Mohandesi, *J. Biomed. Mater. Res., Part B*, 2020, **108**, 2276–2293.
102. G. Zhao, Y. Feng, L. Xue, M. Cui, Q. Zhang, F. Xu, N. Peng, Z. Jiang, D. Gao and X. Zhang, *Acta Biomater.*, 2022, **139**, 190–203.
103. K. Hosoyama, M. Ahumada, C. D. McTiernan, D. R. Davis, F. Variola, M. Ruel, W. Liang, E. J. Suuronen and E. I. Alarcon, *ACS Appl. Mater. Interfaces*, 2018, **10**, 44668–44677.
104. M. Malki, S. Fleischer, A. Shapira and T. Dvir, *Nano Lett.*, 2018, **18**, 4069–4073.
105. V. T. Nkomo, J. M. Gardin, T. N. Skelton, J. S. Gottdiener, C. G. Scott and M. Enriquez-Sarano, *Lancet*, 2006, **368**, 1005–1011.
106. R. El Oakley, P. Kleine and D. S. Bach, *Circulation*, 2008, **117**, 253–256.
107. J. G. Webb, D. A. Wood, J. Ye, R. Gurvitch, J. B. Masson, J. Rodés-Cabau, M. Osten, E. Horlick, O. Wendler, E. Dumont, R. G. Carere, N. Wijesinghe, F. Nietlispach, M. Johnson, C. R. Thompson, R. Moss, J. Leipsic, B. Munt, S. V. Lichtenstein and A. Cheung, *Circulation*, 2010, **121**, 1848–1857.
108. R. O. Bonow, B. Carabello, A. C. de Leon Jr, L. H. Edmunds Jr, B. J. Fedderly, M. D. Freed, W. H. Gaasch, C. R. McKay, R. A. Nishimura, P. T. O'Gara, R. A. O'Rourke, S. H. Rahimtoola, J. L. Ritchie, M. D. Cheitlin, K. A. Eagle, T. J. Gardner, A. Garson Jr, R. J. Gibbons, R. O. Russell, T. J. Ryan and S. C. Smith Jr, *Circulation*, 1998, **98**, 1949–1984.
109. A. Vahanian, H. Baumgartner, J. Bax, E. Butchart, R. Dion, G. Filippatos, F. Flachskampf, R. Hall, B. Iung, J. Kasprzak, P. Nataf, P. Tornos, L. Torracca, A. Wenink, Task Force on the Management of Valvular Hearth Disease of the European Society of Cardiology, and ESC Committee for Practice Guidelines, *Eur. Heart J.*, 2007, **28**, 230–268.
110. B. Kassaï, F. Gueyffier, M. Cucherat and J. P. Boissel, *Cardiac. Surg.*, 2000, **8**, 477–483.
111. R. R. Makkar, G. Fontana, H. Jilaihawi, T. Chakravarty, K. F. Kofoed, O. De Backer, F. M. Asch, C. E. Ruiz, N. T. Olsen, A. Trento, J. Friedman, D. Berman, W. Cheng, M. Kashif, V. Jelnin, C. A. Kliger, H. Guo, A. D. Pichard, N. J. Weissman, S. Kapadia, E. Manasse, D. L. Bhatt, M. B. Leon and L. Søndergaard, *N. Engl. J. Med.*, 2015, **373**, 2015–2024.
112. J. T. Butcher, G. J. Mahler and L. A. Hockaday, *Adv. Drug Delivery Rev.*, 2011, **63**, 242–268.
113. J. Zhou, J. Ding, Z. Zhu, J. Xu, Y. Yi, Y. Li, H. Fan, S. Bai, J. Yang, Y. Tang, X. Dong and N. Dong, *Mater. Sci. Eng., C*, 2019, **97**, 632–643.
114. A. Seyrek, G. Günal and H. M. Aydin, *ACS Appl. Bio Mater.*, 2022, **5**, 3883–3895.
115. W. Yu, X. Zhu, J. Liu and J. Zhou, *Bioengineering*, 2023, **10**, 859.

116. Y. Li, Y. Zhang, J. L. Ding, J. C. Liu, J. J. Xu, Y. H. Tang, Y. P. Yi, W. C. Xu, W. P. Yu, C. Lu, W. Yang, J. S. Yang, Y. Gong and J. L. Zhou, *RSC Adv.*, 2019, **9**, 11882–11893.
117. C. Li, Y. Zhou, S. Liu, R. Guo, C. Lu, D. Yin, Y. Zhang, X. Xu, N. Dong and J. Shi, *ACS Appl. Bio Mater.*, 2022, **5**, 3923–3935.
118. R. Guo, Y. Zhou, S. Liu, C. Li, C. Lu, G. Yang, J. Nie, F. Wang, N. G. Dong and J. Shi, *ACS Appl. Bio Mater.*, 2021, **4**, 2534–2543.
119. C. Hu, R. Luo and Y. Wang, *ACS Appl. Mater. Interfaces*, 2020, **12**, 41113–41126.
120. S. Wu, B. Duan, X. Qin and J. T. Butcher, *Acta Biomater.*, 2017, **51**, 89–100.
121. A. K. Capulli, M. Y. Emmert, F. S. Pasqualini, D. Kehl, E. Caliskan, J. U. Lind, S. P. Sheehy, S. J. Park, S. Ahn, B. Weber, J. A. Goss, S. P. Hoerstrup and K. K. Parker, *Biomaterials*, 2017, **133**, 229–241.
122. S. Wu, Y. Li, C. Zhang, L. Tao, M. Kuss, J. Y. Lim, J. Butcher and B. Duan, *Adv. Healthcare Mater.*, 2022, **11**, e2200053.
123. S. Jana, F. Franchi and A. Lerman, *Biomed. Mater.*, 2019, **15**, 015004.
124. L. Bayarri, *Drugs Today*, 2015, **51**, 505–513.
125. M. Awashra and P. Młynarz, *Nanoscale Adv.*, 2023, **5**, 2674–2723.

CHAPTER 7

Bioceramic-based Cardiovascular Implants

J. KHALIQ*

Northumbria University at Newcastle, NE1 8ST, UK
*E-mail: Jibran.khaliq@northumbria.ac.uk

7.1 Introduction

Cardiovascular implants are important components in the therapeutics against cardiovascular diseases, comprising life-saving devices ranging from stents and heart valves to vascular grafts.[1,2] The importance of these implants lies in their ability to restore normal physiological functions, relieve symptoms, and improve the overall quality of life for individuals suffering from cardiovascular conditions.[3] This chapter sheds light on the area of cardiovascular implants, with a particular focus on the transformative role that bioceramics can play in advancing the field.

Cardiovascular implants have historically faced challenges in the materials traditionally used.[4] While these materials have undoubtedly been influential in various cardiovascular applications, persistent issues, such as biocompatibility, long-term durability, and susceptibility to complications like thrombosis and calcification, have demanded innovative solutions.[5] Bioceramics, due to their distinct material properties, have emerged as a convincing alternative that holds the promise of addressing these challenges and revolutionising cardiovascular implant technology.

Bioceramics, such as alumina,[6] zirconia,[7] hydroxyapatite (HA),[8] and bioactive glasses, exhibit a remarkable synergy of properties, such as

Nanoscience & Nanotechnology Series No. 64
Cardiovascular Nanomedicine
Edited by Roger Narayan and Tanveer Tabish
© The Royal Society of Chemistry 2024
Published by the Royal Society of Chemistry, www.rsc.org

biocompatibility,[9] mechanical strength, and corrosion resistance.[10] The integration of bioceramics holds the promise of not only addressing current challenges but also opening new possibilities in the design and functionality of these life-saving devices. By providing a detailed exploration of bioceramics in the cardiovascular context, this chapter aims to contribute to the broader discourse on advancing medical technology for the treatment of cardiovascular diseases.

This chapter is focused on the significance of bioceramics in the context of cardiovascular implants, emphasising their potential to address the limitations associated with conventional materials. Later in the chapter, the focus is on the intricate interplay between bioceramics and cardiovascular tissues, exploring the biocompatibility of these materials and their capacity for effortless integration. In addition, properties, including the mechanical properties of bioceramics, such as strength, fracture toughness, and fatigue resistance, which are critical attributes to determine their success in dynamic cardiovascular environments, will be discussed.

7.2 Types of Bioceramics

Bioceramics for application in cardiovascular implants are chosen based on their distinct properties and specific applications within the field. This section briefly describes the commonly used bioceramics (also shown in Figure 7.1), with a specific focus on their properties, making them useful for cardiovascular implant technology.

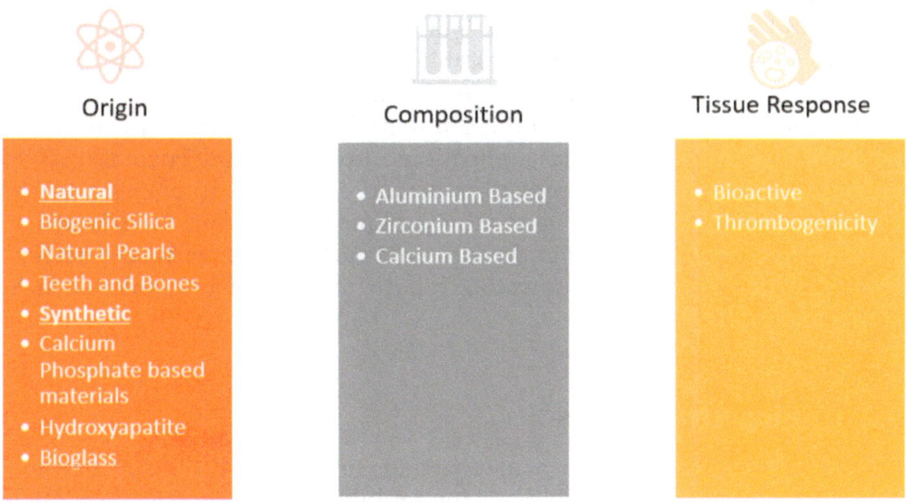

Figure 7.1 Classification of Bioceramics.

7.2.1 Alumina

Alumina, or aluminium oxide (Al_2O_3), is the most commonly used material in cardiovascular implants and is renowned for its exceptional mechanical strength and durability.[6] In cardiovascular implants, alumina is often favoured for its high wear resistance, making it an ideal candidate for applications, such as femoral heads in hip joint prostheses.[11] The robustness of alumina contributes to the longevity of implants, ensuring sustained structural integrity under challenging conditions within the cardiovascular system.

7.2.2 Zirconia

Zirconia, alternatively identified as zirconium dioxide (ZrO_2), occurs naturally in the mineral baddeleyite and can also be synthetically produced from zircon. Zirconia is characterised by its high fracture toughness and biocompatibility, which play a critical role in addressing the challenges associated with wear and fatigue in cardiovascular implants.[12] It is widely used in components, such as dental implants and artificial joints. The biocompatible nature of zirconia promotes tissue integration, while its fracture toughness enhances the overall reliability of implants, making it an important material in cardiovascular applications.

7.2.3 Hydroxyapatite

HA is an inorganic mineral characterised by its standard apatite lattice structure, denoted as $A_{10}(BO_4)_6C_2$, with A, B, and C representing Ca, PO_4, and OH, respectively.[13] In its pure form, HA comprises 39.68% calcium by weight and 18% phosphorus by weight, resulting in a Ca/P molar ratio of 1.67. Hydroxyapatite is a biocompatible ceramic that mimics the mineral component of natural bone and has the ability to promote osseointegration, which makes it a very unique material in the field of cardiovascular implants. In orthopaedic applications, hydroxyapatite-coated surfaces encourage bone ingrowth, providing a secure base for implants.[14] In the case of cardiovascular implants, this unique bioceramic contributes to enhancing biocompatibility and promoting interactions with surrounding tissues, which are crucial indicators for the success of implants, such as stents and grafts.[15]

7.2.4 Bioactive Glasses

Bioactive glasses belong to a category of surface-reactive glass–ceramic biomaterials. Their remarkable biocompatibility and bioactivity have made them useful for applications such as implant devices within the human body, facilitating the repair and replacement of bones affected by disease or damage.[16] These glasses, when in contact with bodily fluids, stimulate the formation of a hydroxyapatite layer, facilitating the integration with the host tissue with their ability to bond with living tissues. In cardiovascular applications, bioactive glasses show promise in promoting endothelialisation, mitigating the risk of thrombosis, and offering an innovative approach to enhancing the biocompatibility of implants.

The above short description has made it evident that each material brings a unique set of properties, which makes them suitable for a particular application. The biocompatibility, corrosion resistance, and mechanical strength inherent in alumina, zirconia, hydroxyapatite, and bioactive glasses contribute to designing a tailored cardiovascular implant, showcasing the dynamic synergy between materials science and medical innovation.

7.3 Biocompatibility and Tissue Integration

Designing cardiovascular implants is very challenging, as one of the most important aspects of the materials is to be biocompatible so the body can accept the foreign implant; hence, the concept of biocompatibility stands as a fundamental pillar for seamless integration of bioceramic materials with the complex environment of cardiovascular tissues.[17] This section highlights the subtle interplay between bioceramics and the biological environment, elucidating the mechanisms that encourage cellular adhesion, proliferation, and tissue regeneration.

Biocompatibility, the cornerstone of successful implantology, refers to the ability of a material to extract an appropriate biological response without causing harmful reactions.[18] In the context of bioceramics, their intrinsic biocompatibility is a result of detailed material selection and engineering. Alumina, zirconia, hydroxyapatite, and bioactive glasses each exhibit characteristics that have excellent compatibility with the physiological environment, reducing the risk of inflammation or rejection, which is the first step in the integration of an implant within the cardiovascular system. The biocompatibility of bioceramics can also be enhanced using advanced manufacturing techniques, such as micropatterning. As shown in Figure 7.2, the micropatterned surface of HA bioceramics showed an enhanced adhesion (S1–S3) compared to the flat sample (S1). A bioceramic implant acts as a scaffold for cellular growth, providing a structured framework that supports the multiplication of cells. This phase is pivotal for the formation of a robust cellular network, laying the foundation for the implant's integration into the host tissue.[20]

A material is said to be biocompatible when, upon being inserted into the human body for a specific use, it does not cause infection and helps in tissue regeneration. Bioceramics, due to their ability to effortlessly integrate with cardiovascular tissues, catalyse the regrowth of the tissues. The

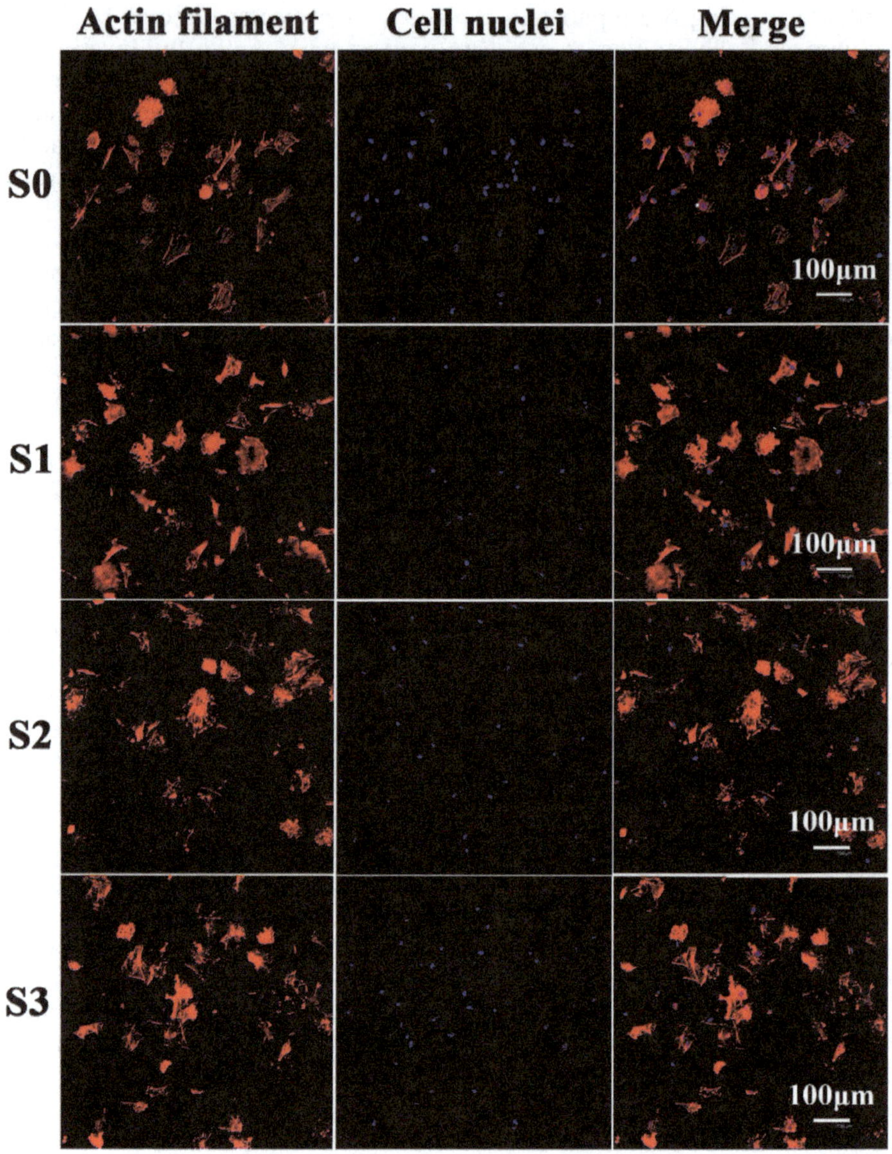

Figure 7.2 Confocal microscopic images of the hydroxyapatite bioceramics with a flat surface (S0) as the control sample and with the micropatterned surfaces (S1–S3) after cell seeding for 6 hours. Reproduced from ref. 19 with permission from the Royal Society of Chemistry.

interaction between the implant and the surrounding biological environment triggers cellular signalling pathways that initiate tissue repair and regeneration processes.

One of the critical steps towards tissue integration starts with the critical phase of cellular adhesion, where the bioceramic surface establishes initial contact with host cells. Bioceramics, owing to their tailored surface properties, facilitate the adhesion of cells, creating a conducive environment for subsequent interactions.[21] This adhesive interaction forms the core for establishing the cellular community on the implant surface, initiating a series of events that orchestrate tissue integration.

In principle, the biocompatibility of bioceramics is a reciprocal relationship with cardiovascular tissues, where the implant becomes an integral part of the biological environment. This intricate phenomenon of cellular adhesion, proliferation, and tissue regeneration not only ensures the successful integration of bioceramic cardiovascular implants but also advances the field of regenerative medicine and implantology. In the subsequent sections, specific applications of these bioceramics in cardiovascular implants will be discussed.

7.4 Mechanical Properties and Design Considerations

The success of cardiovascular implants relies heavily on the mechanical properties of bioceramics, such as strength, fracture toughness, and fatigue resistance. Each of these properties plays a distinct yet coherent role in ensuring the durability and functionality of implants.

The mechanical strength of bioceramics emerges as a foundational element in cardiovascular implant engineering.[22] As these implants move within the intricate and throbbing environment of the cardiovascular system, their ability to withstand forces becomes vital. By discussing the inherent strength of bioceramics, such as alumina and zirconia, information is highlighted on how these materials strengthen cardiovascular constructs, ensuring that the implants maintain structural integrity within the dynamic stresses forced by blood flow.

Fracture toughness is another crucial mechanical property that protects cardiovascular implants against structural vulnerabilities.[23] Materials, such as zirconia, explain the ability of bioceramics to resist crack propagation and fracture initiation.[24] Hydroxyapatite possesses inadequate fracture toughness, which makes it vulnerable to crack, and researchers are trying to improve the fracture toughness through the integration of various additives to enhance the fracture toughness. Moreover, the dynamic nature of cardiovascular environments requires a special focus on the fatigue resistance of bioceramics. Implants, including those made of hydroxyapatite and bioactive glasses, must endure cyclic loading without experiencing structural deterioration.

Beyond intrinsic mechanical properties, the success of cardiovascular implants relies on careful design considerations. Tailoring the shape, size, and surface properties of implants becomes of utmost importance to optimise their interaction with the dynamic cardiovascular environment. The integration of innovative design approaches, including those facilitated by additive manufacturing techniques, further improves the precision required for optimal performance in cardiovascular applications.[25]

7.5 Applications in Stents and Vascular Grafts

Bioceramic-coated stents are promising in reducing restenosis rates, a common complication post stent implantation.[26] The controlled release of bioactive substances from the bioceramic coating, such as drugs or growth factors, contributes to inhibiting excessive tissue proliferation, ensuring the patency of the treated coronary artery. This targeted approach not only mitigates restenosis but also promotes the healing of the arterial wall, elevating the overall success of coronary interventions.

Materials, such as hydroxyapatite, calcium phosphate, and bioactive glasses, exhibit a natural affinity for blood components, fostering an environment conducive to smooth blood flow and reduced clotting.[27] The integration of bioceramics into the design of vascular grafts aims to minimise thrombotic events and enhance the compatibility of these artificial conduits with the intricate vascular network. This works particularly well for patients requiring bypass surgery, where the integration of bioceramics can contribute to the creation of durable and hemocompatible

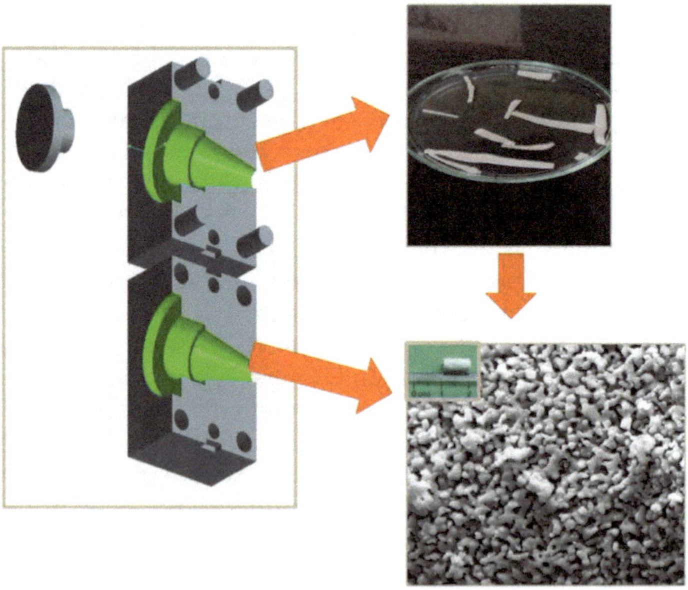

Figure 7.3 Extrusion-based manufacturing technique for bioceramic-based stents. Reproduced from ref. 29 with permission from the Royal Society of Chemistry.

pathways for blood circulation. Moreover, the inherent strength of materials, such as alumina and zirconia, coupled with their resistance to fatigue, positions bioceramic implants as reliable constructs capable of withstanding the continuous mechanical stresses within arterial environments.[28] Currently, metal stents or stents coated with a bioceramic are being used; however, achieving the desired quality is a challenge due to complex manufacturing. Extrusion using calcium phosphate ceramics has recently been proposed as a new technique to produce stents, as shown in Figure 7.3.[29] This characteristic is particularly crucial in arterial applications, where implants must endure the pulsatile forces exerted by blood flow over extended periods.

7.6 Bioceramic Heart Valves

Bioceramic heart valves are different from those made from conventional materials, such as mechanical valves, or animal tissues. The search for alternative heart valves arose from the limitations of traditional materials, such as the gradual onset of calcification and the risk of clot formation (thrombogenicity), which can compromise the structural integrity of the valve over time.[30] Heart valves made of bioceramics are an excellent replacement choice in mitigating these challenges. One of the critical advantages of bioceramic heart valves is their potential to address issues related to calcification.[31] Traditional valves are susceptible to calcium deposits over time, leading to stiffness and reduced functionality. Bioceramic valves, with their inherent resistance to calcification, provide a durable solution that maintains flexibility and functionality over an extended period of time. This feature not only enhances the longevity of the valve but also mitigates the need for early reoperations, which significantly benefits the patients undergoing valve replacement procedures.[32]

Thrombogenicity, the tendency for blood clot formation, is a critical concern in heart valve replacements. Bioceramic heart valves, designed with a focus on hemocompatibility, offer a surface that minimises the risk of blood clotting.[33] The bioinert nature of certain bioceramics, coupled with their ability to promote a harmonious interaction with the bloodstream, reduces the likelihood of thrombotic events. This feature not only enhances the safety of bioceramic heart valves but also eliminates the need for anticoagulant therapy in certain cases, improving the overall quality of life for patients.[34]

These valves not only demonstrate resilience against calcification and thrombogenicity but also offer a material platform that mimics the natural properties of the human heart valve.[35] The biocompatibility of bioceramic valves facilitates seamless integration with the surrounding tissues, promoting a physiological response that contributes to long-term functionality and patient well-being.

7.7 Challenges and Future Directions

As bioceramics continue to emerge as a prime candidate in cardiovascular implant technology, it is a requirement of the current scientific field to acknowledge and address the existing challenges towards future advancements. One of the current challenges in the manufacturing of bioceramic-based cardiovascular implants involves intricate and precise manufacturing techniques. The precision required in manufacturing these implants to meet the stringent health and safety and the demand of the cardiovascular environment often comes with increased complexity. Traditional manufacturing methods pose limitations in achieving the intricate shapes/structures and tailored designs essential for optimal implant performance. Balancing the need for precision with efficient manufacturing processes remains a challenge. Research should focus on streamlining manufacturing techniques and developing innovative methods that maintain precision while enhancing efficiency. The integration of advanced robotics and computer-aided manufacturing holds promise in mitigating this challenge, allowing for the creation of intricate bioceramic structures with accuracy.

Cost consideration is another bottleneck in the widespread adoption of bioceramic-based cardiovascular implants. The unique material properties and manufacturing complexities often translate into higher production costs. Addressing cost barriers is crucial for these implants to make a significant impact. Future research should explore avenues for cost reduction without compromising the quality and efficacy of the implants. Investigating alternative raw materials, optimising manufacturing processes, and encouraging economies of scale are potential strategies to make bioceramic-based cardiovascular implants more accessible to a broader patient population. Collaborations between researchers, manufacturers, and regulatory bodies will be central in establishing a framework that ensures affordability without compromising on safety and performance.

The integration of cutting-edge technologies, such as additive manufacturing, commonly known as 3D printing, presents an exciting opportunity for precise, customised implant manufacturing.[36] This technology allows for the layer-by-layer fabrication of intricate structures, opening new possibilities for personalised implants tailored to individual patient anatomy. Simultaneously, nanotechnology offers the potential to enhance the properties of bioceramics at the molecular level, improving their mechanical strength, biocompatibility, and drug delivery capabilities.[37,38] Research efforts should, therefore, be directed towards optimising these technologies for cardiovascular applications, ensuring seamless integration into clinical practice.

7.8 Conclusion

The intrinsic properties of bioceramics, from biocompatibility and mechanical strength to the potential for tailored design and innovative applications, collectively underline the huge impact these materials can have on cardiovascular care. The key findings of this chapter emphasise the multifaceted nature of bioceramics and their applications in cardiovascular implants. Bioceramics, ranging from alumina and zirconia to hydroxyapatite and bioactive glasses, have exhibited remarkable biocompatibility, providing a compatible interface with cardiovascular tissues. Their mechanical properties, highlighted by strength, fracture toughness, and fatigue resistance, position bioceramics as strong contenders for durable implant applications. Real-world applications in stents, vascular grafts, and heart valves have showcased their potential to mitigate challenges, such as restenosis, thrombogenicity, and calcification, significantly enhancing patient outcomes.

The potential of bioceramics to revolutionise the cardiovascular industry cannot be overstated. These materials not only offer solutions to existing challenges but also hold promise for innovation in personalised medicine. The ability to tailor designs, coupled with advancements in manufacturing techniques and emerging technologies, positions bioceramics at the forefront of a transformative shift in how we approach and execute cardiovascular procedures. The promise of reduced complications, enhanced durability, and improved patient quality of life is a testament to the paradigm shift that bioceramics bring to the field.

References

1. L. Fomundam and J. Lin, in *Design of miniaturized high frequency printed coils for wireless power transfer to biomedical implants*, 2016 Mediterranean Microwave Symposium (MMS), 2016, pp. 1–4.
2. J. S. Ho, A. J. Yeh and E. Neofytou, *et al.*, Wireless power transfer to deep-tissue microimplants, *Proc. Natl. Acad. Sci.*, 2014, **111**, 7974–7979.
3. Y. H. Joung, Development of implantable medical devices: from an engineering perspective, *Int. Neurourol. J.*, 2013, **17**, 98–106.
4. K. Scholten and E. Meng, Materials for microfabricated implantable devices: a review, *Lab Chip*, 2015, **15**, 4256–4272.
5. M. Brovold, J. I. Almeida and I. Pla-Palacín, *et al.*, Naturally-Derived Biomaterials for Tissue Engineering Applications, in *Novel Biomaterials for Regenerative Medicine*, ed. H. J. Chun, K. Park, C.-H. Kim and G. Khang, Springer, Singapore, 2018, pp. 421–449.
6. H. Fischer, M. Luk, B. Oedekoven, R. Telle and K. Mottaghy, Hemocompatibility of high strength oxide ceramic materials: An in vitro study, *J. Biomed. Mater. Res., Part A*, 2007, **81A**, 982–986.

7. Z. V. Parlak, S. Wein and R. Zybała, *et al.*, High-strength ceramics as innovative candidates for cardiovascular implants, *J. Biomater. Appl.*, 2019, **34**, 585–596.
8. G. Anneroth, A. R. Ericsson and L. Zetterqvist, Tissue integration of A12O3-ceramic dental implants (Frialit)—a case report, *Swed. Dent. J.*, 1990, **14**, 63–70.
9. G. Heimke and P. Griss, Ceramic implant materials, *Med. Biol. Eng. Comput.*, 1980, **18**, 503–510.
10. H. Ismail and H. Mohamad, Bioactivity and Biocompatibility Properties of Sustainable Wollastonite Bioceramics from Rice Husk Ash/Rice Straw Ash: A Review, *Materials*, 2021, **14**(18), 5193.
11. L. Vaiani, A. Boccaccio, A. E. Uva, G. Palumbo, A. Piccininni, P. Guglielmi, S. Cantore, L. Santacroce, I. A. Charitos and A. Ballini, Ceramic Materials for Biomedical Applications: An Overview on Properties and Fabrication Processes, *J. Funct. Biomater.*, 2023, **14**(3), 146.
12. P. Kumar, M. Oka and K. Ikeuchi, *et al.*, Low wear rate of UHMWPE against zirconia ceramic (Y-PSZ) in comparison to alumina ceramic and SUS 316L alloy, *J. Biomed. Mater. Res.*, 1991, **25**, 813–828.
13. N. Eliaz and N. Metoki, Calcium Phosphate Bioceramics: A Review of Their History, Structure, Properties, Coating Technologies and Biomedical Applications, *Materials*, 2017, **10**, 334.
14. H. Yoshikawa, N. Tamai, T. Murase and A. Myoui, Interconnected porous hydroxyapatite ceramics for bone tissue engineering, *J. R. Soc., Interface*, 2009, **6**, S341–S348.
15. H. Lu, Y. Zhou, Y. Ma, L. Xiao, W. Ji, Y. Zhang and X. Wang, Current Application of Beta-Tricalcium Phosphate in Bone Repair and Its Mechanism to Regulate Osteogenesis, *Front. Mater.*, 2021, **8**, DOI: 10.3389/fmats.2021.698915.
16. D. U. Tulyaganov, S. Agathopoulos and P. Valerio, *et al.*, Synthesis, bioactivity and preliminary biocompatibility studies of glasses in the system CaO-MgO-SiO2-Na2O-P2O5-CaF2, *J. Mater. Sci.: Mater. Med.*, 2011, **22**, 217–227.
17. M. B. Gorbet and M. V. Sefton, Biomaterial-associated thrombosis: roles of coagulation factors, complement, platelets and leukocytes, *Biomaterials*, 2004, **25**, 5681–5703.
18. C. Werner, M. F. Maitz and C. Sperling, Current strategies towards hemocompatible coatings, *J. Mater. Chem.*, 2007, **17**, 3376–3384.
19. C. Zhao, L. Xia and D. Zhai, *et al.*, Designing ordered micropatterned hydroxyapatite bioceramics to promote the growth and osteogenic differentiation of bone marrow stromal cells, *J. Mater. Chem. B*, 2015, **3**, 968–976.
20. X. Zhang, F. Jiang, T. Groth and K. S. Vecchio, Preparation, characterization and mechanical performance of dense β-TCP ceramics with/without magnesium substitution, *J. Mater. Sci.: Mater. Med.*, 2008, **19**, 3063–3070.

21. J. M. Grunkemeier, W. B. Tsai, C. D. McFarland and T. A. Horbett, The effect of adsorbed fibrinogen, fibronectin, von Willebrand factor and vitronectin on the procoagulant state of adherent platelets, *Biomaterials*, 2000, **21**, 2243–2252.
22. A. S. Rizkalla and D. W. Jones, Mechanical properties of commercial high strength ceramic core materials, *Dent. Mater.*, 2004, **20**, 207–212.
23. R. O. Ritchie, Toughening materials: enhancing resistance to fracture, *Phil. Trans. R. Soc., A*, 2021, **379**, 20200437.
24. G. Maccauro, A. Cittadini and G. Magnani, *et al.*, In Vivo Characterization of Zirconia Toughened Alumina Material: A Comparative Animal Study, *Int. J. Immunopathol. Pharmacol.*, 2010, **23**, 841–846.
25. A. Festas, A. Ramos and J. Davim, Medical devices biomaterials – A review, *Proc. Inst. Mech. Eng., Part L*, 2020, **234**, 218–228.
26. A. Llama-Palacios, M. C. Sánchez and L. A. Díaz, *et al.*, In vitro biofilm formation on different ceramic biomaterial surfaces: Coating with two bactericidal glasses, *Dent. Mater.*, 2019, **35**, 883–892.
27. J. M. Cordell, M. L. Vogl and A. J. Wagoner Johnson, The influence of micropore size on the mechanical properties of bulk hydroxyapatite and hydroxyapatite scaffolds, *J. Mech. Behav. Biomed. Mater.*, 2009, **2**, 560–570.
28. A.-M. Pobloth, M. J. Mersiowsky and L. Kliemt, *et al.*, Bioactive coating of zirconia toughened alumina ceramic implants improves cancellous osseointegration, *Sci. Rep.*, 2019, **9**, 16692.
29. A. Marimuthu, M. Logesh, K. El Mabrouk and A. M. Ballamurugan, In vitro hemocompatibility studies on small-caliber stents for cardiovascular applications, *RSC Adv.*, 2023, **13**, 6793–6799.
30. B. V. S. S. Rao, Application of Bio-ceramics—Carbon Ceramics for Heart Valve Prosthesis, *Trans. Indian Ceram. Soc.*, 1980, **39**, 45–51.
31. B. Taghizadeh, L. Ghavami, H. Derakhshankhah, E. Zangene, M. Razmi, M. Jaymand, P. Zarrintaj, N. Zarghami, M. R. Jaafari, M. Moallem Shahri, A. Moghaddasian, L. Tayebi and Z. Izadi, Biomaterials in Valvular Heart Diseases, *Front. Bioeng. Biotechnol.*, 2020, **8**, DOI: 10.3389/fbioe.2020.529244.
32. Z. Shao, T. Tao and H. Xu, *et al.*, Recent progress in biomaterials for heart valve replacement: Structure, function, and biomimetic design, *VIEW*, 2021, **2**, 20200142.
33. Q. Han, S. M. Shea, T. Arleo, J. Y. Qian and D. N. Ku, Thrombogenicity of biomaterials depends on hemodynamic shear rate, *Artif. Organs*, 2022, **46**, 606–617.
34. C. Sperling, M. Fischer, M. F. Maitz and C. Werner, Blood coagulation on biomaterials requires the combination of distinct activation processes, *Biomaterials*, 2009, **30**, 4447–4456.
35. I. H. Jaffer, J. C. Fredenburgh, J. Hirsh and J. I. Weitz, Medical device-induced thrombosis: what causes it and how can we prevent it?, *J. Thromb. Haemostasis*, 2015, **13**, S72–S81.

36. S. Hassanajili, A. Karami-Pour, A. Oryan and T. Talaei-Khozani, Preparation and characterization of PLA/PCL/HA composite scaffolds using indirect 3D printing for bone tissue engineering, *Mater. Sci. Eng., C*, 2019, **104**, 109960.
37. M. Sternitzke, Structural ceramic nanocomposites, *J. Eur. Ceram. Soc.*, 1997, **17**, 1061–1082.
38. N. Garmendia, B. Olalde and I. Obieta, Biomedical applications of ceramic nanocomposites, in *Ceramic Nanocomposites*, ed. R. Banerjee and I. Mannal, Woodhead Publishing, 2013, ch. 16, pp. 530–547.

CHAPTER 8

Targeted Gold Nanotherapy for the Attenuation of Chemotherapy-associated Cardiotoxicity

F. MELIA[a], TANVEER A. TABISH[b] AND M. Z. I. PRANJOL*[a]

[a]Department of Biochemistry and Biomedicine, School of Life Sciences, University of Sussex, Brighton, UK; [b]Division of Cardiovascular Medicine, Radcliffe Department of Medicine, University of Oxford, Oxford, UK
*E-mail: z.pranjol@sussex.ac.uk

8.1 Introduction

Cancer is the second leading cause of death among noncommunicable diseases worldwide.[1] In 2020, there were approximately 19.3 million new cases and 10 million cancer deaths globally, among which the highest incident and most lethal cancer was lung cancer for males and breast cancer for females.[2] In the UK, there are around 375 000 new cancer cases per year (1000 per day) and 167 000 cancer deaths every year (460 per day). In the last 40 years, cancer survival has doubled, and nowadays, 50% of cancer patients survive for 10 years or more.[3] Despite the improvements in prognosis within the last few decades, the quality of life and life expectancy of patients treated with chemotherapy are endangered by the development of systemic tissue damage and deterioration of major organs, such

Nanoscience & Nanotechnology Series No. 64
Cardiovascular Nanomedicine
Edited by Roger Narayan and Tanveer Tabish
© The Royal Society of Chemistry 2024
Published by the Royal Society of Chemistry, www.rsc.org

as the heart, liver and kidneys.[4] The occurrence of these adverse effects is attributed to the lack of selectivity of chemotherapeutic drugs against cancer, resulting in the systemic action of these cytotoxic drugs.[5] Among chemotherapy adverse effects, cardiotoxicity is of particular importance as it represents one major cause of chemotherapy-associated mortality. For instance, Bradshaw *et al.* showed that breast cancer survivors treated with chemotherapy present increased cardiovascular disease mortality ratios compared to untreated breast cancer survivors.[6] In addition, Armstrong *et al.* showed that the incidence rate of cardiac events in cancer survivors was approximately 10-fold higher than in their healthy counterparts, and cancer survivors have increased risk of developing cardiac events [hazard ratio (HR): 6.9; 95% CI: 5.2–9.1 *versus* HR: 1.0].[7] The development of cardiovascular complications in cancer patients hinders the benefits of cancer therapy. As a result, cardiologists and oncologists have been collaborating for the management of cardio-oncological comorbidities, leading to the establishment of the field named cardio-oncology.[8]

Cardiotoxicity leads to life-threatening complications, such as arrhythmia, dilated cardiomyopathy, myocardial infarction and heart failure.[9,10] These complications are observed in patients treated with anthracyclines (*e.g.* doxorubicin), alkylating agents (*e.g.* cisplatin), antimetabolites (*e.g.* fluorouracil), taxanes (*e.g.* paclitaxel), tyrosine kinase inhibitors (*e.g.* lapatinib) and other drugs.[4,11–13] The mechanism of action that induces cardiac damage differs depending on the chemotherapeutic agent. For example, anthracyclines cause DNA damage, mitochondrial damage and oxidative stress; taxanes disrupt microtubule formation and some antimetabolites inhibit thymidylate synthase.[14] In addition to the health hazard posed on patients, cardiotoxicity also limits treatment options and testing of novel chemotherapeutic drugs. In fact, potentially effective drugs stop at phase 2 of clinical trials because of severe cardiac events occurring in patients (*e.g.* NCT00679029). Certainly, novel therapeutic approaches, such as nanomedicine, are required to minimise the systemic action of chemotherapeutic drugs while maintaining therapeutic efficacy.

Nanomedicine is an emerging therapeutic strategy in cancer therapy, as it offers several advantages over conventional chemotherapy.[15] In fact, the encapsulation of chemotherapeutic drugs in nano-sized carriers, or nanoparticles, restricts the drug cytotoxic action in healthy tissues, resulting in improved tumour selectivity and reduced adverse effects. Over the past decades, several studies have focussed on the nanoparticle competence in eliminating several types of tumours compared to conventional chemotherapy.[16] However, fewer studies have focussed on the ability of nanotherapy to attenuate cardiotoxicity while maintaining optimal therapeutic efficacy. Considering the effects of chemotherapy on the heart as aforementioned, mitigation of cardiotoxicity would improve the therapeutic benefits of cancer therapy by reducing the frequency of cardiac events in cancer patients. In addition, attenuation of cardiotoxicity would allow the testing of a wider range of novel drugs, therefore contributing to expediting the progress of cancer research. Among the several different types of nanoparticles, gold nanoparticles (GNPs) have been shown to effectively reduce chemotherapy-associated cardiotoxicity and enhance tumour cytotoxicity.

This review will discuss the efficacy of gold nanoparticle treatment in the attenuation of chemotherapy-associated cardiotoxicity. In addition, a novel combination therapy known as chemo-photothermal therapy (PTT) will be discussed.

8.2 Principles of Nanomedicine

Nanomedicine is the application of nano-sized carriers (0.1–100 nm), or nanoparticles, in medical imaging, diagnosis, prevention and treatment.[17] Nanoparticles have been used since ancient times. For example, the special optical effects of certain Roman glass artefacts were achieved by the presence of dispersed gold particles within the glass.[18] Nonetheless, nanoparticle usage in medicine for cancer drug delivery only began during the 1990s, and 5 years later, the first nano-formulated drug "Doxil" (liposomal doxorubicin) was approved by the Food and Drug Administration (FDA) for the treatment of Kaposi's sarcoma.[19,20] In the last decade, nanomedicine has shown promising outcomes in overcoming multiple challenges of cancer therapy: tumour selectivity is enhanced through the conjugation of nanoparticles with tumour-specific ligands, such as antibodies, that promote the localisation and release of chemotherapeutic drugs at the tumour site[21]; combination therapy to treat drug-resistant tumours is achieved *via* the encapsulation of multiple drugs within the same nanoparticle[22] and drug biocompatibility and pharmacokinetics are improved *via* enhanced nanoparticle tissue penetration and decreased opsonisation.[16,23]

Nowadays, several materials are used to synthesise nanoparticles. These include polymers [*e.g.* poly(lactic-*co*-glycolic acid) (PLGA)], metals (*e.g.* gold), liposomes, carbon and silica. Nanoparticles can be synthesised either by dividing a bulk material into nanostructures (top-down approach) or by using molecular precursors to construct nanostructures (bottom-up approach) (Figure 8.1A).[24] Each material determines different structural and functional nanoparticle properties.[16] For example, liposomal nanoparticles are excellent platforms for drug delivery due to their ability to encapsulate both hydrophilic and hydrophobic drugs. However, their application is limited due to their rapid clearance from the bloodstream and the induction of a strong immune response.[16,25] On the other hand, silica nanoparticles are suitable for gene delivery but can be toxic to the immune system.[26]

The route of nanoparticle administration depends on the nanoparticle material. For example, gold nanoparticles can be injected intravenously, polymeric nanoparticles are administered orally and liposomal

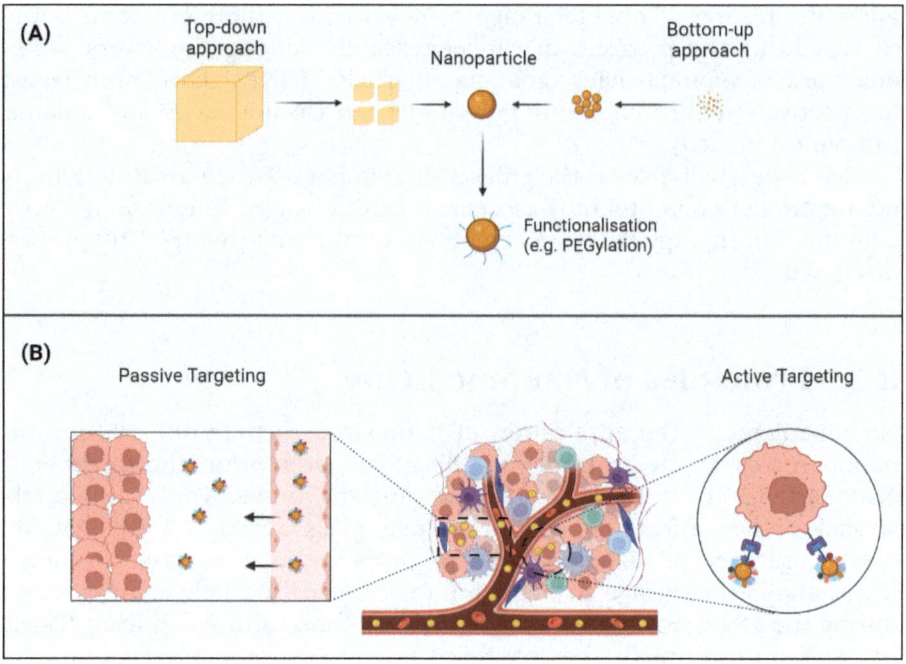

Figure 8.1 Nanoparticle synthesis and tumour-targeting strategies. (A) Nanoparticles are synthesised *via* the top-down or bottom-up approaches. The top-down approach consists of the fragmentation of a starting bulk material into nanoparticles, while the bottom-up approach consists of the generation of nanoparticles from molecular precursors. (B) Nanoparticle tumour targeting is achieved *via* two main strategies, known as passive and active targeting. Passive targeting exploits the leakiness of tumour microvasculature to localise the nanoparticles at the tumour site, while active targeting consists of the ligand-mediated nanoparticle localisation at the tumour site. This figure was created with BioRender.com.

nanoparticles are introduced *via* transdermal administration.[27,28] After the nanoparticles enter the bloodstream, their biological fate depends on the complex interaction with the bloodstream proteins. The set of proteins binding the nanoparticles is known as protein corona. The protein corona is involved in mediating nanoparticle clearance from the immune system, therefore affecting nanoparticle bioavailability. In addition, the protein corona can affect nanoparticle function through its physical destabilisation.[29] For this reason, nanoparticle drug delivery is often enhanced *via* chemical modifications that alter the nanoparticle interaction with the protein corona. This process is known as functionalisation and consists of the addition of functional groups, various peptides and other molecules to the nanoparticle surface.[30] Functionalisation results in improved nanoparticle bioavailability, uptake into the cell and extracellular and intracellular drug release.[31] A common functionalisation is the addition of polyethylene glycol (PEG), which improves systemic circulation time and decreases immunogenicity as well as clearance by macrophages.[32,33]

Nanoparticle tumour targeting is achieved *via* two main strategies, known as passive and active targeting (Figure 8.1B). Passive targeting exploits the characteristics of the tumour vasculature: due to the rapid vascularisation induced by the tumour to cope with the lack of nutrients, the newly formed blood vessels and junctions are undeveloped and become leaky. Thus, the blood vessel permeation is enhanced. The enhanced permeation also determines an increase in the amount of interstitial fluid, which leads to the reduced drainage capability of the lymphatic system. Consequently, the extracellular fluid is retained for a prolonged time in the interstitium. This phenomenon is known as the enhanced permeation and retention (EPR) effect. The EPR effect determines an increase in the nanoparticle amount that reaches the tumour and a prolonged tumour exposure to the nanoparticles. However, because the EPR effect depends on the physical characteristics of the tumour microvasculature, the permeability and retention are different across different types of tumours.[34] Active targeting consists of the conjugation of nanoparticles with a tumour-binding ligand, with the purpose of localising the nanoparticle at the tumour site.[35–37] For example, the transferrin receptor is overexpressed in malignant cells compared to normal cells, and therefore, nanoparticles can be conjugated with transferrin.[38] However, active targeting is limited by the possible expression of the target ligand in non-cancerous cells, which would result in the nanoparticle action on healthy tissues.[39]

8.3 Gold Nanomedicine Attenuates Chemotherapy-associated Cardiotoxicity

GNPs are metallic particles of various morphologies: nanospheres, nanocubes, nanoshells, nanoflowers, nanostars, nanorods and nanocages.[40,41] These are characterised by small size-to-volume ratio, thermal stability, low toxicity and simple synthesis.[42,43] These properties designate GNPs as a versatile tool in biomedicine for both disease diagnosis and treatment. In cancer therapy, GNPs are vastly used as a carrier for drug and gene delivery. This section will evaluate the research that explored the ability of the gold nanoparticle to enhance tumour cell death and reduce cardiotoxicity.

Gold nanotherapy has been shown to treat both liquid and solid tumours, including lymphomas, sarcomas and carcinomas. Arunkumar *et al.* showed that GNPs conjugated with doxorubicin (Dox), referred to as D-GNPs, induce *in vivo* cell death of Dalton lymphoma ascites (DLA) and mitigate cardiotoxicity. Specifically, DLA-bearing Swiss albino mice were treated with either free Dox or GNPs at a concentration of 5 mg kg^{-1} for 15 days. After treatment, the DLA tumour cell viability in both treatment groups was reduced by approximately 2-fold compared to the untreated control ($p < 0.001$). Although the D-GNPs achieved tumour toxicity similar to free Dox administration, the cardiotoxicity in mice treated with D-GNPs was reduced compared to mice treated with free Dox. In fact, heart histological analysis revealed that the heart tissue from mice treated with free Dox presented myofibrillar loss, interstitial oedema and cytoplasmic vacuolation, while the heart tissue from mice treated with D-GNPs presented no visible damage. These observations were also consistent with the mice electrocardiographic evaluation: the electrocardiogram (ECG) of mice treated with free Dox presented decreased heart rate, increased respiratory rate (RR) interval and increased ST interval compared to the untreated control ($p < 0.001$). This suggests that these mice developed arrhythmia and left ventricular dysfunction. On the other hand, the ECG of mice treated with D-GNPs presented values similar to the control, although slightly decreased heart rate as well as slightly increased RR and ST intervals were observed ($p < 0.05$). This minimal ECG difference between the untreated control and the D-GNPs treatment group may suggest that some D-GNPs still localise at the heart. However, because no cardiac damage is observed, it is probable that the number of localised D-GNPs was scarce to impact the heart function.[44]

Dhamecha *et al.* showed that D-GNPs enhance *in vivo* fibrosarcoma cytotoxicity and reduce cardiotoxicity. Specifically, fibrosarcoma-bearing mice were treated with either free Dox or D-GNPs at a concentration of 2.5 mg kg^{-1} for 60 days. After treatment, the percentage of tumour burden in mice treated with D-GNPs was reduced by 2-fold compared to mice treated with free Dox ($p < 0.001$) and by 10-fold compared to the untreated control ($p < 0.001$). Cardiotoxicity in these mice was then assessed *via* measurements of the creatine kinase-myocardial band (CK-MB) levels in blood plasma. CK-MB is a creatine kinase isoenzyme present in the myocytes and released into the bloodstream upon myocardial necrosis. Therefore, CK-MB is a reliable biomarker of cardiac damage.[45,46] Importantly, CK-MB plasma levels in mice treated with free Dox were 4 times higher compared to the untreated control ($p < 0.001$). On the other hand, CK-MB plasma levels in mice treated with GNPs were decreased by 2-fold compared to mice treated with free Dox but slightly more elevated compared to the control ($p < 0.05$).[47] Thus, minor levels of cardiotoxicity are still observed in the D-GNP treatment group, suggesting that some D-GNPs localise in the heart.

Samra *et al.* showed that D-GNPs enhance the *in vivo* cytotoxicity of Ehrlich ascites carcinoma (EAC). Specifically, EAC-bearing Swiss female albino mice were treated with either free Dox or D-GNPs at a concentration of 2 mg kg^{-1} for a 21-day regimen. After treatment, the tumour volume in mice treated with D-GNPs was moderately reduced compared to mice treated with free Dox and decreased by 3-fold compared to the untreated control. Cardiotoxicity in these mice was then evaluated *via* heart histological analysis. The heart tissue from mice treated with free Dox presented severe degeneration and necrosis of the cardiomyocytes, while the heart tissue from mice treated with D-GNPs showed reduced necrosis. Thus, as observed in the previous experiments, minor cardiotoxicity is still observed.[48]

Peng *et al.* showed that rapid renal-clearable D-GNPs are cleared more rapidly and accumulate less in the heart compared to free Dox treatment. Specifically, female CD-1 mice were treated with either free Dox or D-GNPs at a concentration of 5 mg kg^{-1}; 24 hours after treatment, there was minimal Dox accumulation in the heart of mice treated with GNPs, and these levels were reduced by 7-fold compared to the Dox levels in the heart of mice treated with free Dox ($p < 0.01$).[49]

The aforementioned experiments showed encouraging outcomes in the cardio-oncological field. In fact, the attenuation of cardiotoxicity and the enhancement of tumour cell death suggest that gold nanomedicine achieves improved tumour localisation compared to conventional chemotherapy (Figure 8.2). However, there are some limitations that need to be considered. First, minor cardiotoxicity is still observed, suggesting that GNPs are not entirely selective. In addition, the Dox concentration used in these experiments is considerably lower than the Dox human equivalent dose known to cause myocardial damage (300–450 mg m^{-2}).[50–52] Thus, when a higher dosage is required for more aggressive cancers, it is unknown whether the nanoparticle treatment would mitigate cardiotoxicity. An additional limiting factor is immunogenicity: GNPs have been shown to elicit an immune response *via* the interaction with immune cell receptors such as toll-like receptors, resulting in an inflammatory response and potential systemic tissue damage.[53,54] Nanoparticle opsonisation can also occur, leading to nanoparticle aggregation into bigger particles, which can potentially occlude smaller blood vessels and cause ischaemia.[55] Immune system activation by GNPs has been shown to occur in a size-dependent manner, and therefore, immunogenicity represents a limiting factor for the design of bigger nanoparticle formulations.[56] Lastly, enhanced endothelial cell penetration by GNPs may facilitate metastasis by inducing the formation of micrometre-sized gaps in the endothelial wall of blood vessels. Peng *et al.* showed that gold nanoparticles disrupt the integrity of human mammary microvascular endothelial cells (HMMECs), determining the translocation of the molecule serum fluorescein isothiocyanate dextran (FITC-d), which is not able to cross the epithelial barrier in normal physiological conditions.[57,58] Specifically, the levels of the translocated FITC-d across the HMMECs treated with GNPs were increased by 3-fold compared to the control ($p = 0.0006$).[58] The nanoparticle-induced endothelial leakiness is exploited by cancer cells to metastasise to other regions. In fact, the number of MCF-7 cells that migrated across the barrier of human mammary vascular endothelial cells (HMVECs) treated with gold nanoparticles was increased by 3-fold compared to the control ($p < 0.05$). Analogous results are observed *in vivo*: after intravenous injection of MDA-MB-231 cells in female NSG mice, the MDA-MB-231 cell amount in the liver and

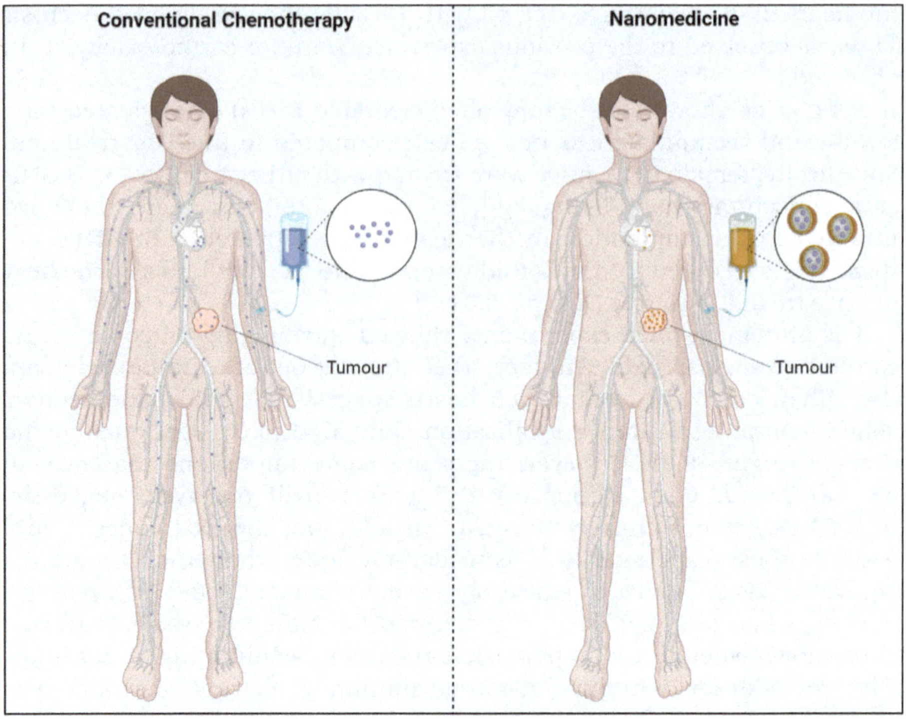

Figure 8.2 Mitigation of chemotherapy-associated cardiotoxicity by gold nanomedicine. Conventional chemotherapy consists of the intravenous administration of non-specific chemotherapeutic drugs, resulting in the systemic action of these drugs. By contrast, gold nanomedicine consists of the encapsulation of the chemotherapeutic drugs in gold nanoparticles, resulting in the accumulation of the drug at the tumour site and reduced accumulation in the heart. This figure was created with BioRender.com.

lungs of mice treated with GNPs was 2-fold higher compared to the control ($p < 0.05$).[58] These results support the idea that nanoparticle treatment favours tumour extravasation. Thus, the attempt to reduce cancer therapy side effects on the heart through the use of GNPs may indirectly favour cancer metastasis. This emphasises an additional level of complexity in cardio-oncological treatment.

8.4 Chemo-photothermal Therapy Enhances Gold Nanotherapy Efficacy

Gold nanotherapy has been shown to achieve improved tumour selectivity and cytotoxicity, reducing Dox accumulation in the heart. However, as previously mentioned, the increase in CK-MB serum levels and the presence of minor heart damage suggest that a portion of GNPs localises in the heart. As a matter of fact, controlling the biological fate of nanoparticles after administration is difficult, and therefore, selective drug delivery is not always achieved.[59] Thus, further selective therapies that improve the control of drug release are required.

PTT is a non-invasive combination therapy proposed for cancer treatment in the early 2000s.[60] PTT consists of the generation of local hyperthermia at the tumour site to induce tumour cell death.[61,62] This therapy consists of two major phases. First, light-sensitive nanoparticles (photothermal agents) are localised at the tumour site *via* ligand-mediated binding (active targeting). The nanoparticles enter the tumour cells *via* endocytosis and evade digestion by cancer cells *via* chemical modifications that favour lysosomal escape.[62–64] Then, electromagnetic radiation at a specific wavelength is applied at the tumour site to generate heat in a process known as photoconversion: upon irradiation, the light-sensitive nanoparticles absorb the photon energy and become excited. This energy is then converted into vibration when the nanoparticles return to their unexcited state. As a result, the kinetic energy of the nanoparticles and the surrounding molecules increases, leading to an increase in temperature locally.[65,66] Hyperthermia has been shown to reduce tumour cell viability around 40–45 °C.[67] However, this temperature alone may not be sufficient to completely eradicate a tumour.[68] As a consequence, in recent years, a novel treatment has been developed to combine PTT and nanoparticle-based chemotherapy in the same treatment. This treatment is known as photothermal chemotherapy or chemo-photothermal therapy (CPTT), and it is achieved through the use of nanoparticles that are both light-sensitive and drug-loaded (Figure 8.3). As a result, radiation can stimulate these nanoparticles to generate heat and release a chemotherapeutic drug within the same tumour cell. Importantly, CPTT introduces a mechanism to control drug release at a specific location: the need for local laser application at the tumour site to release the drug reduces the likelihood of drug release from nanoparticles that may have localised in other sites, such as the heart, therefore reducing potential adverse effects in healthy tissues.

GNPs are reliable photothermal agents: their surface electron cloud oscillations, known as localised surface plasmon, can resonate with the incident light applied to the nanoparticle. This phenomenon is known as surface plasmon resonance. As a result, GNPs have optimal photoconversion efficiency. In addition, modifications to the morphology of GNPs allow the surface plasma to resonate with light at the near-infrared (NIR) frequency.[69,70] The NIR range (750–1700 nm) is the most common type of radiation used in CPTT because it has the highest transmittance in body tissues and is unharmful to body tissues.[71,72] This range is further divided into the bands NIR-I (750–1000 nm) and NIR-II (1000–1700 nm). NIR-II presents reduced interaction with tissues, therefore achieving improved tissue penetration (3–5 cm) compared to NIR-I (1 cm).[72] The radiation power density is variable, but it usually lies between 0 and 35 W cm^{-2}.[73]

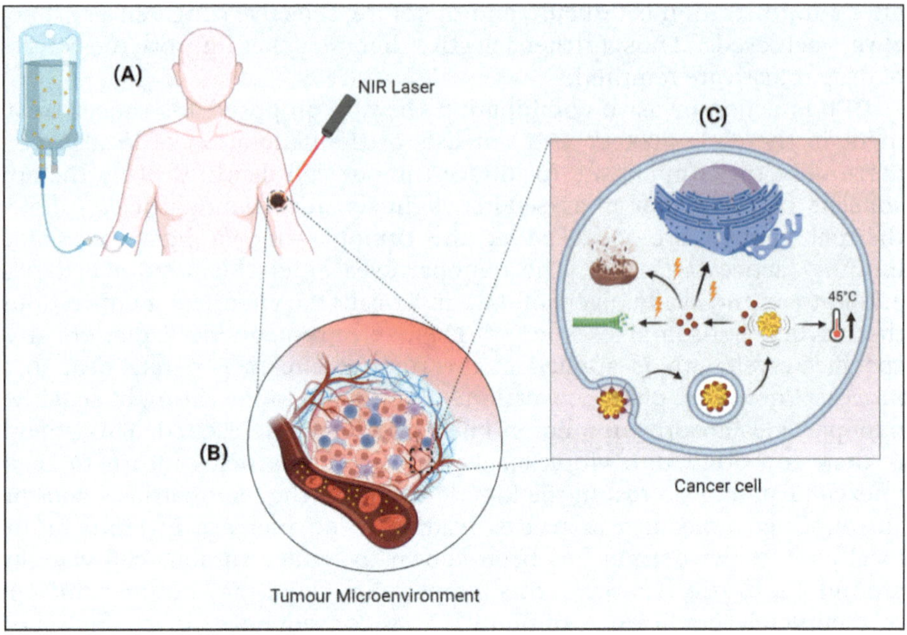

Figure 8.3 Chemo-photothermal therapy. (A) Photo-sensitive and drug-loaded gold nanoparticles are administered intravenously, and subsequently, a near-infrared laser is applied at the tumour site. (B) The injected gold nanoparticles travel in the bloodstream and localise at the tumour site *via* the tumour microvasculature. (C) Gold nanoparticles enter the cancer cell *via* endocytosis and avoid digestion *via* lysosomal escape. The localised surface plasmon of the gold nanoparticles resonates with the near-infrared irradiation surrounding the cell. Surface plasmon resonance is transduced in vibration, which determines an increase in cancer cell temperature to 45 °C or more. Gold nanoparticle vibration also determines the release of the chemotherapeutic drug, which causes microtubule formation disruption, mitochondrial damage and nuclear damage. This figure was created with BioRender.com.

CPTT has shown significant results in the treatment of solid tumours. Xing *et al.* showed that CPTT with liposome-coated GNPs increased toxicity against cervical carcinoma and minimised cardiotoxicity. U14 tumour-bearing Kunming mice were treated with either free Dox or Dox-loaded GNPs for chemo-photothermal therapy (CPTT-GNPs) at a concentration of 10 $\mu g\ mL^{-1}$. The tumour site in these mice was then irradiated at 2 $W\ cm^{-2}$ for 1 minute with an NIR laser (808 nm). After 1 minute, the tumour temperature of the CPTT-GNPs treatment group increased to 54.7 °C, while the tumour temperature of the control increased only up to 38.2 °C. The treatment was repeated every other day for 15 days. After 15 days, the tumour volume of mice treated with CPTT was reduced by 9-fold compared to the free Dox treatment group and by 19-fold compared to the untreated control ($p < 0.01$). Histological analysis revealed that the heart tissue from mice treated with CPTT presented no visible damage.[74]

Zhao *et al.* showed that CPTT with hollow mesoporous carbon-combined GNPs increased breast cancer toxicity and minimised cardiotoxicity. 4T1 tumour-bearing Kunming mice were treated with either free Dox or CPTT-GNPs at a concentration of 10 mg kg^{-1}. The tumour site in these mice was then irradiated at 1 W cm^{-2} for 3 minutes with an NIR laser (808 nm). After 3 minutes, the tumour temperature of the CPTT-GNPs treatment group increased to 46.9 °C, while the tumour temperature of the control increased only up to 37.2 °C. The treatment was repeated every other day for 17 days. After 17 days, the tumour weight of mice treated with CPTT was reduced by 7-fold compared to the free Dox treatment group and by 27-fold compared to the untreated control ($p < 0.01$). Histological analysis in these mice revealed that the heart tissue from mice treated with free Dox presented myocardial haemorrhage, while the heart tissue from mice treated with CPTT showed no evident pathological change.[75]

Nair *et al.* showed that CPTT with hollow GNPs enhanced EAC cytotoxicity and minimised cardiotoxicity. EAC-bearing mice were treated with either free Dox or CPTT-GNPs at a concentration of 4 mg kg^{-1}. The tumour site in these mice was then irradiated at 0.5 W cm^{-2} for 3 minutes with an NIR laser (808 nm). The treatment was repeated approximately every 3 days for a total of 5 irradiations. After 31 days, the tumour weight of mice treated with CPTT was reduced by approximately 2-fold compared to the free Dox treatment group and by 3-fold compared to the untreated control ($p < 0.001$). Histological analysis in these mice revealed that the heart tissue from mice treated with free Dox presented lesions, while the heart tissue from mice treated with CPTT showed no visible damage.[76]

Overall, CPTT showed remarkable results in cardio-oncological nanotherapy due to the ability to selectively target the tumour, control drug release and treat the tumour both thermally and pharmacologically. However, the major constraint of CPTT is the limited NIR wave penetration into body tissues, and therefore, the treatment success depends on the anatomical location of the tumour.[77] With regard to brain tumours, NIR waves have been shown to partly cross the skull and reach the brain, but large GNPs designed as drug carriers may not be able to cross the blood–brain barrier.[78] Thus, further research aiming to improve CPTT efficacy in deep tumours is required.

8.5 Future Perspectives

Further research is required to assess the effect of gold nanotherapy on endothelial dysfunction. Endothelial dysfunction has been associated with several cardiovascular conditions, such as hypertension, atherosclerosis and heart failure.[79] In addition, Franses *et al.* showed that dysfunctional endothelial cells promote cancer inflammation and contribute to the metastatic process.[80] Multiple nanoparticles, such as zinc oxide, silica and carbon, have been shown to cause endothelial dysfunction.[81–83] Although GNPs have been shown to contribute to the normalisation of the endothelial barrier function in mouse atherosclerotic plaques, Liu *et al.* demonstrated that 20 nm bare gold nanoparticles caused a 50% increase in endothelial barrier permeability and slowed the recovery of the barrier function after the removal of GNPs. However, this was not observed with nanoparticles of 5 and 60 nm, suggesting the possibility of a size-dependent interaction of GNPs with the endothelial barrier.[84,85] Because nanoparticles are administered intravenously, the first cells they encounter are endothelial cells. Thus, considering the impact that endothelial dysfunction has on the organism, understanding the action of GNPs on the endothelial barrier is crucial to preventing secondary treatment complications.

An additional area to explore is the application of gold nanomedicine in the detection of tumour extracellular vesicles (EVs). Tumour EVs have been shown to promote immune system suppression, induce fibroblast transformation into cancer-associated fibroblasts, reprogramme cancer metabolism, promote drug resistance, induce tumour invasion and restructure vasculature.[86] Importantly, circulating EVs have been shown to transport molecules, such as growth factors and RNAs, to other sites of the body to prepare a favourable tumour microenvironment for the generation of a metastatic niche.[87] GNPs can be designed to label tumour EVs with the purpose to trace exosome activity and preferential sites of uptake. Lara *et al.* showed that GNP-labelled EVs secreted from B16F10 melanoma cell lines preferentially accumulate in metastatic lung tumours compared to other regions in the body.[88] Therefore, targeting EVs with GNPs would allow the study of the phenotype of different tumours, their modulatory activity on the tumour microenvironment and their preferential metastatic region in the body. In addition, EVs could be used as a vehicle to deliver drug-loaded GNPs to metastatic niches, therefore disrupting metastasis formation.

8.6 Conclusions

Gold nanomedicine attenuates chemotherapy-associated cardiotoxicity and enhances tumour cytotoxicity. Conventional chemotherapy is known to cause cardiotoxicity due to the lack of drug selectivity, therefore hindering cancer treatment benefits. Nanomedicine prevents chemotherapy-associated cardiotoxicity through the encapsulation of chemotherapeutic drugs in nano-sized carriers, restricting the drug action on the heart and resulting in improved tumour selectivity. Among several nanomaterials, gold nanoparticles are a versatile tool for cancer treatment due to their small size-to-volume ratio, thermal stability, low toxicity and simple synthesis. Gold nanotherapy has been shown to mitigate cardiotoxicity and enhance tumour cytotoxicity in both solid and liquid tumours. Further improvements in therapeutic efficacy and cardiotoxicity minimisation are achieved through a novel combination therapy known as chemo-photothermal therapy. This therapy consists of the irradiation of gold nanoparticles with near-infrared light to stimulate the nanoparticles to generate heat and release chemotherapeutic drugs within the same tumour cell. Importantly, chemo-photothermal therapy introduces a mechanism to control drug release at the tumour site, which reduces the risk of drug release in healthy tissues. Chemo-photothermal therapy minimises cardiotoxicity and improves tumour cytotoxicity. However, this therapy is limited by the small NIR wave penetration into the body tissues. Thus, further research aiming to improve chemo-photothermal therapy efficacy in deep tumours and less accessible tumours, such as brain cancer, is required. Further research should also explore the effect of gold nanomedicine on endothelial dysfunction, allowing the prevention of the possibility of secondary treatment complications, such as hypertension and atherosclerosis. Lastly, an additional research area to explore is the application of gold nanomedicine in the detection of tumour extracellular vesicles. Considering that these vesicles are involved in several tumourigenic processes, targeting EVs with GNPs would allow the study of the phenotype of different tumours, their modulatory activity on the tumour microenvironment and their preferential metastatic region in the body.

References

1. W.H.O., *Global health estimates: Leading causes of death*, 2019, Available from: https://www.who.int/data/gho/data/themes/mortality-and-global-health-estimates/ghe-leading-causes-of-death.
2. H. Sung, *et al.*, Global Cancer Statistics 2020: GLOBOCAN Estimates of Incidence and Mortality Worldwide for 36 Cancers in 185 Countries, *Ca-Cancer J. Clin.*, 2021, **71**(3), 209–249.
3. Cancer Research UK, *Cancer Statistics for the UK*, Available from: https://www.cancerresearchuk.org/health-professional/cancer-statistics-for-the-uk#heading-Two.
4. W. M. C. van den Boogaard, D. S. J. Komninos and W. P. Vermeij, Chemotherapy Side-Effects: Not All DNA Damage Is Equal, *Cancers*, 2022, **14**(3), 627.
5. H. Maeda and M. Khatami, Analyses of repeated failures in cancer therapy for solid tumors: poor tumor-selective drug delivery, low therapeutic efficacy and unsustainable costs, *Clin. Transl. Med.*, 2018, **7**(1), 11.
6. P. T. Bradshaw, *et al.*, Cardiovascular Disease Mortality Among Breast Cancer Survivors, *Epidemiology*, 2016, **27**(1), 6–13.
7. G. T. Armstrong, *et al.*, Aging and Risk of Severe, Disabling, Life-Threatening, and Fatal Events in the Childhood Cancer Survivor Study, *J. Clin. Oncol.*, 2014, **32**(12), 1218–1227.
8. P. Lancellotti, *et al.*, Cardio-Oncology Services: rationale, organization, and implementation, *Eur. Heart J.*, 2019, **40**(22), 1756–1763.

9. H. Essa, *et al.*, Chemotherapy-Induced Arrhythmia - Underrecognized and Undertreated, *Am. J. Med.*, 2021, **134**(10), 1224–1231.e1.
10. J. Herrmann, Vascular toxic effects of cancer therapies, *Nat. Rev. Cardiol.*, 2020, **17**(8), 503–522.
11. L. Zhao and B. Zhang, Doxorubicin induces cardiotoxicity through upregulation of death receptors mediated apoptosis in cardiomyocytes, *Sci. Rep.*, 2017, **7**(1), 44735.
12. S. M. Swain, F. S. Whaley and M. S. Ewer, Congestive heart failure in patients treated with doxorubicin, *Cancer*, 2003, **97**(11), 2869–2879.
13. Y. Hu, *et al.*, Cisplatin-induced cardiotoxicity with midrange ejection fraction, *Medicine*, 2018, **97**(52), e13807.
14. V. Y. Chang and J. J. Wang, Pharmacogenetics of Chemotherapy-Induced Cardiotoxicity, *Curr. Oncol. Rep.*, 2018, **20**(7), 52.
15. J. Shi, *et al.*, Cancer nanomedicine: progress, challenges and opportunities, *Nat. Rev. Cancer*, 2017, **17**(1), 20–37.
16. S. Gavas, S. Quazi and T. M. Karpiński, Nanoparticles for Cancer Therapy: Current Progress and Challenges, *Nanoscale. Res. Lett.*, 2021, **16**(1), 173.
17. R. A. Freitas Jr., *Nanomedicine*, Volume I: Basic Capabilities, Landes Bioscience, Georgetown, TX, 1999, http://www.nanomedicine.com/NMI.htm.
18. D. J. Barber, An investigation of the origin of the colour of the Lycurgus Cup by analytical transmission electron microscopy, *Archaeometry*, 1990, **32**(1), 35–45.
19. J. S. James, DOXIL approved for KS, *AIDS Treat. News*, 1995(236), 6.
20. M. Krukemeyer, V. Krenn and F. Huebner, History and possible uses of nanomedicine based on nanoparticles and nanotechnological progress, *J. Nanomed. Nanotechnol.*, 2015, **06**(6), 1–7.
21. K. Sztandera, M. Gorzkiewicz and B. Klajnert-Maculewicz, Gold Nanoparticles in Cancer Treatment, *Mol. Pharmaceutics*, 2019, **16**(1), 1–23.
22. M. Majidinia, *et al.*, Overcoming multidrug resistance in cancer: Recent progress in nanotechnology and new horizons, *IUBMB Life*, 2020, **72**(5), 855–871.
23. S. Ravindran, *et al.*, Pharmacokinetics, Metabolism, Distribution and Permeability of Nanomedicine, *Curr. Drug Metab.*, 2018, **19**(4), 327–334.
24. N. Baig, I. Kammakakam and W. Falath, Nanomaterials: a review of synthesis methods, properties, recent progress, and challenges, *Mater. Adv.*, 2021, **2**(6), 1821–1871.
25. S. C. Semple, *et al.*, Immunogenicity and Rapid Blood Clearance of Liposomes Containing Polyethylene Glycol-Lipid Conjugates and Nucleic Acid, *J. Pharmacol. Exp. Ther.*, 2005, **312**(3), 1020.
26. L. Chen, *et al.*, The toxicity of silica nanoparticles to the immune system, *Nanomedicine*, 2018, **13**(15), 1939–1962.
27. A. S. Thakor, *et al.*, Gold nanoparticles: a revival in precious metal administration to patients, *Nano Lett.*, 2011, **11**(10), 4029–4036.

28. D. Chenthamara, S. Subramaniam, S. G. Ramakrishnan, S. Krishnaswamy, M. M. Essa, F.-H. Lin and M. W. Qoronfleh, Therapeutic efficacy of nanoparticles and routes of administration, *Biomater. Res.*, 2019, **23**(1), 20.
29. R. Rampado, S. Crotti, P. Caliceti, S. Pucciarelli and M. Agostini, Recent Advances in Understanding the Protein Corona of Nanoparticles and in the Formulation of 'Stealthy' Nanomaterials, *Front. Bioeng. Biotechnol.*, 2020, **8**, 166.
30. R. Mout, *et al.*, Surface functionalization of nanoparticles for nanomedicine, *Chem. Soc. Rev.*, 2012, **41**(7), 2539.
31. G. Sanità, B. Carrese and A. Lamberti, Nanoparticle Surface Functionalization: How to Improve Biocompatibility and Cellular Internalization, *Front. Mol. Biosci.*, 2020, **7**, 587012.
32. J. S. Suk, *et al.*, PEGylation as a strategy for improving nanoparticle-based drug and gene delivery, *Adv. Drug Delivery Rev.*, 2016, **99**, 28–51.
33. Y. Qie, *et al.*, Surface modification of nanoparticles enables selective evasion of phagocytic clearance by distinct macrophage phenotypes, *Sci. Rep.*, 2016, **6**(1), 26269.
34. S. M. Narum, *et al.*, Passive targeting in nanomedicine: fundamental concepts, body interactions, and clinical potential, in *Nanoparticles for Biomedical Applications*, ed. E. J. Chung, L. Leon and C. Rinaldi, Elsevier, 2020, ch. 4, pp. 37–53.
35. M. A. Swartz and M. E. Fleury, Interstitial Flow and Its Effects in Soft Tissues, *Annu. Rev. Biomed. Eng.*, 2007, **9**(1), 229–256.
36. T. D. Clemons, *et al.*, Distinction Between Active and Passive Targeting of Nanoparticles Dictate Their Overall Therapeutic Efficacy, *Langmuir*, 2018, **34**(50), 15343–15349.
37. Z. R. Goddard, *et al.*, Active targeting of gold nanoparticles as cancer therapeutics, *Chem. Soc. Rev.*, 2020, **49**(23), 8774–8789.
38. R. Bazak, *et al.*, Cancer active targeting by nanoparticles: a comprehensive review of literature, *J. Cancer Res. Clin. Oncol.*, 2015, **141**(5), 769–784.
39. W. C. Chen, A. X. Zhang and S.-D. Li, Limitations and niches of the active targeting approach for nanoparticle drug delivery, *Eur. J. Nanomed.*, 2012, **4**(2–4), 89–93.
40. X. Hu, Y. Zhang, T. Ding, J. Liu and H. Zhao, Multifunctional Gold Nanoparticles: A Novel Nanomaterial for Various Medical Applications and Biological Activities, *Front. Bioeng. Biotechnol.*, 2020, **8**, 990.
41. L. Fabris, Gold Nanostars in Biology and Medicine: Understanding Physicochemical Properties to Broaden Applicability, *J. Phys. Chem. C*, 2020, **124**(49), 26540–26553.
42. S. J. Amina and B. Guo, A Review on the Synthesis and Functionalization of Gold Nanoparticles as A Drug Delivery Vehicle, *Int. J. Nanomed.*, 2020, **15**, 9823–9857.

43. A.-L. Bailly, *et al.*, In vivo evaluation of safety, biodistribution and pharmacokinetics of laser-synthesized gold nanoparticles, *Sci. Rep.*, 2019, **9**(1), 12890.
44. P. Arunkumar, *et al.*, Near infra-red laser mediated photothermal and antitumor efficacy of doxorubicin conjugated gold nanorods with reduced cardiotoxicity in swiss albino mice, *Nanomedicine*, 2015, **11**(6), 1435–1444.
45. R. Jacob and M. Khan, Cardiac Biomarkers: What Is and What Can Be, *Indian J. Cardiovasc. Dis. Women WINCARS*, 2018, **3**(4), 240–244.
46. C. D. Cabaniss, Creatine Kinase, in *Clinical Methods: The History, Physical, and Laboratory Examinations*, ed. H. K. Walker, W. D. Hall and J. W. Hurst, Butterworth Publishers, a division of Reed Publishing, Boston, 1990.
47. D. Dhamecha, *et al.*, Doxorubicin loaded gold nanoparticles: Implication of passive targeting on anticancer efficacy, *Pharmacol. Res.*, 2016, **113**, 547–556.
48. Y. A. Samra, A. M. Abdelghany and R. A. Zaghloul, Polydatin gold nanoparticles potentiate antitumor effect of doxorubicin in Ehrlich ascites carcinoma-bearing mice, *J. Biochem. Mol. Toxicol.*, 2021, **35**(10), e22869.
49. C. Peng, J. Xu, M. Yu, X. Ning, Y. Huang, B. Du, E. Hernandez, P. Kapur, J. Hsieh and J. Zheng, Tuning the In Vivo Transport of Anticancer Drugs Using Renal - Clearable Gold Nanoparticles, *Angew. Chem.*, 2019, **58**(25), 8479–8483.
50. A. B. Nair and S. Jacob, A simple practice guide for dose conversion between animals and human, *J. Basic Clin. Pharm.*, 2016, **7**(2), 27.
51. Pfizer Medical Information, *DOXORUBICIN Indications and Usage*, Available from: https://www.pfizermedicalinformation.com/en-us/doxorubicin/indications-usage.
52. National Institute for Health and Care Excellence, *Doxorubicin hydrochloride*, 2022, Available from: https://bnf.nice.org.uk/drugs/doxorubicin-hydrochloride.
53. D. Rachmawati, *et al.*, Innate stimulatory capacity of high molecular weight transition metals Au (gold) and Hg (mercury), *Toxicol. In Vitro*, 2015, **29**(2), 363–369.
54. I. Durocher, *et al.*, Evaluation of the in vitro and in vivo proinflammatory activities of gold (+) and gold (-) nanoparticles, *Inflammation Res.*, 2017, **66**(11), 981–992.
55. B. Li and L. A. Lane, Probing the biological obstacles of nanomedicine with gold nanoparticles, *Wiley Interdiscip. Rev.: Nanomed. Nanobiotechnol.*, 2019, **11**(3), e1542.
56. S. Fallarini, *et al.*, Factors affecting T cell responses induced by fully synthetic glyco-gold-nanoparticles, *Nanoscale*, 2013, **5**(1), 390–400.

57. M. F. A. Baxter, *et al.*, Optimizing Fluorescein Isothiocyanate Dextran Measurement As a Biomarker in a 24-h Feed Restriction Model to Induce Gut Permeability in Broiler Chickens, *Front. Vet. Sci.*, 2017, **4**, 56.
58. F. Peng, *et al.*, Nanoparticles promote in vivo breast cancer cell intravasation and extravasation by inducing endothelial leakiness, *Nat. Nanotechnol.*, 2019, **14**(3), 279–286.
59. Y. F. Li and C. Chen, Fate and Toxicity of Metallic and Metal-Containing Nanoparticles for Biomedical Applications, *Small*, 2011, **7**(21), 2965–2980.
60. L. R. Hirsch, *et al.*, in *Nanoshell-mediated near infrared photothermal tumor therapy*, IEEE, 2003.
61. Medical Subject Headings, *Photothermal Therapy*, National Library of Medicine, 2021.
62. L. Zhao, *et al.*, Recent advances in selective photothermal therapy of tumor, *J. Nanobiotechnol.*, 2021, **19**(1), 335.
63. D. Manzanares and V. Ceña, Endocytosis: The Nanoparticle and Submicron Nanocompounds Gateway into the Cell, *Pharmaceutics*, 2020, **12**(4), 371.
64. I. Plaza-Ga, *et al.*, pH-triggered endosomal escape of pore-forming Listeriolysin O toxin-coated gold nanoparticles, *J. Nanobiotechnol.*, 2019, **17**(1), 108.
65. X. Li, *et al.*, Clinical development and potential of photothermal and photodynamic therapies for cancer, *Nat. Rev. Clin. Oncol.*, 2020, **17**(11), 657–674.
66. J. Li, *et al.*, Near infrared photothermal conversion materials: mechanism, preparation, and photothermal cancer therapy applications, *J. Mater. Chem. B*, 2021, **9**(38), 7909–7926.
67. B. K. Bhuyan, *Kinetics of Cell Kill by Hyperthermia1*, *Cancer Res.*, 1979, **39(6_part_2)**, 2277–2284.
68. X. Tang, *et al.*, Cancer cells resist hyperthermia due to its obstructed activation of caspase 3, *Rep. Pract. Oncol. Radiother.*, 2020, **25**(3), 323–326.
69. E. E. Bedford, *et al.*, Surface Plasmon Resonance Biosensors Incorporating Gold Nanoparticles, *Macromol. Biosci.*, 2012, **12**(6), 724–739.
70. W. Yang, *et al.*, Shape effects of gold nanoparticles in photothermal cancer therapy, *Mater. Today Sustainability*, 2021, **13**, 100078.
71. M. Myakishev-Rempel, *et al.*, A Preliminary Study of the Safety of Red Light Phototherapy of Tissues Harboring Cancer, *Photomed. Laser Surg.*, 2012, **30**(9), 551–558.
72. Y. Lyu, J. Li and K. Pu, Second Near - Infrared Absorbing Agents for Photoacoustic Imaging and Photothermal Therapy, *Small Methods*, 2019, **3**(11), 1900553.
73. T. Ahmad, *et al.*, Recent advances in combinatorial cancer therapy via multifunctionalized gold nanoparticles, *Nanomedicine*, 2020, **15**(12), 1221–1237.

74. S. Xing, *et al.*, Doxorubicin/gold nanoparticles coated with liposomes for chemo-photothermal synergetic antitumor therapy, *Nanotechnology*, 2018, **29**(40), 405101.
75. Q. Zhao, *et al.*, Gold nanoparticles modified hollow carbon system for dual-responsive release and chemo-photothermal synergistic therapy of tumor, *J. Colloid Interface Sci.*, 2019, **554**, 239–249.
76. J. B. Nair, *et al.*, Elucidating a Thermoresponsive Multimodal Photo-Chemotherapeutic Nanodelivery Vehicle to Overcome the Barriers of Doxorubicin Therapy, *ACS Appl. Mater. Interfaces*, 2020, **12**(39), 43365–43379.
77. G. Hong, *et al.*, Through-skull fluorescence imaging of the brain in a new near-infrared window, *Nat. Photonics*, 2014, **8**(9), 723–730.
78. V. Sokolova, *et al.*, Transport of ultrasmall gold nanoparticles (2nm) across the blood-brain barrier in a six-cell brain spheroid model, *Sci. Rep.*, 2020, **10**(1), 18033.
79. S. I. Bloom, *et al.*, Mechanisms and consequences of endothelial cell senescence, *Nat. Rev. Cardiol.*, 2023, **20**(1), 38–51.
80. J. W. Franses, *et al.*, Dysfunctional endothelial cells directly stimulate cancer inflammation and metastasis, *Int. J. Cancer*, 2013, **133**(6), 1334–1344.
81. K.-J. Chuang, *et al.*, Effects of zinc oxide nanoparticles on human coronary artery endothelial cells, *Food Chem. Toxicol.*, 2016, **93**, 138–144.
82. L. Feng, *et al.*, Silica nanoparticles trigger the vascular endothelial dysfunction and prethrombotic state via miR-451 directly regulating the IL6R signaling pathway, *Part. Fibre Toxicol.*, 2019, **16**(1), 16.
83. N. Majumder, *et al.*, Oxidized carbon black nanoparticles induce endothelial damage through C-X-C chemokine receptor 3-mediated pathway, *Redox Biol.*, 2021, **47**, 102161.
84. T. J. Beldman, *et al.*, Nanoparticle-Aided Characterization of Arterial Endothelial Architecture during Atherosclerosis Progression and Metabolic Therapy, *ACS Nano*, 2019, **13**(12), 13759–13774.
85. Y. Liu, *et al.*, Nanoparticle size-specific actin rearrangement and barrier dysfunction of endothelial cells, *Nanotoxicology*, 2017, **11**(7), 846–856.
86. Y. Liu, *et al.*, Exosomes and Their Role in Cancer Progression, *Front. Oncol.*, 2021, **11**, 639159.
87. F. Wendler, *et al.*, Extracellular vesicles swarm the cancer microenvironment: from tumor-stroma communication to drug intervention, *Oncogene*, 2017, **36**(7), 877–884.
88. P. Lara, *et al.*, Gold nanoparticle based double-labeling of melanoma extracellular vesicles to determine the specificity of uptake by cells and preferential accumulation in small metastatic lung tumors, *J. Nanobiotechnol.*, 2020, **18**(1), 20.

CHAPTER 9

Extracellular Vesicles for the Treatment of Cardiovascular Diseases

IQRA AZHAR, SANA KARIM, AYESHA ISHTIAQ AND IRAM MURTAZA*

Signal Transduction Laboratory, Quaid-i-Azam University, Islamabad, 45320, Pakistan
*E-mail: irambch@qau.edu.pk

9.1 Classification of EVs

Extracellular vesicles (EVs) are divided into different subtypes, microvesicles (MVs), exosomes, and apoptotic bodies (APoBDs), based on their sizes, content, and composition.[1] Among the different populations of EVs, a remarkable difference in biophysical properties has been reported. Moreover, specific markers of APoBDs, exosomes, and MVs have also been discovered.[2]

9.1.1 Exosomes

Exosomes are small molecules with a size range of 40–150 nm, which are produced in living cells in the intercellular atmosphere by the process of exocytosis from multivesicular bodies.[3] The essential markers of exosomes are ALG-2-interacting protein, tetraspanins (CD63, CD81, and

Nanoscience & Nanotechnology Series No. 64
Cardiovascular Nanomedicine
Edited by Roger Narayan and Tanveer Tabish
© The Royal Society of Chemistry 2024
Published by the Royal Society of Chemistry, www.rsc.org

CD9), and tumor-susceptible gene101. They also express clathrin-dependent heat shock proteins (Hsp60, Hsp90, and Hsp70) and Alix.[4,5] Exosomes are enclosed by the bilayer membrane and function as biological vehicles to transfer the information between the donor and target cells and help their content travel a long distance in tissues.[6] Moreover, exosomes also contain lipids, including ceramides, cholesterol, sphingomyelin, phosphatidylserine, and glycosphingolipids. These lipid moieties are essential for exosome biogenesis, exosome morphology, and homeostasis regulation.[7] They can transfer genetic material, such as messenger RNA (mRNA), circular RNA (circRNA), microRNA (miRNA), and long non-coding RNA (lncRNA).[8] Exosomes interact with their target recipient cells through three pathways. They may transfer their material *via* adhesion at the surface of target cells, facilitated by the receptor–ligand interaction. Exosomes also enter the host cells either through endocytosis or by direct interaction of the exosome's vesicles with the cell membrane. This revealed that exosomes play an important role in cell–cell communication, physiological homeostasis, and pathophysiology.[9–11] Exosomes have considerable attention in clinical applications, such as therapeutic cargo exporters and diagnostic markers.[12] Their small size, reduced immunogenicity, and membrane composition protect the genetic material from degradation, allow them to cross the biological membrane, and make them useful as therapeutic vectors.[13]

9.1.2 Microvesicles

MVs are membrane-derived vesicles, also known as shedding vesicles, released by the budding of the membrane in healthy cells.[9] They have a heterogeneous size ranging from 0.1 to 1 μm in diameter.[14] They comprise phosphatidylserine proteins linked with lipid rafts, ceramide, cholesterol, and sphingomyelin and contain CD40 surface markers. MVs transfer lipids, microRNAs, mRNAs, and proteins and interact with the host recipient cells by receptor–ligand binding.[8] The majority of MVs are derived from blood components, including platelets, leukocytes, and erythrocytes.[15] They are also released from the endothelial cells (EC), which, upon activation, release the MVs representing E-selectin (CD62E).[16] MVs from the endothelial cells enhance angiogenesis *via* proangiogenic molecules. MVs have similar protein content to exosomes, including glycoproteins, integrins, and metalloproteinases.[17,18]

9.1.3 Apoptotic Bodies

APoBDs have a size range of 50–5000 nm and are only released from programmed cell death.[19] These vesicles are produced by different types of cells, including precursor cells, endothelial cells, osteoblasts, stem cells, and immunocytes.[20] Indeed, apoptotic extracellular vesicles (ApoEVs) play a significant role in the clearance of dead cells. Moreover, ApoEVs have "find me" and "eat me" signals for the clearance of dead cells and to prevent secondary necrosis.[21] The universal "eat me" signal is phosphatidylserine, while the "find me" signal is CX3CL1.[22] After the phagocytosis, ApoEVs may also transfer the signaling molecules, *i.e.* RNA, DNA, and protein, to the surrounding cells and control the cell differentiation, cell proliferation, and tissue regeneration. ApoEVs facilitate cell blebbing by coordinating many molecular patterns, such as heat shock protein 90 (Hsp90), mobility group box-1, and interleukin-33 (IL-33).[23] ApoEVs have specific markers (C3b, Tsb, and PS) that promote the function of cells by providing the necessary signals and affect the neighboring tissues through circulation.[24] ApoEVs stimulate the regenerative process and purify skin, kidneys, bones, muscles, and the cardiovascular system. ApoEVs attenuate the inflammation for smooth tissue regeneration by overwhelming the inflammatory cytokines, such as IL-6 and MCP-1, and upregulate the tumor necrosis factor (TNF)-stimulated gene-6 (TSG-6).[25] The most commonly used approaches to purify the ApoEVs are differential centrifugation and fluorescence-activated cell sorting.[26]

9.2 Biogenesis of Extracellular Vesicles

9.2.1 Exosome Biogenesis

Exosomes are released by eukaryotic and prokaryotic cells and control intracellular communication in diseased and healthy cells.[27,28] Exosomes are produced by the endosomal network system, which sorts several intraluminal vesicles and leads them to their proper destination, such as cell surface membranes and lysosomes. During this process, endosomes target innumerable protein and lipid molecules for lysosome breakdown while targeting some other molecules for exocytosis and recycling. Endosomes are divided into three discrete compartments: early, late, and recycling endosomes. Early endosomes incorporate their material with endocytic vesicles and destine them for exocytosis, degradation, and recycling. Recycling materials are dealt with recycling endosomes. Then, residual early endosomes experience several transformations and become late endosomes or multivascular bodies (MVBs).[29] MVB comprises the intraluminal vesicles (ILVs).[30] Successively, MVBs circulate to and fuse with the plasma membrane, discharging their ILV content into outer space.[31,32] After this process, intraluminal vesicles are referred to as exosomes. Several investigations have reported that the biogenesis of exosomes occurs through two pathways: the endosomal sorting complexes required for transport (ESCRT)-dependent pathway and ESCRT-independent pathway, and is highly regulated by multiple signaling cascades.[33–35] ESCRT has four protein complexes: ESCRT-0, ESCRT-I, ESCRT-II, ESCRT-III, and the associated AAA ATPase Vps4 complex.[36] The ESCRT-dependent mechanism originated from the recognition of ubiquitinated proteins on the endosomal membrane through the ESCRT-0 binding. It interacts with the

ESCRT-I and ESCRT-II units to form a complex, which combines with ESCRT-III and involves in the budding process. Lastly, the bud cleaves to produce the ILV.[37] Furthermore, the exosome protein Alix participates in membrane budding and also interacts with Syndecan for cargo selection.[38] Various protein molecules, including tetraspanins, also contribute to exosome synthesis. It is reported that tetraspanin-enriched domains with tetraspanin CD81 significantly categorize the intracellular components and receptors to exosomes.[39,40]

9.2.2 Microvesicle Biogenesis

The mechanism of biogenesis of MVs is less clear in comparison to exosome vesicles. Researchers showed that different mechanisms are responsible for the shedding of MVs. MVs are referred to as ubiquitous vesicles that are formed by the direct budding and fission of the plasma membrane.[41] Several factors participate in MV formation, including the contraction of actin and myosin and the redistribution of phosphatidylserine on the outer membrane.[42] However, the formation of the nascent MVs requires cholesterol, as its depletion affects the shedding of MVs. One of the mechanisms involved in the biosynthesis of MVs is the ESCART machinery. Knockdown of ESCRT proteins (Tsg101, Vsp4, Vps22, Alix, and Chmp1/3) significantly decreased the release of Hedgehog through EVs.[43] An investigation shows the recruitment of the ESCR-I subunit TSG101 on the membrane, where it binds with the tetrapeptide protein of the adaptor protein Arrestin-1 domain-containing protein-1 (ARRCD).[44] The formation of MVs requires the ubiquitination of ARRDC1 and interaction with the vacuolar protein sorting 4 (VPS4) ATPase attached to the E3 ligase (WWP2). As a result, MVs are released with ARRDC1, TSG101, and some other proteins.[44] Another mechanism involved in the synthesis of MVs is controlled by the ADP ribosylation factor 6, which is a member of the Ras GTPase family. They significantly regulate actin–myosin contraction and membrane trafficking. Interestingly, during MV shedding, the activity of ADP-ribosylation factor 6 (ARF6) is required for the activation of the phospholipase D and the extracellular signal-regulated kinase (ERK) recruitment toward the membrane. On the cell membrane, ERK causes the phosphorylation of the myosin light chain kinase (MLCK), which in turn phosphorylates MLCK and regulates the actin–myosin contractility and triggers the MV release.[45,46] These MVs are loaded with vesicle-associated membrane protein 3 (VAMP3), major histocompatibility complex-I (MHC-I), membrane type 1 matrix metalloproteinase (MT1MMP), ARF6, and β1-integrin.[42] Additionally, as a family member, sphingomyelinase (A-SMase), which is also present on the membrane surface, can significantly trigger the EV release through the ceramide generation in red blood cells and glial cells.[47,48] Some extracellular factors, including hypoxic environment, also influence the MV shedding.[49] Hypoxia increases the expression of hypoxia-inducible factors (HIFs). HIFs bind with the 5′-untranslated region of the *RAB22A* gene to upregulate the gene expression of RAB22A and significantly promote the MV release.[50] In addition to membrane dynamics and signaling alterations, MV release is also influenced by calcium homeostasis. Calcium works in cells as a secondary messenger and can highly regulate the MV release. An increased level of calcium mediates the calpain activation, which redistributes cysteine protease calpain to the cell periphery and facilitates the remodeling of the cytoskeleton to increase the MV shedding.[51,52]

9.2.3 Apoptotic Body Biogenesis

APoBDs have a size close to that of platelets and are formed during programmed cell death. The APoBDs are produced from blebbing of the cell membrane.[41] Macrophages phagocytize these bodies and fuse with the lysosomes within macrophages to inhibit the spillover and injury to the neighboring tissues.[53] The clearance is achieved by the interaction between the respective receptors present on the surface of the phagocytes and definite changes, including oxidation of molecules, craft-binding sites for the thrombospondin (Tsp), and complement protein C3b in apoptotic membrane composition. Phosphatidylserine translocates to the outer leaflet and binds with Annexin V. Thrombospondin, Annexin V, and C3b are well-recognized markers of APoBDs.[54] Apoptosis also sends a communication signal to the neighboring tissues to participate in cell survival, remodeling, and apoptosis.[55] Some investigations suggest that ApoBDs also contribute to immune and inflammatory responses in special circumstances.[56] Moreover, recent studies also pointed out that apoptotic vesicles of the dying cells are one of the central controllers of immune regulations.[56–59] During the biogenesis of APoBDs, cells experience several changes, such as cell shrinkage, DNA division, and cytoplasm membrane externalization. These changes lead to the encapsulation of the content, which causes cell death and results in the floating of membranes to produce APoBDs.[58] The ApoEV production is affected by several molecular factors, such as Rho-associated, coiled-coil-containing protein kinase 1 (ROCK1), and myosin light chain kinase (MLCK). The ApoEV production is controlled in a time- and dose-dependent manner.[60–63] An apoptosis pathway involving cysteinase, which is facilitated by Rho-associated kinase-1, triggers the APoBD formation.[64] The functional microtubules facilitate the nuclear content shrinkage, and MLCK participates in the packaging procedure of this nuclear content into ApoEVs.[65] The bleb formation held by the actomyosin increases cell contraction and hydrostatic pressure.[66] A recent investigation also found that the membrane channel pannexin-1 acts as a negative regulator of apoptotic body production, and a pannexin-1 inhibitor (trovafloxacin) disassembles the apoptotic cell.[67] Furthermore, extracellular and intracellular factors are required for the separation of the apoptotic cells into vesicles.

9.3 Composition of Extracellular Vesicles

The components loaded within EVs include proteins, nucleic acids, metabolites, and miRNAs. The packing of different types of molecules within EVs depends on the paternal cells and the packing mechanism. The enrichment of numerous molecules in cargo also relies on the specific cell type being targeted to several stimuli, pathological conditions, and drugs. EVs play an important role in various biological processes, including inflammation, coagulation, communication from cell to cell, immune response regulation, and metastasis. Bioactive lipids, chemokines, growth factors, peptides, nucleotides, and ions are uniquely carried by EVs.[68] EVs are thought to be an effort by the cell to avoid, delay, or reverse death by disabling some segments of the caspase machinery and inhibiting "eat me" signals.[69] EVs operate as a "post-service", moderating and stimulating target cells in an autocrine, endocrine, or paracrine manner.[17]

9.3.1 Protein Content

Comprehensive investigations have been performed on the cargo proteins of EVs, revealing that several proteins are involved in the biogenesis of EVs. Exosomes contain a unique set of glycolytic enzymes, transport proteins (GLUT1 and GLUT4),[70] fusion proteins, and heat shock proteins (Hsp70, Hsp60, and Hsp20).[71] The proteins involved in the exosome biogenesis are glycosylphosphatidylinositol (GPI) anchor proteins, receptor proteins (tumor necrosis factor receptor-1), palmitoylated associated proteins, cholesterol-binding proteins (caveolin), annexins, tetraspanins (CD9, CD81, CD63, and CD82),[72] and flotillins. The exosome Annexin A2 protein plays a significant role in the exosome membrane organization, as it binds to the phosphatidylinositol 4,5-bisphosphate domain, and flotillin 1 and flotillin 2 involve in the cell membrane composition of EVs and cargo sorting.[73] The cholesterol and caveolins also play a significant role in EV synthesis, as they control multiple cellular processes, such as shape maintenance, exocytosis, and endocytosis.[74]

Proteomic analysis also revealed that metalloproteinases, such as disintegrin and metalloproteinase (ADAMs), are enriched in EVs. ADAMs are multi-purpose proteins that contribute to cellular communication, cell adhesion, cell migration, and cytokine and growth factor regulation.[75] Several lysosome-associated proteins (LAMPs) protect the vesicle membrane from enzymatic degradation.[76] Other typical proteins are antigen-presenting proteins (MHC-I and MHC-II), extracellular matrix (ECM) proteins (collagen and fibronectin),[77] cell motility proteins (myosin, actin, and tubulin), ESCRT complex proteins (Alix and Tsg101),[78] metabolic enzymes (pyruvate kinase, beta-enolase, peroxidases, and fatty acid synthase), signaling proteins (Hras, GTPase, RAP1B, and RhoA), and membrane fusion proteins (Rab family proteins and AFR6), which are also involved in exosome cargo composition.[79] An investigation also reports the presence of transcription factors, including Wnt and Notch, inside the EVs.[80] MVs are enriched in several repertoires of protein molecules, such as integrins, P-selectin, and glycoprotein Ib.[81] MVs have more post-transcriptionally modified proteins, including phosphoprotein or glycoprotein, compared to exosomes.[82] ARF6 (a GTP-binding protein) controls the cytoskeleton alteration and also facilitates the shedding of vesicles.[83] APoBDs also comprise DNA-binding histone proteins.[79]

9.3.2 Lipid Content

The EV activity not only depends on the protein and nucleic acids but also relies on lipids. Exosome vesicles are enriched in cholesterol, phosphatidylserine, arachidonic acid, sphingomyelin, phosphatidic acid,[84,85] and some other fatty acids, including leukotrienes and prostaglandins, which contribute to structural rigidity and stability.[86] Furthermore, several lipolytic enzymes are also observed in exosomes, which produce the bioactive lipids independently. Several lipid raft proteins are involved in the biogenesis of exosomes.[87] Lipid rafts are mostly composed of sphingolipids and cholesterol and are located in the plasma membrane. They are recognized as microdomains and found in the detergent-resistant membrane of the apparent cover of vesicles.[88] The lipid content is cell-specific and conserved in exosome vesicles. Lipids are essential for exosome biogenesis, preserving exosome shape, and regulating homeostasis in the inheritor cells.[89–91] Ceramide is highly enriched in the inner membrane of the exosomes and helps in membrane invagination and preserves the vesicle structure and shape.[84,87]

Cholesterol is abundantly present in the exosome membranes, leading to the release of exosomes enriched with flotillins. A high-density lipid (lysobisphosphatidic acid) present in MVBs facilitates ILV formation and, consequently, exosome formation.[89,92] Bone morphogenetic proteins (BMPs), sphingomyelin, and phosphatidylcholine differentiate several categories of vesicles. Several types of MVs have similar concentrations of phosphatidylcholine and sphingomyelin, whereas sphingomyelin has a higher exosome concentration.[29,93] Moreover, the MV lipid composition is similar to that of the donor cells. They are enriched with phosphatidylserine (PS) and polyunsaturated glycerophosphoserine.[94,95] Investigations also reported the presence of saccharides (mannose, *N*-linked glycan, poly-lactosamine, and α-2,6-sialic acid) on the external surface of the exosomes.[96] EVs are also enriched with the enzymes that are responsible for lipid metabolisms, such as phospholipases A2 and D. The activity of these enzymes is initiated by the addition of GTP.[97] Moreover, under hypoxic conditions, adipocytes release EVs with a high level of *de novo* lipogenesis enzymes (fatty acid synthase).[98]

9.3.3 Nucleic Acids

EVs also enclose various forms of nucleic acids, including DNAs, mRNAs, and non-coding RNAs (miRNAs and lncRNAs). EVs transfer the nuclear content horizontally between cells and bring phenotypic changes in receptor cells.[99] EV membranes protect their internal materials from degradation by proteases and nucleases and also from the fluctuation of the microenvironment (osmolality and pH).[100] Researchers have recognized that several exosome RNAs are tissue-specific, and various exosome RNA molecules are present regardless of their origin.[101] EV mRNAs are shorter than 1 kb[102] and are translated into proteins in target cells.[103] The mRNAs enriched in EV are the same as expressed in the cell of origin and thus represent their parental cells.[104] It has also been established that the relative amount of each mRNA is not associated.[104] The enrichment of mRNA into EV is explained *via* the zip code mechanism. Zip codes are core sequences found at the 3′UTRs of the target messenger RNA, which intervene in the binding of the ribonuclear–protein complex and facilitate the mRNA localization in vesicles.[105] The zip code is also found in miRNAs and is known as EXO-motif, with a sequence of "GCAAG".[106] These zip code sequences in both miRNA and mRNAs regulate their loading into EVs.

miRNAs are small (21–25 nucleotides), double-stranded, highly conserved RNA molecules found in both plants and animals, with the basic function to post-transcriptionally control the gene expression of the target mRNA by binding at the open reading frame (ORF) and 3-untranslated region.[107,108] The finding of exosomal miRNA suggests that miRNAs can be directly delivered into the targeted cells through exosomes and highlights the functional variations in targeted mRNA. miRNAs play a significant role in cell communication *via* exosomes, as they are transferred from the donor cells.[109] Researchers also find that miRNAs do not incorporate into the exosome vesicles randomly, but particular miRNAs enter into exosomes selectively. The selective miRNA loading into the exosomes is mostly carried out by four different mechanisms: (1) the sphingomyelinase-related pathway, (2) the RNA-induced silencing complex (RISC)-dependent pathway, (3) the 3′-miRNA sequence-dependent path, and (4) the ribonucleoprotein-dependent pathway. Exosomal miRNAs contribute to basic cell functions, such as immune regulation, apoptosis, metastasis, proliferation, and cytokine production, through endocrine and paracrine signaling.[110,111] In the cardiovascular system, exosomal miRNAs also play a significant role as cell communicators. A recent investigation identified that exosomal miRNA plays a significant role in myocardial infarction (MI) injury. Exosome miR-126 decreases apoptosis, improves cardiac function, and enhances cell survival in cardiomyocytes when treated with $CoCl_2$ and hydrogen peroxide.[112] The level of miRNAs (miR-451, miR142-3p, and miR-150) is significantly high in derived exosomes compared to their parent cells.[113] The highly abundant non-coding RNA in EVs followed by miRNA is tRNA, which is significantly involved in the regulation of immunity, epigenetics, and gene expression. The EV tRNA fragment is entirely dependent on the source cell state.[114] EV-associated tRNA fragment acts as a potential biomarker in human disorders, including osteoporosis, male infertility, chronic kidney disease (CDK), gastric carcinoma, liver cancer, and even some infectious diseases.[115]

Another non-coding RNA loaded in the EV is small-nuclear RNA molecules (snoRNAs). The snoRNA's main function is to chemically modify the other RNA molecules. By comparing the levels of snoRNAs in EVs and source cells, it is found that they are highly enriched in EVs. In recipient cells, they directly shifted into the nucleus and are also affected by the biological condition of the source cell.[116] In addition, lncRNAs (200 bp) are also present in EVs and are involved in cell communication and signal transduction. They are steadily found in body fluids and remain unaffected by the degradation by the RNAase enzyme. Moreover, exosome-derived lncRNAs can regulate angiogenesis, apoptosis, migration, and proliferation.[117,118] Recent investigations reported that cardiomyocytes, under hypoxic condition, release the EVs with an abundance of lncRNA NEAT1 and persuade the pro-fibrotic gene expression program.[119]

Besides RNA, genomic DNA has also been present in EVs. Unlike RNA, the incorporation of genomic DNA into EVs has not yet been understood completely. One of the theories proposes the idea that during mitosis, the nuclear envelope breaks down, releasing genomic sequence fragments into the cytosol, which are then transported to particular packing locations.[120] In EVs, the presence of mitochondrial DNA (mtDNA), double-stranded DNA (dsDNA), single-stranded DNA, and oncogene amplifications (*i.e.* c-Myc) was observed.[121,122] EVs also play a role in the spread of numerous pathologies by providing a way for altered mtDNA to penetrate the other nearby cells.[123] DNA from tumor EVs, including oncogene amplification, represents the genetic condition of the tumor.[121,124,125] EVs were also used to transport DNA into specific fibroblasts, where EVs were labeled for DNA found in the cytoplasm and even the fibroblasts' nuclei.[126] Additionally, it was demonstrated that various EV subgroups transported various DNA cargos.[127] However, it is still unclear if the EVs' DNA has a significant role in cardiovascular disease (CVD).[128]

9.4 Sources of EVs

9.4.1 Mesenchymal Stem Cell-derived EVs

Mesenchymal stem cells (MSCs) play a significant role in restoration and regeneration due to their multi-lineage differentiation ability. MSCs have the potential to differentiate into the vascular system, skeletal muscle cells, and connective tissues.[129] The main origins of MSCs are the embryo, umbilical cord, bone marrow, and adipose tissues. In addition, some rare sources are also included, such as dental MSCs and menstrual blood-derived MSCs.[130–132] MSC-secreted EVs, such as exosomes, MVs, and apoptotic vesicles, have a unique tendency to treat some diseases. MSC-derived EVs promote angiogenesis *via* miRNAs-210 in infarcted hearts.[133] MSC-secreted EVs decrease the infarct size, facilitate vessel formation, and improve the cardiac function in ischemic injury.[134] A recent investigation also found that MSC-secreted ApoEVs significantly improve the cardiac function and enhance angiogenesis in myocardial infarction, which is due to endothelial cell autophagy.[135] Bone marrow MSC-derived exosomes (BMSC-EXO) contain the miRNAs that regulate the microenvironment of the targeted cells in cardiac regeneration. BMSC-EXO reduces cardiac fibrosis and infarct size *via* miR-22 by directly targeting the methyl-binding protein.[136] MSC-derived EVs contribute to cardiac cell functions, such as decreasing cardiac fibrosis, restoring cardiac survival, renovating cardiac function, and increasing capillary density, suggesting that MSC-derived EVs can act as a vehicle and a stimulator for the migration and proliferation of the cardiac cells.[137]

9.4.2 Cardiac Progenitor Cell-derived EVs

Progenitor cells exist in between the stem cells and mature cells; these cells are also recognized as precursor cells because they can differentiate into cardiac cells.[71] Cardiac progenitor cells (CPCs) are a heterogeneous group of cells that show a response to cardiac injuries. To recover the function of damaged cardiac cells, cardiac progenitor cells release certain membrane-bound nano-sized particles known as EVs (exosomes).[138,139] CPC stimulates cardiac cell survival and exerts an anti-fibrotic effect through the transfer of EVs.[140] An investigation shows that after myocardial infarction, intra-myocardial injection of CPC-derived exosomes significantly enhanced the cardiac function and considerably reduced the scars of myocardial injury by overexpressing chemokine receptors (CXCR).[141] Dougherty *et al.*'s findings demonstrated that human cardiac progenitor exosomes have therapeutic potential for cardiac repair and angiogenesis in the ischemic heart, and this potential can be increased by culturing CPCs under hypoxic conditions, which constitute the 5% O_2 microenvironment. Furthermore, this study also revealed that the human CPCs cultured in a physiological physoxic 5% O_2 (low oxygen) microenvironment were capable to retain the manifestation of cardiac markers and the morphology of cardiac cells. This finding elucidated the robustness of the human CPCs under low oxygen.[142] CPC exosomes also protect the heart from oxidative stress and hypoxia-induced cardiac injury. An *in vivo* experiment on rat models shows that doxorubicin (Dox) and trastuzumab (Trz) produce oxidative stress in cardiomyocytes, which induces cardiac toxicity. CPC exosomes, which are derived from the ischemic heart, were enriched with miR-146a-5p, and CPC exosomes effectively alleviate oxidative stress.[143] In another study, hypoxia-induced, CPC-derived exosomes were enriched with the miRNA, enhancing the tube formation in EC, protecting the heart from ischemia-induced reperfusion injury, and improving cardiac function.[144] Under oxidative stress, CPC-derived exosomes are enriched with miR-21, which targets the programmed cell death protein 4 (PDCD4) and protects the cardiac cell from apoptosis in the ischemic injured rat model.[145] All these investigations show that CPC exosomes could be the best source for cardiac repair, as they have the properties of enhancing cardiomyocyte survival and proliferation, exerting anti-apoptotic effect, promoting angiogenesis, reducing the infarct size, and improving cardiac function.

9.4.3 Adipocyte-derived EVs

Under pathological and physiological conditions, adipocytes also secrete the exosomes enriched with mRNA, proteins, and miRNA.[146] Adipose-derived stem cells released exosomes that are enriched with miRNA-126 and have a myocardial protective effect against hypoxia-induced acute myocardial infarction (AMI). The results demonstrated that, under hypoxia conditions, adipose-derived stem cell (ADSC)-exosome miR-126 facilitates the migration of endothelial progenitor cells and promoted angiogenesis by targeting and suppressing the inhibitors (P13KR2 and Spred1) of angiogenesis. These miR-126 exosomes not only facilitated the angiogenesis but also alleviated the apoptosis, fibrosis, and inflammatory factor expressions (IL-1β, TNF- α, and IL-6) in myocardial ischemia.[147] Ischemia–reperfusion (I/R) injury leads to hypertrophy and apoptosis, which causes further damage to the heart. ADSC-EXOs are abundant in miR-221/222, which protects the heart from hypertrophy and apoptosis by targeting the p53 upregulated modulator of apoptosis (PUMA) and ETS-1 (hypertrophic marker) and achieves cardioprotection by activating the Ak strain transforming (AKT) signaling and reducing expressions. Furthermore, studies showed that ADSC-CM enriched with miR221/222 decreased the phosphorylation of the nuclear factor kappa B (NF-κB), which leads to a reduction in PUMA and ETS-1 expressions and protects the ischemic myocardium.[148,149] miR-93-5p-loaded ADSC exosomes protected the heart from acute myocardial ischemic injury by targeting the toll-like receptor 4 (TLR4) and Agt7 and reduced the hypoxia-induced inflammatory cytokines and autophagy. Further studies also demonstrated that ADSC-EXO attenuated the I/R injury *via* the activation of Wnt/β-catenin signaling and alleviated apoptosis and promoted neovascularization after ischemia–reperfusion.[150–152] Under oxidative stress, ADSC-EXO ameliorates the ischemia–reperfusion injury by alleviating apoptosis. ADSC-EXO activates the SK1/S1PR1/S1P signaling pathway, promotes the macrophage (M_2) polarization to suppress the inflammatory factors, and attenuates the activation of TGF-β1 and NF-κβ-p65 and reverses the fibrosis of myocardial cells in AMI.[153,154] ADSC-EXO can be used as a therapeutic potential, as it protected the damaged myocardium by inhibiting fibrosis, cardiac cell apoptosis, hypertrophy, and myocardium autophagy; alleviating the inflammatory cytokines; and promoting angiogenesis.

9.4.4 Cardiac Telocyte-derived EVs

In 2010, new interstitial cells were discovered in the heart (myocardium and epicardium) *via* transmission electron microscopy, known as cardiac telocytes (TCs or CT). They are not only present in the heart but also in all other organs and tissues. In cardioprotection, they are involved in angiogenesis after the acute myocardial infarction. CT releases exosomes, which contain miRNAs, lipids, nucleic acids, and specific proteins. These cells contact with the endothelial tubes for angiogenesis, and CT-EXO interacts with neighboring cells for functional regulation and involvement in the intracellular signal transduction.[155–158] In myocardial infarction, CT-derived exosomes improve angiogenesis and cardiac function. They increase the propagation, migration, and EC tube formation and decrease cardiac fibrosis.[159]

A recent study investigated the communication between endothelial cells and cardiac telocytes to inhibit apoptosis and enhance the persistence of the endothelial cells to stimulate angiogenesis in myocardial infarction. In myocardial infarction, CT-exosomes were enriched with miRNA-21-5p, which improved the survival of the endothelial cells by targeting and silencing the cell death-inducing p53 target 1 gene and downregulating the caspase 3 to inhibit the apoptosis and promoted the angiogenesis of cardiac cells. This finding provides new opportunities for cardiac therapies to promote regeneration in the ischemic heart.[160]

9.5 EVs in the Homeostasis of the Cardiovascular System

Recent research has shown that cells associated with the cardiovascular system, including cardiomyocytes, fibroblasts, endotheliocytes, smooth muscle cells (SMCs), platelets, monocytes, leukocytes, and macrophages, can release EVs.[161] Under physiological circumstances, EVs play a biological role in maintaining normal heart structure and function; however, under pathological circumstances, EVs alter their composition (*i.e.* tissue factors and phosphatidyl serine exposure on the cell surface), thus facilitating CVDs.[162] EVs, therefore, hold vast potential for monitoring as well as treating CVDs. Increasing evidence indicates that the miRNAs and proteins that EVs transfer to target cells are mostly responsible for the impacts of EVs on those cells.[163] Depending on the condition of the source cells, EVs containing miRNAs and proteins have been shown to regulate a multitude of diverse functions in target cells, maintaining cardiovascular balance and health or inducing pathological changes in CVDs.[164] A significant capacity for interventions by raising or lowering EVs that contain miRNAs is also suggested by the fact that EVs are crucial to the remodeling of cardiac muscle following MI.[164] Different types of bodily fluids have been shown to contain EVs, including blood (plasma/serum), saliva, urine, breast milk,[165] nasal secretions,[166] cerebrospinal fluid,[167] semen,[168] and bronchoalveolar lavage fluid.[169] Although it is impossible to pinpoint the precise tissue from which circulating EVs originate, it is mostly accepted that they constitute a combination of EVs from various cell types.[170,171] It should be noted that the mRNA of the liver was discovered in the exosomes of the blood of normal rats, proving that EVs derived from hepatocytes are discharged into circulation under normal circumstances.[172]

EVs can control and take part in the maintenance of stem cells, tissue healing, and coagulation.[173] For instance, one of the major roles of blood platelets during the coagulation process is to create a membrane layer that speeds up blood coagulation and encourages the production of the fibrin matrix in the slot of hemostasis.[174] In terms of tissue repair, EVs can aid in the network formation of a blood vessel during the physiological wound-healing process. Indeed, endothelial cell-derived EVs comprise metalloproteinase proteins (MMP-9 and MMP-2) to enhance matrix disintegration, hence boosting blood vessel development.[175] EVs stimulate effective stem cell differentiation, which helps to maintain the niche of the stem cells and heal damaged tissue. EVs control the bilateral transfer of genetic data between stromal and stem cells. In addition to changing the phenotype of stem cells, they alter the characteristics of damaged tissue cells. EVs may also provoke de-differentiation of cells that have survived an injury, enabling tissue regeneration.[173]

9.6 Potential Use of EVs as Drug Delivery Carriers for Therapeutic Purposes

EVs have been seen as desirable drug delivery systems with some extra inherent qualities that further increase their value.[176–178] The structure, size, and lipid membrane of EVs resemble natural liposomes, *e.g.* nanovesicle-based synthetic drug delivery systems that are used in clinical settings and sold commercially. The local accumulation and pharmacological impact of a drug molecule at the targeted location should be maximized by an optimum drug delivery system while minimizing adverse effects. EVs contain naturally occurring phospholipid bilayer-like liposomes, which keep cargos within the sac-like structure.[177] The EV cargo delivery potential is increased when specific tissue-targeted surface proteins are integrated, which makes it possible for target cells to recognize and absorb them, increasing the effectiveness of drug delivery.

Research on EVs has been attempting to boost their targeting ability to the targeted locations in CVDs through various modifications of EV surface (5–10) and by combining EVs with synthetic drug delivery systems.[179] The therapeutic effect of EVs has been observed in CVDs, such as atherosclerosis, myocardial infarction, and cardiac inflammation. Atherosclerosis is a long-lasting inflammatory condition that causes plaques made of lipids to build up on the artery wall. The effective siRNA delivery into the primary aortic endothelial cells utilizing EVs derived from endothelium[180] demonstrated EVs' ability to transfer therapeutic cargos to the endothelium with inflammation in atherosclerotic conditions, resulting in reduced inflammation.

Research demonstrates that tRNAs in exosomes, one type of EV, are modified after exosome-specific modifications and are comparatively present more in exosomes than in cells.[181] These and other RNA modifications are crucial in both healthy and unhealthy circumstances and may suggest that RNA alterations have a role in RNA cargo sorting inside EVs.[182] In addition, similar research revealed that the RNA-binding protein Y box-binding protein 1 (YBX1) was secreted to sort miRNAs into the intraluminal vesicles of cells using exosome multivesicular structures exported from HEK293T cells through unusual secretion methods.[183]

Endothelial-derived EVs are reported to transfer miR-143 and miR-145 to the atherosclerotic plaque and inhibit the development of atherosclerotic injury.[184] Jansen *et al.* reported the administration of miRNA-126 through EVs produced by endothelial cells undergoing cell death to induce vascular re-endothelialization *in vivo*.[185] These results imply that therapeutic miRNA delivery through EVs is a potential strategy to treat atherosclerosis. Moreover, drug and nucleic acid delivery and imaging agents may also be packed in EVs to provide a therapeutic and diagnostic platform for atherosclerosis, for example, iron oxide and superparamagnetic iron oxide nanoparticles.[186–188] An EV-based therapeutics was reported for atherosclerosis as well as other CVDs due to the inherent targetability of EVs to atherosclerotic lesions[180,184,185,189] and the viability of using superparamagnetic iron oxide nanoparticles for magnetic resonance imaging (MRI) of plaques under atherosclerotic condition in humans.[190]

EVs are widely reported in cases of inflammation, which is one of the important occurrences after myocardial infarction. Inflammation in a damaged heart elevates vascular permeability similarly to aggressive tumors, which may boost the diffusion and retention of therapeutic delivery of drugs. EV, as nanosize encapsulation, is reported to target cardiac inflammation through permeability and retention effect.[191] Several studies using experimental MI models have shown that spontaneously released EVs are a viable method for delivering miRNAs to the damaged heart.[147,151,192] Previous studies demonstrated that enhanced permeability and retention (EPR) effect is the only characteristic inherent to the targetability of EVs, thus influencing their accumulation at the inflamed sites.

Ischemia and cardiomyocyte necrosis in MI necessitate regenerative assistance. Numerous investigations have shown that EVs containing cargo can enhance cardiac functionality and hence serve as a cell-free therapeutic substitute. Exosomes abundant in miR-21-5p and miR-210-3p from induced pluripotent stem cells (iPSs) generated from mouse cardiac fibroblasts inhibited the regulation of NANOG transcription factor and hypoxia-inducible factor in H9C2, leading to cell death.[140] In addition, miR-210, which was generated from CPC, downregulates levels of ephrin and protein-tyrosine phosphatase 1B (PTP1b), which prevents cardiomyocyte death. In another study, RasGAP-p120 was reported to be downregulated by miRNA-132, which improves endothelial cell tube development.[193] Exosomes containing pro-angiogenic factors that are released by human CD34+ cells demonstrated the ability of new blood vessel formation in ischemic hindlimbs.[194,195]

9.7 Clinical Translation: Challenges and Limitations

The EV interventional clinical application is continuously growing in cancer therapy.[196,197] The interventional clinical application of EVs (exosomes and MV), both recruiting and not yet recruiting, can be assessed at https://www.clinicaltrials.gov. Most of the clinical trials on exosomes are occurring in Europe and the United States. However, several clinical therapeutic applications have been implemented for cardiovascular diseases or acute ischemic stroke, and no adverse events were witnessed in clinical applications. The cardiovascular interventions are (1) Role of Exosomes Derived from Epicardial Fat in Atrial Fibrillation held in Sheba Medical Center, Israel, ClinicalTrials.gov Identifier: NCT03478410; (2) Antiplatelet Therapy Effect on EVS in Acute Myocardial Infarction in Medical University of Warsaw, Poland; University of Amsterdam, the Netherlands, identifier: NCT02931045; and (3) Allogenic Mesenchymal Stem Cell-Derived Exosome in Patients with Acute Ischemic Stroke in Isfahan University of Medical Sciences, Iran, identifier: NCT03384433. The EV can easily be stored and thawed for instant use in patients with no loss of biological function.[198] However, EV's clinical applications still face some challenges. Advancements in technology are required for the isolation, characterization, purification, and storage of EVs.[199] A noteworthy challenge lies in the development of exosome-dependent therapies in clinical trials is developing reproducible, accessible, efficient, and sterile approaches for the clinical-grade exosome construction with insignificant batch variations.[200,201] However, the achievability of clinical practice-grade exosomes was demonstrated by using the bio-rector culture system. This purification of exosomes was performed by ultracentrifugation and filtration by using the semiclosed and closed systems, which were followed by flow cytometry and NanoSight analysis. Then, exosomes were tested for sterility and endotoxin to confirm the clinical potential of exosomes.[202] Similar upscale approaches are developing for the treatment of cardiovascular diseases.[203] EVs are heterogeneous groups of vesicles. There is no particular method or marker that can differentiate the exosomes, sub-exosomes, and MV groups. This limitation restricts the application of EVs for clinical therapy.

Moreover, in circulation, EVs essentially evade the degradation in the kidneys, liver, immune cells, and lungs.[204] The therapeutic application of EVs also requires authorized cell sources in the cardiovascular field. EVs can be generated from exogenous and autologous sources. However, their immune compatibility potential makes it difficult for on-demand production and makes them impossible to regulate their manufacturing.[205] One of the main challenges is the non-specific delivery of the EVs at off-target tissues, which also hinders the therapeutic potential of EVs. To obtain the effective dose level for the preferred therapeutic results, an adequate dose of EVs must be reached at the disease site.[206,207] However, there is a shortage of evidence to estimate the toxicity at a higher dose of EVs.

9.8 EVs as a Tool for CVD Theranostics

9.8.1 EVs as Diagnostic Tool

Most of the biological fluids, including blood, urine, saliva, seminal and amniotic fluid, milk, bile, semen, and malignant ascites, contain EVs.[208–212] EVs' presence in several biofluids makes them attractive diagnostic markers for CVDs and other diseases.[213] Additionally, the distinct composition of EVs derived from cells has prognostic and diagnostic values. In one research, 181 participants with clinically stable coronary artery disease (CAD) over a period of six years indicated increased levels of miR-126 and miR-199a in circulating MVs, which potentially decreased the risk of a future adverse cardiac event.[214]

9.8.2 EVs as Diagnostics Markers: Limitations and Challenges

Due to more complicated exosome separation and quantification techniques, exosomes have shown comparably less potential to function as an effective diagnostic marker.[213] Because of their tiny size, exosomes often go unnoticed by commonly used instruments such as flow cytometers, necessitating specialist equipment for quantification, such as Dynamic Light Scattering (DLS) machines or Nanoparticles Tracking Analysis (NTA) monitors.[215] Despite these limitations, there are only a few studies that show how well exosomes predict the development of cardiovascular diseases and other disorders. For instance, miR-1 and miR-133a levels in EVs from individuals with acute coronary syndrome (ACS) and MI were increased.[216,217] Additionally, there were noticeably more exosomes in the plasma of ovarian cancer patients than that in healthy persons.[218,219] These results show the potential of exosomes as biomarkers and the consequent need for more straightforward exosome extraction and quantification techniques.

The potential use of EVs in diagnostics and prognostics is now hindered by the lack of consistency in sample collection, extraction, and measurement. The effects of different sample handling on EVs' physical and molecular characteristics are evidenced by data.[220,221] Additionally, it might be challenging to separate EVs from non-vesicular protein constituents when dealing with plasma or serum samples that is high in proteins.[222] It is also known that the time of day, the sample obtained, and the number of physical activities performed before collection affect the EV levels.[221] It is still difficult to separate tissue-specific EVs from human tissues.

When compared to EVs extracted from cell culture supernatant, EVs from physiological fluids exhibit more variability.[223,224] For example, various exosomes released by several types of cells are mixed together in human plasma.[223–225] Furthermore, the recovery of EV biomarkers relevant to a specific disease is complicated by the presence of EVs released by both sick and healthy cells in biological fluids.[223–225] The difficulty of extracting EVs from human plasma is made more difficult by the absence of capture reagents with enough specificity for certain exosome markers.[223]

9.8.3 Native EVs as a Therapeutic Tool

The first research outlining the potential therapeutic application of EVs for cardiovascular disorders was published in 2010.[226] Early in the new millennium, research from a number of organizations showed that cardiac repair after myocardial infarction was enhanced by the transplantation of several cell types, such as mesenchymal stem cells[227] and hematopoietic progenitor or stem cells.[228] Later, various studies show that the beneficial outcomes of stem cells or progenitor cells were instead mediated by paracrine factors,[229,230] namely EVs[226,231] released by the cells, rather than by the direct benefaction of the engrafted cells. Since these investigations, several research teams have shown that EVs released by stem cells, also by somatic cells, possess the ability to promote tissue regeneration in a variety of conditions, including myocardial infarction,[232–235] ischemic limb,[236] and wounds.[237]

Although there have been many studies on cardiovascular conditions, such as atherosclerosis[238] and stroke,[239,240] the majority of research on EV has focused on ischemic heart disease and MI. Additional research revealed that plasma EVs stimulate the cardioprotective pathway *via* the TLR4-activated heat shock protein (HPS), protecting the heart against I/R damage.[241]

Several cell sources, including MSCs,[242–247] putative CPCs,[144,193,233,248–254] cells derived from iPS,[129,140,234,255–258] and somatic cells[237,241,259,260] produce EVs with cardiovascular effectiveness and therapeutic properties. MSC-derived EVs encourage cardiac healing and enhance cardiac function in non-reperfusion MI animal models by increasing cardiac cell survival, new blood vessel formation, and decreasing cardiac fibrosis.[244,261] Khan *et al.* reported that MSC-derived EVs improve cardiac progenitor cell survival and proliferation, hence enhancing endogenous heart repair and regeneration capacity.[261] According to one of the studies, miRNAs in EVs derived from MSC, *e.g.* miR-182, regulate macrophage polarization toward an anti-inflammatory M2 phenotype, which is thought to be the mechanism through which EVs exert cardioprotective effects.[262] Additionally, EVs derived from MSCs improve cardiac function and repair by increasing cardiac cell survival, new blood vessel formation, and decreasing cardiac fibrosis.[244,261]

Cardiomyocyte and endothelial cell proliferation are both greatly increased and infarct size is significantly decreased by human cardiac progenitor cell (hCPC)-derived EVs.[191] Alexandru *et al.* reported the therapeutic potential of EVs through transferring of microRNAs, including miRNA-10a, miRNA-21, miRNA-126, miRNA-146a, and miRNA-223, to endothelial progenitor cells (EPCs), which results in improvement of vascular repair and atherosclerotic plaque stabilization in a hamster model.[263] EVs released by cardiosphere-derived cells decrease infarction size, alleviate ventricular hypertrophy, and enhance cardiac function by suppressing cardiac cell death and improving new blood vessel formation in the heart.[248] Alexandru *et al.* analyzed the effect of EPC-derived EVs against atherosclerosis in CVD development. They showed the therapeutic effect by transferring bioactive microRNAs, including miRNA-223, miRNA-21, miRNA-126, and miRNA-146a, which results in the repair of vascular function and decreases atherosclerotic vascular disease development.[264]

EVs produced from induced pluripotent stem cells (iPSCs) possess cardioprotective microRNAs that control the NANOG transcription factor expression and hypoxia-inducible factor 1-alpha, regulate hypoxia, and inhibit cardiomyocyte death in I/R damage.[140] Harane *et al.* reported EVs to be effective in the treatment of chronic heart failure by the transfer of specific miRNAs and stimulation of cardioprotective pathways. These iPSC-derived cardiovascular progenitors improve cardiac functionality by promoting cell survival, proliferation, and endothelial cell migration.[256]

Several studies reveal the effects of allogenic EVs after intravenous[243,265,266] and intramyocardial[136,144,193,233,242,248,250,251,253,267] administration in mice,[136,250,254,268] rats,[144,233,252,253,269] and pigs.[248,251] Gallet *et al.* reported EVs to retain for longer in the heart when administrated *via* intramyocardial injection than intracoronary or intravenous injection.[248] EVs may enhance ejection fraction (EF) up to 1.3-fold[144,242,249,253] compared to the control group. According to pre-clinical evidence gathered so far, they decrease infarct size up to 3-fold[136,251,252,268] relative to the control group. Research has reported the therapeutic effects of EVs and donor cells and shows that EVs are just as effective as donor cells in the setting of MI.[136,248]

Bodily fluid-derived, including serum and plasma, EVs have a potential therapeutic effect on MI.[241] EVs extracted from rat or human plasma were capable of cardioprotection both *in vitro* and *in vivo*. In another investigation, it was shown that EVs extracted from mouse serum may prevent programmed cell death in MI by regulating the miRNA-21/PDCD4 pathway.[270] Overall, these results imply that EVs produced from diverse cells or tissues have the ability to lessen both the short- and long-term effects of ischemia–reperfusion injury, MI, and cardiac remodeling.

Several studies have reported the cardioprotective properties of MSCs and MSC-derived EVs by decreasing the infarct size and improving cardiac functionality in ischemia–reperfusion injury.[262,271,272] Studies showed that MSC-derived EV possesses therapeutic properties and inhibits apoptosis and oxidative stress in ischemia–reperfusion injury.[129,257]

9.8.4 Engineered EVs as Therapeutic Modalities in Treating CVDs

Different techniques are implemented to do various modifications in EVs, including culturing them in stressful environments (such as hypoxia,[134,144,260] serum starvation media,[260,273] or inflammation[274] or transfecting them with bioactive substances such as nucleic acids, particularly miRNAs),[113,275,276] miRNA antagonists,[277] and DNA plasmid.[269,278–280] For example, Gray *et al.* identified miRNA clusters, including miR-292, miR-210, miR-103, miR-17, miR-199a, miR-20a, and miR-15b, in EVs from CPCs cultured under hypoxia, which increased the ability of cardiac endothelial cells to form tube formation *in vitro*, decreased the pro-fibrotic genes expression in cardiac fibroblasts cultured *in vitro*, and enhanced the function of the infarcted heart compared to EVs in culture with normal conditions.[144]

Altering the growth medium of cells secreting EVs is another way to modulate EVs. EVs obtained from adipose-derived stem cells cultured in an endothelial differentiation medium display an increase in miRNA-31 levels. The resulting EVs with elevated miRNA-31 levels showed enhanced endothelial cell migration, tube formation, and aortic ring outgrowth in contrast to EVs obtained from adipose-derived stem cells grown in the normal medium and thus not enriched for miRNA-31.[281]

Extracted cellular vesicles are modified by combining them with other cellular vesicles using methods, such as extrusion, sonication, and stirring, to create engineered hybrid EVs.[51] The acquired hybrid EVs display cardiac repair abilities and inherited characteristics of parent cells.[48] Research shows that platelet membranes fused to MSC-derived exosomes *via* a membrane fusion technique have an advantage due to platelets' ability to target injury and boost EVs' pro-angiogenic potential.[48,55] Furthermore, MSC-derived exosomes were reported to be coated with the monocyte cell membrane, which increases specific targeting to injury and encourages cardiac healing.[56]

MSC-derived EVs transfected with miR-181a boosted the pro-reparative state of blood mononuclear cells, serving as an example of EVs modified by external agents. EVs with increased levels of miR-181a enhanced ejection fraction after being administered to the infarcted area of the heart in mice models.[275] In order to customize the enrichment of EVs with certain proteins and RNAs that are of interest, a number of cellular platforms have been established; however, these platforms have not yet been tested for cardiovascular applications.

Genetic alteration of parent cells and natural EVs can also change several aspects of EVs.[51] Transgenes are often transferred *via* lentiviral vectors to alter the expression of genes in cells.[64] For instance, the fusion of lysosome-associated membrane protein 2b to a targeting peptide in MSCs might produce exosomes with greater affinity to the ischemic heart.[65] Furthermore, it has been shown that atorvastatin-activated MSC exosomes increase neovascularization and cardioprotective effectiveness through upregulating long non-coding RNA (H19) in MSCs.[66] Studies reported that EVs produced from MSCs overexpressing miR-126 stimulate neovascularization and support vascular regeneration.[67,68] In one study, lentivirus was utilized to overexpress CD47 on the membrane of MSCs and extract CD47 EVs; the miR-21a was then electroporated into CD47 EVs. The findings showed that CD47 EVs were collected in the cardiac injury and had increased retention duration in an *in vivo* model. Genetic editing may be used in concert with other techniques to achieve accurate membrane engineering, which is especially helpful in the treatment of CVDs.

Additionally, only a few studies showed engineered EVs with certain epitopes capable of targeting the heart[282] by using cellular machinery. Engineering EVs *via* hijacking the cellular machinery has the benefit of maintaining the biophysical characteristics of EVs; but, it has some limitations, such as the possibility that overexpressing a particular molecule in a cell may have unanticipated biological effects that interact with EV formation. EV loading, targeting, and distribution may all be effectively controlled using modification techniques based on post-isolation alteration of EVs, independent of the cell that produced them. Post-isolation techniques could conceal or damage endogenous EV characteristics, which would impede EV's biological activity.

Covering artificial nanoparticles with natural cell membranes allows artificial nanoparticles to take on some characteristics of living cells.[60,61] Wang *et al.* created rapamycin-loaded poly(lactic-*co*-glycolic acid) (PLGA) nanoparticles with a macrophage membrane coating, which efficiently inhibited macrophage phagocytosis and targeted active ECs while exhibiting good drug release kinetics.[283]

In order to reduce cardiac fibrosis, control inflammatory responses, and promote new blood vessel formation, a number of EV formulations have been enhanced utilizing transfection agents or membrane-permeabilizing techniques. For instance, Kang *et al.* utilized miRNA-21-5p to regulate multiple gene targets, such as SMAD 7, sprout receptor tyrosine kinase signaling antagonist, and phosphatase and tensin homolog/protein tyrosine phosphatase,[284] and this plays a crucial role in the formation of fibrosis following myocardial infarction.

9.9 Enhancing the Targeting Abilities of EVs to Heart

Although EVs offer a significant ability for drug delivery, numerous limitations, such as quick liver clearance and low drug-loading efficacy, have been preventing their widespread use as a drug delivery system.[285] Various surface modifications are performed in order to increase the efficacy of EVs as a drug delivery system; *e.g.* Zahid *et al.* reported a peptide that precisely targets the cardiac tissue when administered intravenously using a phage peptide library.[286] Without relying on the permeability and retention effect, tagging this targeting peptide to the surface of EVs boosted EVs' targeting efficiency to the cardiac tissue by fifteen percent.[287] This method has a possible risk of adverse side effects on the healthy myocardium. It is better to use a different peptide that precisely targets damaged myocardium.

A few research teams have suggested that surface modification might increase the targetability of EVs. Kanki *et al.* reported the cardiac homing peptide using a similar phage peptide technique. The peptide with a sequence of CSTSMLKAC selectively binds to the ischemic cardiac muscle tissue.[288] Cardiosphere stem cells (CSCs) were modified by attaching homing peptides to enhance the cardio-targeting ability of their extracellular vesicles.[206]

In ischemia/reperfusion injury in the rat model, the cardiac homing peptide-conjugated EVs improved therapy effectiveness, as shown by lowered infarct size, decreased fibrosis, and improved cardiac function.[206] In a mouse model of MI, Wang *et al.* increased the therapeutic efficiency of murine EV-derived MSCs by genetically tagging cardiac homing peptides onto their surface.[207] The cardiac homing peptide-conjugated EVs accumulated effectively and arbitrarily in the damaged cardiac location.[207] The cardiac homing peptide was the first peptide to target EVs in the damaged myocardium.

Functional ligands to membranes of EVs are incorporated to increase their therapeutic potential[289]; for instance, using functional ligands, dibenzocyclooctynesulfo-*N*-hydroxysuccinimidyl ester, to attach cyclo peptide (c(RGDyK)) onto the surface of EVs derived from MSCs displayed excellent affinity to integrin αvβ3 in cerebral vascular ECs after the occurrence of ischemia. In a similar manner, DBCO-sulfo-NHS on the EV surface was used to attach an ischemia myocardium-targeted peptide, which increases the homing specificity and therapeutic effectiveness of exosomes derived from MSCs in a myocardial infarction model.[290] Exosomes from cardiosphere-derived stem cells can target infarcted hearts and increase cell proliferation and new blood vessel formation.[244] These exosomes were engineered with a targeting peptide conjugation using dioleoylphosphatidylethanolamine *N*-hydroxysuccinimide (DOPE-NHS).

9.10 Conclusion and Future Perspectives

Conclusively, EVs have excellent properties, such as low immunogenicity, good biocompatibility, and biostability. These characteristics make them highly desirable for the development of novel therapeutic delivery methods. EVs derived from different cell sources participate in neovascularization, reduce infarct size, inhibit apoptosis, enhance cardiac function, improve cardiac cell survival, and inhibit fibrosis and anti-inflammatory effects in cardiovascular diseases. In cardiovascular diseases, EVs made progress in various areas, such as clinical transformation, diagnosis, and therapy, and act as drug delivery vehicles. EVs play a significant role in detecting the progression of cardiovascular diseases by reflecting the changes in the biological composition of vesicles and estimating the level of circulatory EVs with specific markers and content. Furthermore, recently, EVs have been actively used in regenerative medicine that directs the therapeutic potential of EVs in cardiovascular therapy. Finally, EVs are a novel and promising approach to facilitate therapeutic results in CVD patients. However, for future research, it will be essential to determine the EV's composition at the individual vesicle level and also necessary to separate the vesicle synthesis at the individual level for the appropriate engineering and designing of vesicles in cardiovascular therapeutic applications. To stipulate the target for EVs in diseases, we could fuse the protein content engaged in EVs with various specific ligands/receptors. In addition, we could also enhance the biogenesis of EVs by genetically modifying the synthesis pathways.

Acknowledgements

All authors acknowledge Department of Biochemistry, Quaid-i-Azam University, Islamabad and Higher Education Commission of Pakistan.

References

1. P. Pfeifer, N. Werner and F. Jansen, Role and function of microRNAs in extracellular vesicles in cardiovascular biology, *BioMed Res. Int.*, 2015, **2015**(1), 161393.
2. E. van der Pol, *et al.*, Recent developments in the nomenclature, presence, isolation, detection and clinical impact of extracellular vesicles, *J. Thromb. Haemostasis*, 2016, **14**(1), 48–56.
3. M. Simons and G. Raposo, Exosomes–vesicular carriers for intercellular communication, *Curr. Opin. Cell Biol.*, 2009, **21**(4), 575–581.
4. N. P. Hessvik and A. Llorente, Current knowledge on exosome biogenesis and release, *Cell. Mol. Life Sci.*, 2018, **75**(2), 193–208.

5. M. Yáñez-Mó, *et al.*, Biological properties of extracellular vesicles and their physiological functions, *J. Extracell. Vesicles*, 2015, **4**(1), 27066.
6. D. Maisano, *et al.*, Uncovering the exosomes diversity: A window of opportunity for tumor progression monitoring, *Pharmaceuticals*, 2020, **13**(8), 180.
7. B. Yue, *et al.*, Exosome biogenesis, secretion and function of exosomal miRNAs in skeletal muscle myogenesis, *Cell Proliferation*, 2020, **53**(7), e12857.
8. G. Raposo and W. Stoorvogel, Extracellular vesicles: exosomes, microvesicles, and friends, *J. Cell Biol.*, 2013, **200**(4), 373–383.
9. V. B. R. Konala, *et al.*, The current landscape of the mesenchymal stromal cell secretome: a new paradigm for cell-free regeneration, *Cytotherapy*, 2016, **18**(1), 13–24.
10. D. Van der Pol, *et al.*, Classification, functions, and clinical relevance of extracellular vesicles, *Pharmacol. Rev.*, 2012, **64**(3), 676–705.
11. G. Camussi, *et al.*, Exosomes/microvesicles as a mechanism of cell-to-cell communication, *Kidney Int.*, 2010, **78**(9), 838–848.
12. Y. Zhang, *et al.*, Exosomes: biogenesis, biologic function and clinical potential, *Cell Biosci.*, 2019, **9**(1), 1–18.
13. M. J. Haney, *et al.*, Exosomes as drug delivery vehicles for Parkinson's disease therapy, *J. Controlled Release*, 2015, **207**, 18–30.
14. Y. Chen, G. Li and M.-L. Liu, Microvesicles as emerging biomarkers and therapeutic targets in cardiometabolic diseases, *Genomics, Proteomics Bioinf.*, 2018, **16**(1), 50–62.
15. F. Jansen, G. Nickenig and N. Werner, Extracellular vesicles in cardiovascular disease: potential applications in diagnosis, prognosis, and epidemiology, *Circ. Res.*, 2017, **120**(10), 1649–1657.
16. F. Deng, *et al.*, Endothelial microvesicles in hypoxic hypoxia diseases, *J. Cell. Mol. Med.*, 2018, **22**(8), 3708–3718.
17. A. Hafiane and S. S. Daskalopoulou, Extracellular vesicles characteristics and emerging roles in atherosclerotic cardiovascular disease, *Metabolism*, 2018, **85**, 213–222.
18. H. Kalra, G. P.Drummen and S. Mathivanan, Focus on extracellular vesicles: introducing the next small big thing, *Int. J. Mol. Sci.*, 2016, **17**(2), 170.
19. M. Ainola, *et al.*, Activation of plasmacytoid dendritic cells by apoptotic particles–mechanism for the loss of immunological tolerance in Sjögren's syndrome, *Clin. Exp. Immunol.*, 2018, **191**(3), 301–310.
20. L. Jiang, *et al.*, Determining the contents and cell origins of apoptotic bodies by flow cytometry, *Sci. Rep.*, 2017, **7**(1), 1–12.
21. H. Ando, *et al.*, Dermal fibroblasts internalize phosphatidylserine-exposed secretory melanosome clusters and apoptotic melanocytes, *Int. J. Mol. Sci.*, 2020, **21**(16), 5789.

22. A. Hochreiter-Hufford and K. S. Ravichandran, Clearing the dead: apoptotic cell sensing, recognition, engulfment, and digestion, *Cold Spring Harbor Perspect. Biol.*, 2013, **5**(1), a008748.
23. I. K. Poon, *et al.*, Apoptotic cell clearance: basic biology and therapeutic potential, *Nat. Rev. Immunol.*, 2014, **14**(3), 166–180.
24. M. Li, L. Liao and W. Tian, Extracellular vesicles derived from apoptotic cells: an essential link between death and regeneration, *Front. Cell Dev. Biol.*, 2020, 1063.
25. S. Liu, *et al.*, Mesenchymal stem cells prevent hypertrophic scar formation via inflammatory regulation when undergoing apoptosis, *J. Invest. Dermatol.*, 2014, **134**(10), 2648–2657.
26. T. K. Phan, I. K. Poon and G. K. Atkin-Smith, Detection and isolation of apoptotic bodies to high purity, *J. Visualized Exp.*, 2018(138), e58317.
27. B. L. Deatherage and B. T. Cookson, Membrane vesicle release in bacteria, eukaryotes, and archaea: a conserved yet underappreciated aspect of microbial life, *Infect. Immun.*, 2012, **80**(6), 1948–1957.
28. L. M. Doyle and M. Z. Wang, Overview of extracellular vesicles, their origin, composition, purpose, and methods for exosome isolation and analysis, *Cells*, 2019, **8**(7), 727.
29. J. Huotari and A. Helenius, Endosome maturation, *EMBO J.*, 2011, **30**(17), 3481–3500.
30. R. C. Piper and D. J. Katzmann, Biogenesis and function of multivesicular bodies, *Annu. Rev. Cell Dev. Biol.*, 2007, **23**, 519.
31. A. Bobrie, *et al.*, Exosome secretion: molecular mechanisms and roles in immune responses, *Traffic*, 2011, **12**(12), 1659–1668.
32. J. Kowal, M. Tkach and C. Théry, Biogenesis and secretion of exosomes, *Curr. Opin. Cell Biol.*, 2014, **29**, 116–125.
33. J. H. Hurley and G. Odorizzi, Get on the exosome bus with ALIX, *Nat. Cell Biol.*, 2012, **14**(7), 654–655.
34. D. J. Katzmann, M. Babst and S. D. Emr, Ubiquitin-dependent sorting into the multivesicular body pathway requires the function of a conserved endosomal protein sorting complex, ESCRT-I, *Cell*, 2001, **106**(2), 145–155.
35. J. H. Hurley, ESCRTs are everywhere, *EMBO J.*, 2015, **34**(19), 2398–2407.
36. W. M. Henne, H. Stenmark and S. D. Emr, Molecular mechanisms of the membrane sculpting ESCRT pathway, *Cold Spring Harbor Perspect. Biol.*, 2013, **5**(9), a016766.
37. W. M. Henne, N. J. Buchkovich and S. D. Emr, The ESCRT pathway, *Dev. Cell*, 2011, **21**(1), 77–91.
38. C. Villarroya-Beltri, *et al.*, Sorting it out: regulation of exosome loading, in *Seminars in cancer biology*, Elsevier, 2014.
39. D. Perez-Hernandez, *et al.*, The intracellular interactome of tetraspanin-enriched microdomains reveals their function as sorting

machineries toward exosomes, *J. Biol. Chem.*, 2013, **288**(17), 11649–11661.

40. J. G. van den Boorn, *et al.*, Exosomes as nucleic acid nanocarriers, *Adv. Drug Delivery Rev.*, 2013, **65**(3), 331–335.
41. C. Théry, *et al.*, Minimal information for studies of extracellular vesicles 2018 (MISEV2018): a position statement of the International Society for Extracellular Vesicles and update of the MISEV2014 guidelines, *J. Extracell. Vesicles*, 2018, **7**(1), 1535750.
42. J. C. Akers, *et al.*, Biogenesis of extracellular vesicles (EV): exosomes, microvesicles, retrovirus-like vesicles, and apoptotic bodies, *J. Neuro-Oncol.*, 2013, **113**(1), 1–11.
43. T. Matusek, *et al.*, The ESCRT machinery regulates the secretion and long-range activity of Hedgehog, *Nature*, 2014, **516**(7529), 99–103.
44. J. F. Nabhan, *et al.*, Formation and release of arrestin domain-containing protein 1-mediated microvesicles (ARMMs) at plasma membrane by recruitment of TSG101 protein, *Proc. Natl. Acad. Sci.*, 2012, **109**(11), 4146–4151.
45. V. Muralidharan-Chari, *et al.*, ARF6-regulated shedding of tumor cell-derived plasma membrane microvesicles, *Curr. Biol.*, 2009, **19**(22), 1875–1885.
46. A. Sedgwick, *et al.*, Extracellular microvesicles and invadopodia mediate non-overlapping modes of tumor cell invasion, *Sci. Rep.*, 2015, **5**, 14748.
47. F. Bianco, *et al.*, Acid sphingomyelinase activity triggers microparticle release from glial cells, *EMBO J.*, 2009, **28**(8), 1043–1054.
48. A. O. Awojoodu, *et al.*, Acid sphingomyelinase is activated in sickle cell erythrocytes and contributes to inflammatory microparticle generation in SCD, *Blood*, 2014, **124**(12), 1941–1950.
49. T. Wang, *et al.*, Hypoxia-inducible factors and RAB22A mediate formation of microvesicles that stimulate breast cancer invasion and metastasis, *Proc. Natl. Acad. Sci.*, 2014, **111**(31), E3234–E3242.
50. L. G. Lima, *et al.*, Malignant transformation in melanocytes is associated with increased production of procoagulant microvesicles, *Thromb. Haemostasis*, 2011, **106**(10), 712–723.
51. J. Taylor, R. Jaiswal and M. Bebawy, Calcium-calpain dependent pathways regulate vesiculation in malignant breast cells, *Curr. Cancer Drug Targets*, 2017, **17**(5), 486–494.
52. J. Taylor, *et al.*, Ca^{2+} mediates extracellular vesicle biogenesis through alternate pathways in malignancy, *J. Extracell. Vesicles*, 2020, **9**(1), 1734326.
53. D. Wlodkowic, *et al.*, Apoptosis and beyond: cytometry in studies of programmed cell death, *Methods Cell Biol.*, 2011, **103**, 55–98.
54. M. Van Engeland, *et al.*, Annexin V-Affinity assay: a review on an apoptosis detection system based on phosphatidylserine exposure, *Cytometry*, 1998, **31**(1), 1–9.

55. A. Pérez-Garijo and H. Steller, Spreading the word: non-autonomous effects of apoptosis during development, regeneration and disease, *Development*, 2015, **142**(19), 3253–3262.
56. V. D. Joshi, D. V. Kalvakolanu and A. S. Cross, Simultaneous activation of apoptosis and inflammation in pathogenesis of septic shock: a hypothesis, *FEBS Lett.*, 2003, **555**(2), 180–184.
57. S. J. Park, *et al.*, Molecular mechanisms of biogenesis of apoptotic exosome-like vesicles and their roles as damage-associated molecular patterns, *Proc. Natl. Acad. Sci.*, 2018, **115**(50), E11721–E11730.
58. S. Caruso and I. K.Poon, Apoptotic cell-derived extracellular vesicles: more than just debris, *Front. Immunol.*, 2018, **9**, 1486.
59. S. Faouzi, *et al.*, Anti-Fas induces hepatic chemokines and promotes inflammation by an NF-kappa B-independent, caspase-3-dependent pathway, *J. Biol. Chem.*, 2001, **276**(52), 49077–49082.
60. M. L. Coleman, *et al.*, Membrane blebbing during apoptosis results from caspase-mediated activation of ROCK I, *Nat. Cell Biol.*, 2001, **3**(4), 339–345.
61. C. D. Gregory and I. Dransfield, Apoptotic tumor cell-derived extracellular vesicles as important regulators of the onco-regenerative niche, *Front. Immunol.*, 2018, **9**, 1111.
62. K. Aoki, *et al.*, Coordinated changes in cell membrane and cytoplasm during maturation of apoptotic bleb, *Mol. Biol. Cell*, 2020, **31**(8), 833–844.
63. J. C. Mills, *et al.*, Apoptotic membrane blebbing is regulated by myosin light chain phosphorylation, *J. Cell Biol.*, 1998, **140**(3), 627–636.
64. R. Tixeira, *et al.*, ROCK1 but not LIMK1 or PAK2 is a key regulator of apoptotic membrane blebbing and cell disassembly, *Cell Death Differ.*, 2020, **27**(1), 102–116.
65. M. Zirngibl, *et al.*, Loading of nuclear autoantigens prototypically recognized by systemic lupus erythematosus sera into late apoptotic vesicles requires intact microtubules and myosin light chain kinase activity, *Clin. Exp. Immunol.*, 2015, **179**(1), 39–49.
66. K. A. Orlando, N. L. Stone and R. N. Pittman, Rho kinase regulates fragmentation and phagocytosis of apoptotic cells, *Exp. Cell Res.*, 2006, **312**(1), 5–15.
67. I. K. Poon, *et al.*, Unexpected link between an antibiotic, pannexin channels and apoptosis, *Nature*, 2014, **507**(7492), 329–334.
68. T. H. Lee, E. D'Asti, N. Magnus, K. Al-Nedawi, B. Meehan and J. Rak, Microvesicles as mediators of intercellular communication in cancer —the emerging science of cellular 'debris', *Semin. Immunopathol.*, 2011, **33**, 455–467.
69. S.-Y. Park and I.-S. Kim, Engulfment signals and the phagocytic machinery for apoptotic cell clearance, *Exp. Mol. Med.*, 2017, **49**(5), e331–e331.

70. N. A. Garcia, *et al.*, Cardiomyocyte exosomes regulate glycolytic flux in endothelium by direct transfer of GLUT transporters and glycolytic enzymes, *Cardiovasc. Res.*, 2016, **109**(3), 397–408.
71. M.-y. Xu, *et al.*, Differences in the cargos and functions of exosomes derived from six cardiac cell types: a systematic review, *Stem Cell Res. Ther.*, 2019, **10**(1), 1–11.
72. E. Morita, *et al.*, Human ESCRT and ALIX proteins interact with proteins of the midbody and function in cytokinesis, *EMBO J.*, 2007, **26**(19), 4215–4227.
73. E. B. Babiychuk and A. Draeger, Annexins in cell membrane dynamics. Ca(2+)-regulated association of lipid microdomains, *J. Cell Biol.*, 2000, **150**(5), 1113–1124.
74. K. Ni, *et al.*, The evolving role of Caveolin-1: a critical regulator of extracellular vesicles, *Med. Sci.*, 2020, **8**(4), 46.
75. M. Shimoda and R. Khokha, Metalloproteinases in extracellular vesicles, *Biochim. Biophys. Acta, Mol. Cell Res.*, 2017, **1864**(11), 1989–2000.
76. D. A. Leone, *et al.*, Surface LAMP-2 is an endocytic receptor that diverts antigen internalized by human dendritic cells into highly immunogenic exosomes, *J. Immunol.*, 2017, **199**(2), 531–546.
77. J. Cosme, *et al.*, Hypoxia-induced changes in the fibroblast secretome, exosome, and whole-cell proteome using cultured, cardiac-derived cells isolated from neonatal mice, *J. Proteome Res.*, 2017, **16**(8), 2836–2847.
78. H. Pollet, *et al.*, Plasma membrane lipid domains as platforms for vesicle biogenesis and shedding?, *Biomolecules*, 2018, **8**(3), 94.
79. C. Théry, *et al.*, Proteomic analysis of dendritic cell-derived exosomes: a secreted subcellular compartment distinct from apoptotic vesicles, *J. Immunol.*, 2001, **166**(12), 7309–7318.
80. H. Kalra, *et al.*, Vesiclepedia: a compendium for extracellular vesicles with continuous community annotation, *PLoS Biol.*, 2012, **10**(12), e1001450.
81. H. F. Heijnen, *et al.*, Activated Platelets Release Two Types of Membrane Vesicles: Microvesicles by Surface Shedding and Exosomes Derived From Exocytosis of Multivesicular Bodies and Alpha-Granules, *Blood*, 1999, **94**(11), 3791–3799.
82. G. Palmisano, *et al.*, Characterization of membrane-shed microvesicles from cytokine-stimulated β-cells using proteomics strategies, *Mol. Cell. Proteomics*, 2012, **11**(8), 230–243.
83. D. Di Vizio, Large oncosomes in human prostate cancer tissues and in the circulation of mice with metastatic disease, *Am. J. Pathol.*, 2012, **181**(5), 1573–1584.
84. K. Trajkovic, *et al.*, Ceramide triggers budding of exosome vesicles into multivesicular endosomes, *Science*, 2008, **319**(5867), 1244–1247.

85. A. Llorente, *et al.*, Molecular lipidomics of exosomes released by PC-3 prostate cancer cells, *Biochim. Biophys. Acta, Mol. Cell Biol. Lipids*, 2013, **1831**(7), 1302–1309.
86. M. Record, *et al.*, Exosomes as new vesicular lipid transporters involved in cell–cell communication and various pathophysiologies, *Biochim. Biophys. Acta, Mol. Cell Biol. Lipids*, 2014, **1841**(1), 108–120.
87. G. Skryabin, *et al.*, Lipid rafts in exosome biogenesis, *Biochemistry*, 2020, **85**(2), 177–191.
88. A. de Gassart, *et al.*, Lipid raft-associated protein sorting in exosomes, *Blood*, 2003, **102**(13), 4336–4344.
89. Z. Chu, D. P. Witte and X. Qi, Saposin C–LBPA interaction in late-endosomes/lysosomes, *Exp. Cell Res.*, 2005, **303**, 300–307.
90. M. Vidal, *et al.*, Asymmetric distribution of phospholipids in the membrane of vesicles released during in vitro maturation of guinea pig reticulocytes: evidence precluding a role for "aminophospholipid translocase", *J. Cell. Physiol.*, 1989, **140**(3), 455–462.
91. V. R. Minciacchi, M. R. Freeman and D. Di Vizio, Extracellular vesicles in cancer: exosomes, microvesicles and the emerging role of large oncosomes, *Semin. Cell Dev. Biol.*, 2015, **40**, 41–51.
92. C. Bissig, *et al.*, Viral infection controlled by a calcium-dependent lipid-binding module in ALIX, *Dev. Cell*, 2013, **25**(4), 364–373.
93. D. D. Taylor, S. Akyol and C. Gercel-Taylor, Pregnancy-associated exosomes and their modulation of T cell signaling, *J. Immunol.*, 2006, **176**(3), 1534–1542.
94. B. Bicalho, J. L. Holovati and J. P. Acker, Phospholipidomics reveals differences in glycerophosphoserine profiles of hypothermically stored red blood cells and microvesicles, *Biochim. Biophys. Acta, Biomembr.*, 2013, **1828**(2), 317–326.
95. M. P. Zaborowski, *et al.*, Extracellular vesicles: composition, biological relevance, and methods of study, *Bioscience*, 2015, **65**(8), 783–797.
96. B. S. Batista, *et al.*, Identification of a conserved glycan signature for microvesicles, *J. Proteome Res.*, 2011, **10**(10), 4624–4633.
97. C. Subra, *et al.*, Exosomes account for vesicle-mediated transcellular transport of activatable phospholipases and prostaglandins, *J. Lipid Res.*, 2010, **51**(8), 2105–2120.
98. S. Sano, *et al.*, Lipid synthesis is promoted by hypoxic adipocyte-derived exosomes in 3T3-L1 cells, *Biochem. Biophys. Res. Commun.*, 2014, **445**(2), 327–333.
99. M. Mittelbrunn, *et al.*, Unidirectional transfer of microRNA-loaded exosomes from T cells to antigen-presenting cells, *Nat. Commun.*, 2011, **2**(1), 1–10.
100. K. M. Kim, *et al.*, RNA in extracellular vesicles, *Wiley Interdiscip. Rev.: RNA*, 2017, **8**(4), e1413.

101. M. Ragusa, *et al.*, Asymmetric RNA distribution among cells and their secreted exosomes: biomedical meaning and considerations on diagnostic applications, *Front. Mol. Biosci.*, 2017, **4**, 66.
102. Z. Wei, *et al.*, Coding and noncoding landscape of extracellular RNA released by human glioma stem cells, *Nat. Commun.*, 2017, **8**(1), 1–15.
103. A. O. Batagov and I. V. Kurochkin, Exosomes secreted by human cells transport largely mRNA fragments that are enriched in the 3′-untranslated regions, *Biol. Direct*, 2013, **8**(1), 1–8.
104. P. Jenjaroenpun, *et al.*, Characterization of RNA in exosomes secreted by human breast cancer cell lines using next-generation sequencing, *PeerJ*, 2013, **1**, e201.
105. K. C. Martin and A. Ephrussi, mRNA localization: gene expression in the spatial dimension, *Cell*, 2009, **136**(4), 719–730.
106. C. Villarroya-Beltri, *et al.*, Sumoylated hnRNPA2B1 controls the sorting of miRNAs into exosomes through binding to specific motifs, *Nat. Commun.*, 2013, **4**(1), 1–10.
107. M. Pu, *et al.*, Regulatory network of miRNA on its target: coordination between transcriptional and post-transcriptional regulation of gene expression, *Cell. Mol. Life Sci.*, 2019, **76**(3), 441–451.
108. J. T. Mendell, MicroRNAs: critical regulators of development, cellular physiology and malignancy, *Cell Cycle*, 2005, **4**(9), 1179–1184.
109. M. Alexander, *et al.*, Exosome-delivered microRNAs modulate the inflammatory response to endotoxin, *Nat. Commun.*, 2015, **6**(1), 1–16.
110. J. Sadri Nahand, *et al.*, Pathogenic role of exosomes and microRNAs in HPV-mediated inflammation and cervical cancer: a review, *Int. J. Cancer*, 2020, **146**(2), 305–320.
111. I. Huang-Doran, C.-Y. Zhang and A. Vidal-Puig, Extracellular vesicles: novel mediators of cell communication in metabolic disease, *Trends Endocrinol. Metab.*, 2017, **28**(1), 3–18.
112. W. Wang, *et al.*, Exosomes derived miR-126 attenuates oxidative stress and apoptosis from ischemia and reperfusion injury by targeting ERRFI1, *Gene*, 2019, **690**, 75–80.
113. J. Guduric-Fuchs, *et al.*, Selective extracellular vesicle-mediated export of an overlapping set of microRNAs from multiple cell types, *BMC Genomics*, 2012, **13**(1), 1–14.
114. N.-T. Chiou, R. Kageyama and K. M. Ansel, Selective export into extracellular vesicles and function of tRNA fragments during T cell activation, *Cell Rep.*, 2018, **25**(12), 3356–3370.e4.
115. Q. Weng, *et al.*, Extracellular vesicles-associated tRNA-derived fragments (tRFs): biogenesis, biological functions, and their role as potential biomarkers in human diseases, *J. Mol. Med.*, 2022, 1–17.
116. T. A. Driedonks, *et al.*, Immune stimuli shape the small non-coding transcriptome of extracellular vesicles released by dendritic cells, *Cell. Mol. Life Sci.*, 2018, **75**(20), 3857–3875.

117. P. Zhang, *et al.*, Exosome-mediated delivery of MALAT1 induces cell proliferation in breast cancer, *OncoTargets Ther.*, 2018, **11**, 291.
118. A. Shabbir, *et al.*, Mesenchymal stem cell exosomes induce proliferation and migration of normal and chronic wound fibroblasts, and enhance angiogenesis in vitro, *Stem Cells Dev.*, 2015, **24**(14), 1635–1647.
119. F. Kenneweg, *et al.*, Long noncoding RNA-enriched vesicles secreted by hypoxic cardiomyocytes drive cardiac fibrosis, *Mol. Ther.—Nucleic Acids*, 2019, **18**, 363–374.
120. C. Brandimarte, *et al.*, Relapsing polychondritis: A case with anticartilage, anticardiolipin and anticentromere antibodies, *Recenti Prog. Med.*, 1988, **79**(2), 67–69.
121. L. Balaj, *et al.*, Tumour microvesicles contain retrotransposon elements and amplified oncogene sequences, *Nat. Commun.*, 2011, **2**(1), 1–9.
122. T. H. Lee, *et al.*, Oncogenic ras-driven cancer cell vesiculation leads to emission of double-stranded DNA capable of interacting with target cells, *Biochem. Biophys. Res. Commun.*, 2014, **451**(2), 295–301.
123. M. Guescini, *et al.*, Astrocytes and Glioblastoma cells release exosomes carrying mtDNA, *J. Neural Transm.*, 2010, **117**(1), 1–4.
124. B. K. Thakur, *et al.*, Double-stranded DNA in exosomes: a novel biomarker in cancer detection, *Cell Res.*, 2014, **24**(6), 766–769.
125. D. C. García-Olmo, *et al.*, Cell-free nucleic acids circulating in the plasma of colorectal cancer patients induce the oncogenic transformation of susceptible cultured cells, *Cancer Res.*, 2010, **70**(2), 560–567.
126. A. Waldenström, *et al.*, Cardiomyocyte microvesicles contain DNA/RNA and convey biological messages to target cells, *PLoS One*, 2012, **7**(4), e34653.
127. E. Lázaro-Ibáñez, *et al.*, Different gDNA content in the subpopulations of prostate cancer extracellular vesicles: apoptotic bodies, microvesicles, and exosomes, *Prostate*, 2014, **74**(14), 1379–1390.
128. Y. Kawamura, *et al.*, Extracellular vesicles as trans-genomic agents: Emerging roles in disease and evolution, *Cancer Sci.*, 2017, **108**(5), 824–830.
129. R. C. Lai, *et al.*, Exosome secreted by MSC reduces myocardial ischemia/reperfusion injury, *Stem Cell Res.*, 2010, **4**(3), 214–222.
130. R. Dalirfardouei, *et al.*, Promising effects of exosomes isolated from menstrual blood-derived mesenchymal stem cell on wound-healing process in diabetic mouse model, *J. Tissue Eng. Regener. Med.*, 2019, **13**(4), 555–568.
131. M. A. Lopez-Verrilli, *et al.*, Mesenchymal stem cell-derived exosomes from different sources selectively promote neuritic outgrowth, *Neuroscience*, 2016, **320**, 129–139.
132. P. Stanko, *et al.*, Dental mesenchymal stem/stromal cells and their exosomes, *Stem Cells Int.*, 2018, **2018**(1), 8973613.

133. N. Wang, *et al.*, Mesenchymal stem cells-derived extracellular vesicles, via miR-210, improve infarcted cardiac function by promotion of angiogenesis, *Biochim. Biophys. Acta, Mol. Basis Dis.*, 2017, **1863**(8), 2085–2092.
134. S. Bian, *et al.*, Extracellular vesicles derived from human bone marrow mesenchymal stem cells promote angiogenesis in a rat myocardial infarction model, *J. Mol. Med.*, 2014, **92**(4), 387–397.
135. H. Liu, *et al.*, Donor MSCs release apoptotic bodies to improve myocardial infarction via autophagy regulation in recipient cells, *Autophagy*, 2020, **16**(12), 2140–2155.
136. Y. Feng, *et al.*, Ischemic preconditioning potentiates the protective effect of stem cells through secretion of exosomes by targeting Mecp2 via miR-22, *PLoS One*, 2014, **9**(2), e88685.
137. Z. Zhang, *et al.*, Pretreatment of cardiac stem cells with exosomes derived from mesenchymal stem cells enhances myocardial repair, *J. Am. Heart Assoc.*, 2016, **5**(1), e002856.
138. J. P. Sluijter and E. Van Rooij, Exosomal microRNA clusters are important for the therapeutic effect of cardiac progenitor cells, *Circ. Res.*, 2015, 219–221.
139. C. Castaldo and I. Chimenti, Cardiac progenitor cells: the matrix has you, *Stem Cells Transl. Med.*, 2018, **7**(7), 506–510.
140. Y. Wang, *et al.*, Exosomes/microvesicles from induced pluripotent stem cells deliver cardioprotective miRNAs and prevent cardiomyocyte apoptosis in the ischemic myocardium, *Int. J. Cardiol.*, 2015, **192**, 61–69.
141. A. Ciullo, *et al.*, Exosomal expression of CXCR4 targets cardioprotective vesicles to myocardial infarction and improves outcome after systemic administration, *Int. J. Mol. Sci.*, 2019, **20**(3), 468.
142. J. A. Dougherty, *et al.*, Human cardiac progenitor cells enhance exosome release and promote angiogenesis under physoxia, *Front. Cell Dev. Biol.*, 2020, **8**, 130.
143. G. Milano, *et al.*, Intravenous administration of cardiac progenitor cell-derived exosomes protects against doxorubicin/trastuzumab-induced cardiac toxicity, *Cardiovasc. Res.*, 2020, **116**(2), 383–392.
144. W. D. Gray, *et al.*, Identification of therapeutic covariant microRNA clusters in hypoxia-treated cardiac progenitor cell exosomes using systems biology, *Circ. Res.*, 2015, **116**(2), 255–263.
145. J. Xiao, *et al.*, Cardiac progenitor cell-derived exosomes prevent cardiomyocytes apoptosis through exosomal miR-21 by targeting PDCD4, *Cell Death Dis.*, 2016, **7**(6), e2277.
146. Y. Zhang, M. Yu and W. Tian, Physiological and pathological impact of exosomes of adipose tissue, *Cell Proliferation*, 2016, **49**(1), 3–13.
147. Q. Luo, *et al.*, Exosomes from MiR-126-overexpressing adscs are therapeutic in relieving acute myocardial ischaemic injury, *Cell. Physiol. Biochem.*, 2017, **44**(6), 2105–2116.

148. T.-L. Lee, *et al.*, Conditioned medium from adipose-derived stem cells attenuates ischemia/reperfusion-induced cardiac injury through the microRNA-221/222/PUMA/ETS-1 pathway, *Theranostics*, 2021, **11**(7), 3131.
149. T.-C. Lai, *et al.*, MicroRNA-221/222 mediates ADSC-exosome-induced cardioprotection against ischemia/reperfusion by targeting PUMA and ETS-1, *Front. Cell Dev. Biol.*, 2020, **8**, 1475.
150. X. Cui, *et al.*, Exosomes from adipose-derived mesenchymal stem cells protect the myocardium against ischemia/reperfusion injury through Wnt/β-catenin signaling pathway, *J. Cardiovasc. Pharmacol.*, 2017, **70**(4), 225.
151. J. Liu, *et al.*, miR-93-5p-containing exosomes treatment attenuates acute myocardial infarction-induced myocardial damage, *Mol. Ther.--Nucleic Acids*, 2018, **11**, 103–115.
152. Y. Bai, *et al.*, Adipose mesenchymal stem cell-derived exosomes stimulated by hydrogen peroxide enhanced skin flap recovery in ischemia-reperfusion injury, *Biochem. Biophys. Res. Commun.*, 2018, **500**(2), 310–317.
153. Z. Liu, *et al.*, Exosomes from adipose-derived mesenchymal stem cells prevent cardiomyocyte apoptosis induced by oxidative stress, *Cell Death Discovery*, 2019, **5**(1), 1–7.
154. S. Deng, *et al.*, Exosomes from adipose-derived mesenchymal stem cells ameliorate cardiac damage after myocardial infarction by activating S1P/SK1/S1PR1 signaling and promoting macrophage M2 polarization, *Int. J. Biochem. Cell Biol.*, 2019, **114**, 105564.
155. L. M. Popescu, *et al.*, Telocytes in human epicardium, *J. Cell Mol. Med.*, 2010, **14**(8), 2085–2093.
156. M. Marini, L. Ibba-Manneschi and M. Manetti, Cardiac telocyte-derived exosomes and their possible implications in cardiovascular pathophysiology, in *Exosomes in Cardiovascular Diseases*, 2017, pp. 237–254.
157. Y. Bei, *et al.*, Telocytes in cardiac regeneration and repair, *Semin. Cell Dev. Biol.*, 2016, **55**, 14–21.
158. C. Manole, *et al.*, Experimental acute myocardial infarction: telocytes involvement in neo-angiogenesis, *J. Cell. Mol. Med.*, 2011, **15**(11), 2284–2296.
159. J. Yang, *et al.*, Exosomes derived from cardiac telocytes exert positive effects on endothelial cells, *Am. J. Transl. Res.*, 2017, **9**(12), 5375.
160. Z. Liao, *et al.*, Cardiac telocytes inhibit cardiac microvascular endothelial cell apoptosis through exosomal miRNA-21-5p-targeted cdip1 silencing to improve angiogenesis following myocardial infarction, *Theranostics*, 2021, **11**(1), 268.
161. G. Sarlon-Bartoli, *et al.*, Plasmatic level of leukocyte-derived microparticles is associated with unstable plaque in asymptomatic patients

with high-grade carotid stenosis, *J. Am. Coll. Cardiol.*, 2013, **62**(16), 1436–1441.
162. P.-E. Rautou, *et al.*, Microparticles, vascular function, and atherothrombosis, *Circ. Res.*, 2011, **109**(5), 593–606.
163. P. Diehl, *et al.*, Microparticles: major transport vehicles for distinct microRNAs in circulation, *Cardiovasc. Res.*, 2012, **93**(4), 633–644.
164. S. Fu, *et al.*, Extracellular vesicles in cardiovascular diseases, *Cell Death Discovery*, 2020, **6**(1), 1–9.
165. C. Lässer, *et al.*, Human saliva, plasma and breast milk exosomes contain RNA: uptake by macrophages, *J. Transl. Med.*, 2011, **9**, 9.
166. C. Lässer, *et al.*, RNA-containing exosomes in human nasal secretions, *Am. J. Rhinol. Allergy*, 2011, **25**(2), 89–93.
167. J. M. Street, *et al.*, Identification and proteomic profiling of exosomes in human cerebrospinal fluid, *J. Transl. Med.*, 2012, **10**(1), 1–7.
168. A. Poliakov, *et al.*, Structural heterogeneity and protein composition of exosome-like vesicles (prostasomes) in human semen, *Prostate*, 2009, **69**(2), 159–167.
169. B. Levänen, *et al.*, Altered microRNA profiles in bronchoalveolar lavage fluid exosomes in asthmatic patients, *J. Allergy Clin. Immunol.*, 2013, **131**(3), 894–903.e8.
170. K. M. de Vooght, *et al.*, Extracellular vesicles in the circulation: are erythrocyte microvesicles a confounder in the plasma haemoglobin assay?, *Biochem. Soc. Trans.*, 2013, **41**(1), 288–292.
171. X. Huang, *et al.*, Characterization of human plasma-derived exosomal RNAs by deep sequencing, *BMC Genomics*, 2013, **14**(1), 1–14.
172. B. A. Wetmore, *et al.*, Quantitative analyses and transcriptomic profiling of circulating messenger RNAs as biomarkers of rat liver injury, *Hepatology*, 2010, **51**(6), 2127–2139.
173. A. Gaceb, M. C. Martinez and R. Andriantsitohaina, Extracellular vesicles: new players in cardiovascular diseases, *Int. J. Biochem. Cell Biol.*, 2014, **50**, 24–28.
174. G. Tans, *et al.*, Comparison of anticoagulant and procoagulant activities of stimulated platelets and platelet-derived microparticles, *Blood*, 1991, **77**(12), 2641–2648.
175. G. Taraboletti, *et al.*, Shedding of the matrix metalloproteinases MMP-2, MMP-9, and MT1-MMP as membrane vesicle-associated components by endothelial cells, *Am. J. Pathol.*, 2002, **160**(2), 673–680.
176. S. Y. Chong, *et al.*, Extracellular vesicles in cardiovascular diseases: alternative biomarker sources, therapeutic agents, and drug delivery carriers, *Int. J. Mol. Sci.*, 2019, **20**(13), 3272.
177. G. Fuhrmann, I. K. Herrmann and M. M. Stevens, Cell-derived vesicles for drug therapy and diagnostics: opportunities and challenges, *Nano Today*, 2015, **10**(3), 397–409.

178. S. T. Yurkin and Z. Wang, Cell membrane-derived nanoparticles: emerging clinical opportunities for targeted drug delivery, *Nanomedicine*, 2017, **12**(16), 2007–2019.
179. W. J. Goh, *et al.*, EXOPLEXs: chimeric drug delivery platform from the fusion of cell-derived nanovesicles and liposomes, *Biomacromolecules*, 2018, **19**(1), 22–30.
180. A. B. Banizs, *et al.*, In vitro evaluation of endothelial exosomes as carriers for small interfering ribonucleic acid delivery, *Int. J. Nanomed.*, 2014, **9**, 4223.
181. M. J. Shurtleff, *et al.*, Broad role for YBX1 in defining the small non-coding RNA composition of exosomes, *Proc. Natl. Acad. Sci.*, 2017, **114**(43), E8987–E8995.
182. P. Mathiyalagan, *et al.*, FTO-dependent N6-methyladenosine regulates cardiac function during remodeling and repair, *Circulation*, 2019, **139**(4), 518–532.
183. M. J. Shurtleff, *et al.*, Y-box protein 1 is required to sort microRNAs into exosomes in cells and in a cell-free reaction, *eLife*, 2016, **5**, e19276.
184. E. Hergenreider, *et al.*, Atheroprotective communication between endothelial cells and smooth muscle cells through miRNAs, *Nat. Cell Biol.*, 2012, **14**(3), 249–256.
185. F. Jansen, *et al.*, Endothelial microparticle–mediated transfer of microRNA-126 promotes vascular endothelial cell repair via SPRED1 and is abrogated in glucose-damaged endothelial microparticles, *Circulation*, 2013, **128**(18), 2026–2038.
186. S. S. Kelkar and T. M. Reineke, Theranostics: combining imaging and therapy, *Bioconjugate Chem.*, 2011, **22**(10), 1879–1903.
187. C. He, *et al.*, Exosome theranostics: biology and translational medicine, *Theranostics*, 2018, **8**(1), 237.
188. B. Yang, Y. Chen and J. Shi, Exosome biochemistry and advanced nanotechnology for next-generation theranostic platforms, *Adv. Mater.*, 2019, **31**(2), 1802896.
189. M. E. Lobatto, *et al.*, Perspectives and opportunities for nanomedicine in the management of atherosclerosis, *Nat. Rev. Drug Discovery*, 2011, **10**(11), 835–852.
190. T. Y. Tang, *et al.*, The ATHEROMA (Atorvastatin Therapy: Effects on Reduction of Macrophage Activity) Study: evaluation using ultrasmall superparamagnetic iron oxide-enhanced magnetic resonance imaging in carotid disease, *J. Am. Coll. Cardiol.*, 2009, **53**(22), 2039–2050.
191. J. A. Maring, *et al.*, Cardiac progenitor cell–derived extracellular vesicles reduce infarct size and associate with increased cardiovascular cell proliferation, *J. Cardiovasc. Transl. Res.*, 2019, **12**(1), 5–17.
192. J. Pan, *et al.*, Exosomes derived from miR-146a-modified adipose-derived stem cells attenuate acute myocardial infarction– induced

myocardial damage via downregulation of early growth response factor 1, *J. Cell. Biochem.*, 2019, **120**(3), 4433–4443.
193. L. Barile, *et al.*, Extracellular vesicles from human cardiac progenitor cells inhibit cardiomyocyte apoptosis and improve cardiac function after myocardial infarction, *Cardiovasc. Res.*, 2014, **103**(4), 530–541.
194. S. Sahoo, *et al.*, Exosomes from human CD34+ stem cells mediate their proangiogenic paracrine activity, *Circ. Res.*, 2011, **109**(7), 724–728.
195. P. Mathiyalagan, *et al.*, Angiogenic mechanisms of human CD34+ stem cell exosomes in the repair of ischemic hindlimb, *Circ. Res.*, 2017, **120**(9), 1466–1476.
196. E. Eitan, *et al.*, In a randomized trial in prostate cancer patients, dietary protein restriction modifies markers of leptin and insulin signaling in plasma extracellular vesicles, *Aging Cell*, 2017, **16**(6), 1430–1433.
197. E. Kontopoulou, *et al.*, Evaluation of dsDNA from extracellular vesicles (EVs) in pediatric AML diagnostics, *Ann. Hematol.*, 2020, **99**(3), 459–475.
198. Y. Cheng, *et al.*, Effect of pH, temperature and freezing-thawing on quantity changes and cellular uptake of exosomes, *Protein Cell*, 2019, **10**(4), 295–299.
199. Y. Hao, *et al.*, Promotion or inhibition of extracellular vesicle release: Emerging therapeutic opportunities, *J. Controlled Release*, 2021, **340**, 136–148.
200. A. T. Reiner, *et al.*, Concise review: developing best-practice models for the therapeutic use of extracellular vesicles, *Stem Cells Transl. Med.*, 2017, **6**(8), 1730–1739.
201. L. Ayers, *et al.*, Clinical requirements for extracellular vesicle assays, *J. Extracell. Vesicles*, 2019, **8**(1), 1593755.
202. M. Mendt, S. Kamerkar, H. Sugimoto, K. M. McAndrews, C.-C. Wu, M. Gagea, S. Yang, *et al.*, Generation and testing of clinical-grade exosomes for pancreatic cancer, *JCI Insight*, 2018, **3**(8), e99263.
203. G. Andriolo, *et al.*, Exosomes from human cardiac progenitor cells for therapeutic applications: development of a GMP-grade manufacturing method, *Front. Physiol.*, 2018, **9**, 1169.
204. I. K. Herrmann, M. J. A. Wood and G. Fuhrmann, Extracellular vesicles as a next-generation drug delivery platform, *Nat. Nanotechnol.*, 2021, **16**(7), 748–759.
205. R. C. de Abreu, *et al.*, Native and bioengineered extracellular vesicles for cardiovascular therapeutics, *Nat. Rev. Cardiol.*, 2020, **17**(11), 685–697.
206. A. Vandergriff, *et al.*, Targeting regenerative exosomes to myocardial infarction using cardiac homing peptide, *Theranostics*, 2018, **8**(7), 1869.

207. X. Wang, *et al.*, Engineered exosomes with ischemic myocardium-targeting peptide for targeted therapy in myocardial infarction, *J. Am. Heart Assoc.*, 2018, **7**(15), e008737.
208. S. L. Maas, X. O. Breakefield and A. M. Weaver, Extracellular vesicles: unique intercellular delivery vehicles, *Trends Cell Biol.*, 2017, **27**(3), 172–188.
209. G. Raposo and W. Stoorvogel, Extracellular vesicles: exosomes, microvesicles, and friends, *J. Cell Biol.*, 2013, **200**(4), 373–383.
210. S. Sahoo, *et al.*, Therapeutic and diagnostic translation of extracellular vesicles in cardiovascular diseases: roadmap to the clinic, *Circulation*, 2021, **143**(14), 1426–1449.
211. S. Rahmati, *et al.*, An overview of current knowledge in biological functions and potential theragnostic applications of exosomes, *Chem. Phys. Lipids*, 2020, **226**, 104836.
212. B. Tancini, *et al.*, Insight into the role of extracellular vesicles in lysosomal storage disorders, *Genes*, 2019, **10**(7), 510.
213. F. Jansen, G. Nickenig and D. Werner, Extracellular vesicles in cardiovascular disease: potential applications in diagnosis, prognosis, and epidemiology, *Circ. Res.*, 2017, **120**(10), 1649–1657.
214. F. Jansen, *et al.*, MicroRNA Expression in Circulating Microvesicles Predicts Cardiovascular Events in Patients With Coronary Artery Disease, *J. Am. Heart Assoc.*, 2014, **3**(6), e001249.
215. D. M. Yellon and S. M. Davidson, Exosomes: nanoparticles involved in cardioprotection?, *Circ. Res.*, 2014, **114**(2), 325–332.
216. Y. Kuwabara, *et al.*, Increased microRNA-1 and microRNA-133a levels in serum of patients with cardiovascular disease indicate myocardial damage, *Circ.: Cardiovasc. Genet.*, 2011, **4**(4), 446–454.
217. P. Zamani, *et al.*, The therapeutic and diagnostic role of exosomes in cardiovascular diseases, *Trends Cardiovasc. Med.*, 2019, **29**(6), 313–323.
218. M. Szajnik and *et al.*, Exosomes in plasma of patients with ovarian carcinoma: potential biomarkers of tumor progression and response to therapy, *Gynecol. Obstet.*, 2013, 3.
219. M. K. S. Tang and A. S. T. Wong, Exosomes: Emerging biomarkers and targets for ovarian cancer, *Cancer Lett.*, 2015, **367**(1), 26–33.
220. T. L. Whiteside, The potential of tumor-derived exosomes for noninvasive cancer monitoring, *Expert Rev. Mol. Diagn.*, 2015, **15**(10), 1293–1310.
221. R. E. Lane, *et al.*, Extracellular vesicles as circulating cancer biomarkers: opportunities and challenges, *Clin. Transl. Med.*, 2018, **7**(1), 1–11.
222. A. Abramowicz, P. Widlak and M. Pietrowska, Proteomic analysis of exosomal cargo: the challenge of high purity vesicle isolation, *Mol. BioSyst.*, 2016, **12**(5), 1407–1419.
223. T. L. Whiteside, Extracellular vesicles isolation and their biomarker potential: are we ready for testing?, *Ann. Transl. Med.*, 2017, **5**(3), 54.

224. C.-S. Hong, *et al.*, Isolation of biologically active and morphologically intact exosomes from plasma of patients with cancer, *J. Extracell. Vesicles*, 2016, **5**(1), 29289.
225. A. Yekula, *et al.*, From laboratory to clinic: Translation of extracellular vesicle based cancer biomarkers, *Methods*, 2020, **177**, 58–66.
226. R. C. Lai, *et al.*, Exosome secreted by MSC reduces myocardial ischemia/reperfusion injury, *Stem Cell Res.*, 2010, **4**(3), 214–222.
227. A. A. Mangi, *et al.*, Mesenchymal stem cells modified with Akt prevent remodeling and restore performance of infarcted hearts, *Nat. Med.*, 2003, **9**(9), 1195–1201.
228. A. Kawamoto, *et al.*, CD34-positive cells exhibit increased potency and safety for therapeutic neovascularization after myocardial infarction compared with total mononuclear cells, *Circulation*, 2006, **114**(20), 2163–2169.
229. L. Timmers, *et al.*, Reduction of myocardial infarct size by human mesenchymal stem cell conditioned medium, *Stem Cell Res.*, 2008, **1**(2), 129–137.
230. M. Gnecchi, *et al.*, Paracrine action accounts for marked protection of ischemic heart by Akt-modified mesenchymal stem cells, *Nat. Med.*, 2005, **11**(4), 367–368.
231. S. Sahoo, *et al.*, Exosomes from human CD34(+) stem cells mediate their proangiogenic paracrine activity, *Circ. Res.*, 2011, **109**(7), 724–728.
232. A. G.-E. Ibrahim, K. Cheng and E. Marbán, Exosomes as critical agents of cardiac regeneration triggered by cell therapy, *Stem Cell Rep.*, 2014, **2**(5), 606–619.
233. U. Agarwal, *et al.*, Experimental, systems, and computational approaches to understanding the microRNA-mediated reparative potential of cardiac progenitor cell–derived exosomes from pediatric patients, *Circ. Res.*, 2017, **120**(4), 701–712.
234. B. Liu, *et al.*, Cardiac recovery via extended cell-free delivery of extracellular vesicles secreted by cardiomyocytes derived from induced pluripotent stem cells, *Nat. Biomed. Eng.*, 2018, **2**(5), 293–303.
235. P. Saha, *et al.*, Circulating exosomes derived from transplanted progenitor cells aid the functional recovery of ischemic myocardium, *Sci. Transl. Med.*, 2019, **11**(493), eaau1168.
236. P. Mathiyalagan, *et al.*, Angiogenic mechanisms of human CD34+ stem cell exosomes in the repair of ischemic hindlimb, *Circ. Res.*, 2017, **120**(9), 1466–1476.
237. H. Henriques-Antunes, *et al.*, The kinetics of small extracellular vesicle delivery impacts skin tissue regeneration, *ACS Nano*, 2019, **13**(8), 8694–8707.
238. L. Chen, *et al.*, Exosomal lncRNA GAS5 regulates the apoptosis of macrophages and vascular endothelial cells in atherosclerosis, *PLoS One*, 2017, **12**(9), e0185406.

239. P. A. Lapchak, *et al.*, Intravenous xenogeneic human cardiosphere-derived cell extracellular vesicles (exosomes) improves behavioral function in small-clot embolized rabbits, *Exp. Neurol.*, 2018, **307**, 109–117.
240. J. Yang, *et al.*, Exosome mediated delivery of miR-124 promotes neurogenesis after ischemia, *Mol. Ther.--Nucleic Acids*, 2017, **7**, 278–287.
241. J. M. Vicencio, *et al.*, Plasma exosomes protect the myocardium from ischemia-reperfusion injury, *J. Am. Coll. Cardiol.*, 2015, **65**(15), 1525–1536.
242. B. Yu, *et al.*, Exosomes secreted from GATA-4 overexpressing mesenchymal stem cells serve as a reservoir of anti-apoptotic microRNAs for cardioprotection, *Int. J. Cardiol.*, 2015, **182**, 349–360.
243. J. Ma, et al, Exosomes derived from Akt-modified human umbilical cord mesenchymal stem cells improve cardiac regeneration and promote angiogenesis via activating platelet-derived growth factor D, *Stem Cells Transl. Med.*, 2017, **6**(1), 51–59.
244. S. Bian, *et al.*, Extracellular vesicles derived from human bone marrow mesenchymal stem cells promote angiogenesis in a rat myocardial infarction model, *J. Mol. Med.*, 2014, **92**(4), 387–397.
245. Y. Feng, *et al.*, Ischemic preconditioning potentiates the protective effect of stem cells through secretion of exosomes by targeting Mecp2 via miR-22, *PLoS One*, 2014, **9**(2), e88685.
246. Y. Zhao, X. Sun, W. Cao, J. Ma, L. Sun, H. Qian, W. Zhu and W. Xu, Exosomes derived from human umbilical cord mesenchymal stem cells relieve acute myocardial ischemic injury, *Stem Cells Int.*, 2015, **2015**(1), 761643.
247. X. Teng, *et al.*, Mesenchymal stem cell-derived exosomes improve the microenvironment of infarcted myocardium contributing to angiogenesis and anti-inflammation, *Cell. Physiol. Biochem.*, 2015, **37**(6), 2415–2424.
248. R. Gallet, *et al.*, Exosomes secreted by cardiosphere-derived cells reduce scarring, attenuate adverse remodelling, and improve function in acute and chronic porcine myocardial infarction, *Eur. Heart J.*, 2017, **38**(3), 201–211.
249. L. Barile, *et al.*, Cardioprotection by cardiac progenitor cell-secreted exosomes: role of pregnancy-associated plasma protein-A, *Cardiovasc. Res.*, 2018, **114**(7), 992–1005.
250. A. G.-E. Ibrahim, K. Cheng and E. Marbán, Exosomes as critical agents of cardiac regeneration triggered by cell therapy, *Stem Cell Rep.*, 2014, **2**(5), 606–619.
251. G. de Couto, *et al.*, Exosomal microRNA transfer into macrophages mediates cellular postconditioning, *Circulation*, 2017, **136**(2), 200–214.

252. L. Cambier, *et al.*, Y RNA fragment in extracellular vesicles confers cardioprotection via modulation of IL-10 expression and secretion, *EMBO Mol. Med.*, 2017, **9**(3), 337–352.
253. S. Sharma, *et al.*, A deep proteome analysis identifies the complete secretome as the functional unit of human cardiac progenitor cells, *Circ. Res.*, 2017, **120**(5), 816–834.
254. A. Kervadec, *et al.*, Cardiovascular progenitor-derived extracellular vesicles recapitulate the beneficial effects of their parent cells in the treatment of chronic heart failure, *J. Heart Lung Transplant.*, 2016, **35**(6), 795–807.
255. M. Adamiak, *et al.*, Induced pluripotent stem cell (iPSC)–derived extracellular vesicles are safer and more effective for cardiac repair than iPSCs, *Circ. Res.*, 2018, **122**(2), 296–309.
256. N. El Harane, *et al.*, Acellular therapeutic approach for heart failure: in vitro production of extracellular vesicles from human cardiovascular progenitors, *Eur. Heart J.*, 2018, **39**(20), 1835–1847.
257. F. Arslan, *et al.*, Mesenchymal stem cell-derived exosomes increase ATP levels, decrease oxidative stress and activate PI3K/Akt pathway to enhance myocardial viability and prevent adverse remodeling after myocardial ischemia/reperfusion injury, *Stem Cell Res.*, 2013, **10**(3), 301–312.
258. M. Khan, *et al.*, Embryonic stem cell-derived exosomes promote endogenous repair mechanisms and enhance cardiac function following myocardial infarction, *Circ. Res.*, 2015, **117**(1), 52–64.
259. H. Liu, *et al.*, Exosomes derived from dendritic cells improve cardiac function via activation of CD4+ T lymphocytes after myocardial infarction, *J. Mol. Cell. Cardiol.*, 2016, **91**, 123–133.
260. T. M. Ribeiro-Rodrigues, *et al.*, Exosomes secreted by cardiomyocytes subjected to ischaemia promote cardiac angiogenesis, *Cardiovasc. Res.*, 2017, **113**(11), 1338–1350.
261. M. Khan, *et al.*, Embryonic stem cell-derived exosomes promote endogenous repair mechanisms and enhance cardiac function following myocardial infarction, *Circ. Res.*, 2015, **117**(1), 52–64.
262. J. Zhao, *et al.*, Mesenchymal stromal cell-derived exosomes attenuate myocardial ischaemia-reperfusion injury through miR-182-regulated macrophage polarization, *Cardiovasc. Res.*, 2019, **115**(7), 1205–1216.
263. N. Alexandru, *et al.*, Microparticles of healthy origins improve endothelial progenitor cell dysfunction via microRNA transfer in an atherosclerotic hamster model, *Acta Physiol.*, 2017, **221**(4), 230–249.
264. N. Alexandru, *et al.*, Intravenous administration of allogenic cell-derived microvesicles of healthy origins defends against atherosclerotic cardiovascular disease development by a direct action on endothelial progenitor cells, *Cells*, 2020, **9**(2), 423.

265. Y. Zhao, *et al.*, Exosomes derived from human umbilical cord mesenchymal stem cells relieve acute myocardial ischemic injury, *Stem Cells Int.*, 2015, **2015**, 761643.
266. R. G. Rogers, M. Fournier, L. Sanchez, A. G. Ibrahim, M. A. Aminzadeh, M. I. Lewis and E. Marbán, Disease-modifying bioactivity of intravenous cardiosphere-derived cells and exosomes in mdx mice, *JCI Insight*, 2019, **4**(7), e125754.
267. X. Teng, *et al.*, Mesenchymal stem cell-derived exosomes improve the microenvironment of infarcted myocardium contributing to angiogenesis and anti-inflammation, *Cell. Physiol. Biochem.*, 2015, **37**(6), 2415–2424.
268. K. M. Luther, *et al.*, Exosomal miR-21a-5p mediates cardioprotection by mesenchymal stem cells, *J. Mol. Cell. Cardiol.*, 2018, **119**, 125–137.
269. J. Ma, *et al.*, Exosomes derived from Akt-modified human umbilical cord mesenchymal stem cells improve cardiac regeneration and promote angiogenesis via activating platelet-derived growth factor D, *Stem Cells Transl. Med.*, 2017, **6**(1), 51–59.
270. H. Gu, *et al.*, Serum-derived extracellular vesicles protect against acute myocardial infarction by regulating miR-21/PDCD4 signaling pathway, *Front. Physiol.*, 2018, **9**, 348.
271. L. Timmers, *et al.*, Reduction of myocardial infarct size by human mesenchymal stem cell conditioned medium, *Stem cell res.*, 2008, **1**(2), 129–137.
272. L. Shao, Y. Zhang, B. Lan, J. Wang, Z. Zhang, L. Zhang, P. Xiao, *et al.*, MiRNA-sequence indicates that mesenchymal stem cells and exosomes have similar mechanism to enhance cardiac repair, *BioMed Res. Int.*, 2017, **2017**(1), 4150705.
273. H. Henriques-Antunes, *et al.*, The kinetics of small extracellular vesicle delivery impacts skin tissue regeneration, *ACS Nano*, 2019, **13**(8), 8694–8707.
274. Y. Zhang, *et al.*, Secreted monocytic miR-150 enhances targeted endothelial cell migration, *Mol. Cell*, 2010, **39**(1), 133–144.
275. Z. Wei, *et al.*, miRNA-181a over-expression in mesenchymal stem cell-derived exosomes influenced inflammatory response after myocardial ischemia-reperfusion injury, *Life Sci.*, 2019, **232**, 116632.
276. L. Shao, *et al.*, Knockout of beta-2 microglobulin enhances cardiac repair by modulating exosome imprinting and inhibiting stem cell-induced immune rejection, *Cell. Mol. Life Sci.*, 2020, **77**(5), 937–952.
277. B. W. van Balkom, *et al.*, Endothelial cells require miR-214 to secrete exosomes that suppress senescence and induce angiogenesis in human and mouse endothelial cells, *Blood*, 2013, **121**(19), 3997–4006.
278. B. Yu, *et al.*, Exosomes secreted from GATA-4 overexpressing mesenchymal stem cells serve as a reservoir of anti-apoptotic microRNAs for cardioprotection, *Int. J. Cardiol.*, 2015, **182**, 349–360.

279. M. Kanada, *et al.*, Differential fates of biomolecules delivered to target cells via extracellular vesicles, *Proc. Natl. Acad. Sci.*, 2015, **112**(12), E1433–E1442.
280. A. G. Ibrahim, *et al.*, Augmenting canonical Wnt signalling in therapeutically inert cells converts them into therapeutically potent exosome factories, *Nat. Biomed. Eng.*, 2019, **3**(9), 695–705.
281. L.-P. Zhu, *et al.*, Hypoxia-elicited mesenchymal stem cell-derived exosomes facilitates cardiac repair through miR-125b-mediated prevention of cell death in myocardial infarction, *Theranostics*, 2018, **8**(22), 6163.
282. T. J. Antes, *et al.*, Targeting extracellular vesicles to injured tissue using membrane cloaking and surface display, *J. Nanobiotechnol.*, 2018, **16**(1), 1–15.
283. Y. Wang, *et al.*, Macrophage membrane functionalized biomimetic nanoparticles for targeted anti-atherosclerosis applications, *Theranostics*, 2021, **11**(1), 164–180.
284. J. Y. Kang, *et al.*, Human peripheral blood-derived exosomes for microRNA delivery, *Int. J. Mol. Med.*, 2019, **43**(6), 2319–2328.
285. T. Imai, *et al.*, Macrophage-dependent clearance of systemically administered B16BL6-derived exosomes from the blood circulation in mice, *J. Extracell. Vesicles*, 2015, **4**(1), 26238.
286. M. Zahid, *et al.*, Identification of a cardiac specific protein transduction domain by in vivo biopanning using a M13 phage peptide display library in mice, *PLoS One*, 2010, **5**(8), e12252.
287. H. Kim, *et al.*, Cardiac-specific delivery by cardiac tissue-targeting peptide-expressing exosomes, *Biochem. Biophys. Res. Commun.*, 2018, **499**(4), 803–808.
288. S. Kanki, *et al.*, Identification of targeting peptides for ischemic myocardium by in vivo phage display, *J. Mol. Cell. Cardiol.*, 2011, **50**(5), 841–848.
289. U. Agarwal, *et al.*, Experimental, Systems, and Computational Approaches to Understanding the MicroRNA-Mediated Reparative Potential of Cardiac Progenitor Cell–Derived Exosomes From Pediatric Patients, *Circ. Res.*, 2017, **120**(4), 701–712.
290. P. Mathiyalagan, *et al.*, Angiogenic Mechanisms of Human $CD34^+$ Stem Cell Exosomes in the Repair of Ischemic Hindlimb, *Circ. Res.*, 2017, **120**(9), 1466–1476.

Subject Index